T0320605

MANIFOLDS, TENSORS, AND FORMS

Providing a succinct yet comprehensive treatment of the essentials of modern differential geometry and topology, this book's clear prose and informal style make it accessible to advanced undergraduate and graduate students in mathematics and the physical sciences.

The text covers the basics of multilinear algebra, differentiation and integration on manifolds, Lie groups and Lie algebras, homotopy and de Rham cohomology, homology, vector bundles, Riemannian and pseudo-Riemannian geometry, and degree theory. It also features over 250 detailed exercises, and a variety of applications revealing fundamental connections to classical mechanics, electromagnetism (including circuit theory), general relativity, and gauge theory. Solutions to the problems are available for instructors at www.cambridge.org/9781107042193.

PAUL RENTELN is Professor of Physics in the Department of Physics, California State University San Bernardino, where he has taught a wide range of courses in physics. He is also Visiting Associate in Mathematics at the California Institute of Technology, where he conducts research into combinatorics.

MANIFOLDS, TENSORS, AND FORMS

An Introduction for Mathematicians and Physicists

PAUL RENTELN

California State University San Bernardino
and
California Institute of Technology

CAMBRIDGE
UNIVERSITY PRESS

CAMBRIDGE
UNIVERSITY PRESS

University Printing House, Cambridge CB2 8BS, United Kingdom

One Liberty Plaza, 20th Floor, New York, NY 10006, USA

477 Williamstown Road, Port Melbourne, VIC 3207, Australia

4843/24, 2nd Floor, Ansari Road, Daryaganj, Delhi - 110002, India

79 Anson Road, #06-04/06, Singapore 079906

Cambridge University Press is part of the University of Cambridge.

It furthers the University's mission by disseminating knowledge in the pursuit of
education, learning and research at the highest international levels of excellence.

www.cambridge.org
Information on this title: www.cambridge.org/9781107042193

© P. Renteln 2014

This publication is in copyright. Subject to statutory exception
and to the provisions of relevant collective licensing agreements,
no reproduction of any part may take place without the written
permission of Cambridge University Press.

First published 2014

A catalogue record for this publication is available from the British Library

Library of Congress Cataloging in Publication data
Renteln, Paul, 1959– author.
Manifolds, tensors, and forms : an introduction for mathematicians and physicists /
Paul Renteln.
pages cm
Includes bibliographical references and index.
ISBN 978-1-107-04219-3 (alk. paper)
1. Geometry, Differential – Textbooks. 2. Manifolds (Mathematics) – Textbooks.
3. Calculus of tensors – Textbooks. 4. Forms (Mathematics) – Textbooks. I. Title.
QA641.R46 2013
516.3′6–dc23
2013036056

ISBN 978-1-107-04219-3 Hardback

Additional resources for this publication at www.cambridge.org/9781107042193

Cambridge University Press has no responsibility for the persistence or
accuracy of URLs for external or third-party internet websites referred to in
this publication, and does not guarantee that any content on such websites is,
or will remain, accurate or appropriate.

Contents

Preface

Q: What's the difference between an argument and a proof? A: An argument will convince a reasonable person, but a proof is needed to convince an unreasonable one.

Anon.

Die Mathematiker sind eine Art Franzosen: Redet man zu ihnen, so bersetzen sie es in ihre Sprache, und dann ist es alsbald ganz etwas anderes. (Mathematicians are like Frenchmen: whatever you say to them they translate into their own language and forthwith it is something entirely different.)

Johann Wolfgang von Goethe

This book offers a concise overview of some of the main topics in differential geometry and topology and is suitable for upper-level undergraduates and beginning graduate students in mathematics and the sciences. It evolved from a set of lecture notes on these topics given to senior-year students in physics based on the marvelous little book by Flanders [25], whose stylistic and substantive imprint can be recognized throughout. The other primary sources used are listed in the references.

By intent the book is akin to a whirlwind tour of many mathematical countries, passing many treasures along the way and only stopping to admire a few in detail. Like any good tour, it supplies all the essentials needed for individual exploration after the tour is over. But, unlike many tours, it also provides language instruction. Not surprisingly, most books on differential geometry are written by mathematicians. This one is written by a mathematically inclined physicist, one who has lived and worked on both sides of the linguistic and formalistic divide that often separates pure and applied mathematics. It is this language barrier that often causes the beginner so much trouble when approaching the subject for the first time. Consequently, the book has been written with a conscious attempt to explain as much as possible from both a "high brow" and a "low brow" viewpoint,[1] particularly in the early chapters.

For many mathematicians, mathematics is the art of avoiding computation. Similarly, physicists will often say that you should never begin a computation unless you know what the answer will be. This may be so, but, more often than not, what happens is that a person works out the answer by ugly computation, and then reworks and publishes the answer in a way that hides all the gory details and makes it seem as though he or she knew the answer all along from pure abstract thought. Still, it is true that there are times when an answer can be obtained much more easily by means of a powerful abstract tool. For this reason, both approaches are given their due here. The result is a compromise between highly theoretical approaches and concrete calculational tools.

This compromise is evident in the use of proofs throughout the book. The one thing that unites mathematicians and scientists is the desire to know, not just *what* is true, but *why* it is true. For this reason, the book contains both proofs and computations. But, in the spirit of the above quotation, arguments sometimes substitute for formal proofs and many long, tedious proofs have been omitted to promote the flow of the exposition. The book therefore risks being not mathematically rigorous enough for some readers and too much so for others, but its virtue is that it is neither encyclopedic nor overly pedantic. It is my hope that the presentation will appeal to readers of all backgrounds and interests.

The pace of this work is quick, hitting only the highlights. Although the writing is deliberately terse the tone of the book is for the most part informal, so as to facilitate its use for self-study. Exercises are liberally sprinkled throughout the text and sometimes referred to in later sections; additional problems are placed at the end of most chapters.[2] Although it is not necessary to do all of them, it is certainly advisable to do some; in any case you should read them all, as they provide flesh for the bare bones. After working through this book a student should have acquired all the tools needed to use these concepts in scientific applications. Of course, many topics are omitted and every major topic treated here has many books devoted to it alone. Students wishing to fill in the gaps with more detailed investigations are encouraged to seek out some of the many fine works in the reference list at the end of the book.

---

[1] The playful epithets are an allusion, of course, to modern humans (abstract thinkers) and Neanderthals (concrete thinkers).

[2] A solutions manual is available to instructors at www.cambridge.org/9781107042193.

The prerequisites for this book are solid first courses in linear algebra, multivariable calculus, and differential equations. Some exposure to point set topology and modern algebra would be nice, but it is not necessary. To help bring students up to speed and to avoid the necessity of looking elsewhere for certain definitions, a mathematics primer is included as Appendix A. Also, the beginning chapter contains all the linear algebra facts employed elsewhere in the book, including a discussion of the correct placement and use of indices.[3] This is followed by a chapter on tensors and multilinear algebra in preparation for the study of tensor analysis and differential forms on smooth manifolds. The de Rham cohomology leads naturally into the topology of smooth manifolds, and from there to a rather brief chapter on the homology of continuous manifolds. The tools introduced there provide a nice way to understand integration on manifolds and, in particular, Stokes' theorem, which is afforded two kinds of treatment. Next we consider vector bundles, connections, and covariant derivatives and then manifolds with metrics. The last chapter offers a very brief introduction to degree theory and some of its uses. This is followed by several appendices providing background material, calculations too long for the main body of the work, or else further applications of the theory.

Originally the book was intended to serve as the basis for a rapid, one-quarter introduction to these topics. But inevitably, as with many such projects, it began to suffer from mission creep, so that covering all the material in ten weeks would probably be a bad idea. Instructors laboring under a short deadline can, of course, simply choose to omit some topics. For example, to get to integration more quickly one could skip Chapter 5 altogether, then discuss only version two of Stokes' theorem. Instructors having the luxury of a semester system should be able to cover everything. Starred sections can be (or perhaps ought to be) skimmed or omitted on a first reading.

This is an expository work drawing freely from many different sources (most of which are listed in the references section), so none of the material harbors any pretense of originality. It is heavily influenced by lectures of Bott and Chern, whose classes I was fortunate enough to take. It also owes a debt to the expository work of many other writers, whose contributions are hereby acknowledged. My sincere apologies to anyone I may have inadvertently missed in my attributions. The manuscript itself was originally typeset using Knuth's astonishingly versatile TEXprogram (and its offspring, LATEX), and the figures were made using Timothy Van Zandt's wonderful graphics tool `pstricks`, enhanced by the three-dimensional drawing packages `pst-3dplot` and `pst-solides3d`.

---

[3] Stephen Hawking jokes in the introduction to his *Brief History of Time* that the publisher warned him his sales would be halved if he included even one equation. Analogously, sales of this book may be halved by the presence of indices, as many pure mathematicians will do anything to avoid them.

It goes without saying that all writers owe a debt to their teachers. In my case I am fortunate to have learned much from Abhay Ashtekar, Raoul Bott, Shing Shen Chern, Stanley Deser, Doug Eardley, Chris Isham, Karel Kuchař, Robert Lazarsfeld, Rainer Kurt Sachs, Ted Shifrin, Lee Smolin, and Philip Yasskin, who naturally bear all responsibility for any errors contained herein . . . I also owe a special debt to Laurens Gunnarsen for having encouraged me to take Chern's class when we were students together at Berkeley and for other very helpful advice. I am grateful to Rick Wilson and the Mathematics Department at the California Institute of Technology for their kind hospitality over the course of many years and for providing such a stimulating research environment in which to nourish my other life in combinatorics. Special thanks go to Nicholas Gibbons, Lindsay Barnes, and Jessica Murphy at Cambridge University Press, for their support and guidance throughout the course of this project, and to Susan Parkinson, whose remarkable editing skills resulted in many substantial improvements to the book. Most importantly, this work would not exist without the love and affection of my wife, Alison, and our sons David and Michael.

1

Linear algebra

Eighty percent of mathematics is linear algebra.

Raoul Bott

This chapter offers a rapid review of some of the essential concepts of linear algebra that are used in the rest of the book. Even if you had a good course in linear algebra, you are encouraged to skim the chapter to make sure all the concepts and notations are familiar, then revisit it as needed.

1.1 Vector spaces

The standard example of a vector space is $\mathbb{R}^n$, which is the Cartesian product of $\mathbb{R}$ with itself n times: $\mathbb{R}^n = \mathbb{R} \times \cdots \times \mathbb{R}$. A vector v in $\mathbb{R}^n$ is an n-tuple $(a_1, a_2, \ldots, a_n)$ of real numbers with scalar multiplication and vector addition defined as follows[†]:

$$c(a_1, a_2, \ldots, a_n) := (ca_1, ca_2, \ldots, ca_n) \tag{1.1}$$

and

$$(a_1, a_2, \ldots, a_n) + (b_1, b_2, \ldots, b_n) := (a_1 + b_1, a_2 + b_2, \ldots, a_n + b_n). \tag{1.2}$$

The zero vector is $(0, 0, \ldots, 0)$.

More generally, a **vector space** V over a field $\mathbb{F}$ is a set $\{u, v, w, \ldots\}$ of objects called **vectors**, together with a set $\{a, b, c, \ldots\}$ of elements in $\mathbb{F}$ called **scalars**, that is closed under the taking of linear combinations:

$$u, v \in V \text{ and } a, b \in \mathbb{F} \Rightarrow au + bv \in V, \tag{1.3}$$

and where $0v = 0$ and $1v = v$. (For the full definition, see Appendix A.)

[†] The notation A := B means that A is defined by B.

A **subspace** of V is a subset of V that is also a vector space. An **affine subspace** of V is a translate of a subspace of V.[1] The vector space V is the **direct sum** of two subspaces U and W, written $V = U \oplus W$, if $U \cap W = 0$ (the only vector in common is the zero vector) and every vector $v \in V$ can be written uniquely as $v = u + w$ for some $u \in U$ and $w \in W$.

A set $\{v_i\}$ of vectors[2] is **linearly independent** (over the field $\mathbb{F}$) if, for any collection of scalars $\{c_i\} \subset \mathbb{F}$,

$$\sum_i c_i v_i = 0 \qquad \text{implies} \qquad c_i = 0 \text{ for all } i. \tag{1.4}$$

Essentially this means that no member of a set of linearly independent vectors may be expressed as a linear combination of the others.

> **EXERCISE 1.1** Prove that the vectors $(1, 1)$ and $(2, 1)$ in $\mathbb{R}^2$ are linearly independent over $\mathbb{R}$ whereas the vectors $(1, 1)$ and $(2, 2)$ in $\mathbb{R}^2$ are linearly dependent over $\mathbb{R}$.

A set B of vectors is a **spanning set** for V (or, more simply, **spans** V) if every vector in V can be written as a linear combination of vectors from B. A spanning set of linearly independent vectors is called a **basis** for the vector space. The cardinality of a basis for V is called the **dimension** of the space, written $\dim V$. Vector spaces have many different bases, and they all have the same cardinality. (For the most part we consider only finite dimensional vector spaces.)

> **Example 1.1** The vector space $\mathbb{R}^n$ is n-dimensional over $\mathbb{R}$. The **standard basis** is the set of n vectors $\{e_1 = (1, 0, \dots, 0), e_2 = (0, 1, \dots, 0), \dots, e_n = (0, 0, \dots, 1)\}$.

Pick a basis $\{e_i\}$ for the vector space V. By definition we may write

$$v = \sum_i v_i e_i \tag{1.5}$$

for any vector $v \in V$, where the v_i are elements of the field $\mathbb{F}$ and are called the **components** of v with respect to the basis $\{e_i\}$. We get a different set of components for the same vector v depending upon the basis we choose.

> **EXERCISE 1.2** Show that the components of a vector are unique, that is, if $v = \sum_i v_i e_i = \sum_i v_i' e_i$ then $v_i = v_i'$.

> **EXERCISE 1.3** Show that $\dim(U \oplus W) = \dim U + \dim W$.

---

[1] By definition, W is a **translate** of U if, for some fixed $v \in V$ with $v \neq 0$, $W = \{u + v : u \in U\}$. An affine subspace is like a subspace without the zero vector.

[2] To avoid cluttering the formulae, the index range will often be left unspecified. In some cases this is because the range is arbitrary, while in other cases it is because the range is obvious.

EXERCISE 1.4 Let W be a subspace of V. Show that we can always **complete a basis** of W to obtain one for V. In other words, if $\dim W = m$ and $\dim V = n$, and if $\{f_1, \ldots, f_m\}$ is a basis for W, show there exists a basis for V of the form $\{f_1, \ldots, f_m, g_1, \ldots, g_{n-m}\}$. Equivalently, show that a basis of a finite-dimensional vector space is just a maximal set of linearly independent vectors.

1.2 Linear maps

Let V and W be vector spaces. A map $T : V \to W$ is **linear** (or a **homomorphism**) if, for $v_1, v_2 \in V$ and $a_1, a_2 \in \mathbb{F}$,

$$T(a_1 v_1 + a_2 v_2) = a_1 T v_1 + a_2 T v_2. \tag{1.6}$$

We will write either $T(v)$ or Tv for the action of the linear map T on a vector v.

EXERCISE 1.5 Show that two linear maps that agree on a basis agree everywhere.

Given a finite subset $U := \{u_1, u_2, \ldots\}$ of vectors in V, any map $T : U \to W$ induces a linear map $T : V \to W$ according to the rule $T(\sum_i a_i u_i) := \sum_i a_i T u_i$. The original map is said to have been **extended by linearity**.

The set of all $v \in V$ such that $Tv = 0$ is called the **kernel** (or **null space**) of T, written $\ker T$; $\dim \ker T$ is sometimes called the **nullity** of T. The set of all $w \in W$ for which there exists a $v \in V$ with $Tv = w$ is called the **image** (or **range**) of T, written $\operatorname{im} T$. The **rank** of T, $\operatorname{rk} T$, is defined as $\dim \operatorname{im} T$.

EXERCISE 1.6 Show that $\ker T$ is a subspace of V and $\operatorname{im} T$ is a subspace of W.

EXERCISE 1.7 Show that T is injective if and only if the kernel of T consists of the zero vector alone.

If T is bijective it is called an **isomorphism**, in which case V and W are said to be **isomorphic**; this is written as $V \cong W$ or, sloppily, $V = W$. Isomorphic vector spaces are not necessarily identical, but they behave as if they were.

Theorem 1.1 *All finite-dimensional vector spaces of the same dimension are isomorphic.*

EXERCISE 1.8 Prove Theorem 1.1.

A linear map from a vector space to itself is called an **endomorphism**, and if it is a bijection it is called an **automorphism**.[3]

---

[3] Physicists tend to call an endomorphism a **linear operator**.

EXERCISE 1.9 A linear map T is **idempotent** if $T^2 = T$. An idempotent endomorphism $\pi : V \to V$ is called a **projection (operator)**. *Remark:* This is not to be confused with an orthogonal projection, which requires an inner product for its definition.

(a) Show that $V = \operatorname{im} \pi \oplus \ker \pi$.
(b) Suppose W is a subspace of V. Show that there exists a projection operator $\pi : V \to V$ that restricts to the identity map on W. (Note that the projection operator is not unique.) *Hint:* Complete a basis of W so that it becomes a basis of V.

EXERCISE 1.10 Show that if $T : V \to V$ is an automorphism then the inverse map T^{-1} is also linear.

EXERCISE 1.11 Show that the set Aut V of all automorphisms of V is a group. (For information about groups, consult Appendix A.)

1.3 Exact sequences

Suppose that you are given a sequence of vector spaces V_i and linear maps $\varphi_i : V_i \to V_{i+1}$ connecting them, as illustrated below:

$$\cdots \longrightarrow V_{i-1} \xrightarrow{\varphi_{i-1}} V_i \xrightarrow{\varphi_i} V_{i+1} \xrightarrow{\varphi_{i+1}} \cdots .$$

The maps are said to be **exact at** V_i if $\operatorname{im} \varphi_{i-1} = \ker \varphi_i$, i.e., the image of φ_{i-1} equals the kernel of φ_i. The sequence is called an **exact sequence** if the maps are exact at V_i for all i. Exact sequences of vector spaces show up everywhere and satisfy some particularly nice properties, so it worth exploring them a bit.

If V_1, V_2, and V_3 are three vector spaces, and if the sequence

$$0 \xrightarrow{\varphi_0} V_1 \xrightarrow{\varphi_1} V_2 \xrightarrow{\varphi_2} V_3 \xrightarrow{\varphi_3} 0 \tag{1.7}$$

is exact, it is called a **short exact sequence**. In this diagram "0" represents the zero-dimensional vector space, whose only element is the zero vector. The linear map φ_0 sends 0 to the zero vector of V_1, while φ_3 sends everything in V_3 to the zero vector.

EXERCISE 1.12 Show that the existence of the short exact sequence (1.7) is equivalent to the statement "φ_1 is injective and φ_2 is surjective." In particular, if

$$0 \longrightarrow V \xrightarrow{\varphi} W \longrightarrow 0 \tag{1.8}$$

is exact, V and W must be isomorphic.

It follows that if $T : V \to W$ is surjective then

$$0 \longrightarrow \ker T \overset{\iota}{\longrightarrow} V \overset{T}{\longrightarrow} W \longrightarrow 0 \qquad (1.9)$$

is a short exact sequence, where ι is the **inclusion map**.[4] By virtue of the above discussion, all short exact sequences are of this form.

Theorem 1.2 *Given the short exact sequence* (1.9), *there exists a linear map* $S : W \to V$ *such that* $T \circ S = 1$. *We say that the exact sequence* (1.9) **splits**.

Proof Let $\{f_i\}$ be a basis of W. By the surjectivity of T, for each i there exists an $e_i \in V$ such that $T(e_i) = f_i$. Let S be the map $f_i \mapsto e_i$ extended by linearity. The composition of linear maps is linear, so $T \circ S = 1$. (Note that S must therefore be injective.) □

Remark The map S is called a **section** of T.

Theorem 1.3 *Let the short exact sequence* (1.9) *be given, and let S be a section of T. Then*

$$V = \ker T \oplus S(W).$$

In particular, $\dim V = \dim \ker T + \dim W$.

Proof Let $v \in V$, and define $w := S(T(v))$ and $u := v - w$. Then $T(u) = T(v - w) = T(v) - T(S(T(v))) = 0$, so $v = u + w$ with $u \in \ker T$ and $w \in S(W)$. Now suppose that $x \in \ker T \cap S(W)$. The map S is injective, so $W \cong S(W)$. In particular, there exists a unique $w \in W$ such that $S(w) = x$. But then $w = T(S(w)) = T(x) = 0$, so, by the linearity of S, $x = 0$. Hence $V = \ker T \oplus S(W)$. By Exercise 1.3, $\dim V = \dim \ker T + \dim S(W) = \dim \ker T + \dim W$. □

Remark The conclusion of Theorem 1.3 is commonly referred to as the **rank–nullity theorem**, which states that the rank plus the nullity of any linear map equals the dimension of the domain.

EXERCISE 1.13 Show that if $T : V \to W$ is an injective linear map between spaces of the same dimension then T must be an isomorphism.

EXERCISE 1.14 Let S and T be two endomorphisms of V. Show that $\mathrm{rk}(ST) \leq \min\{\mathrm{rk}\, S, \mathrm{rk}\, T\}$.

---

[4] It is almost silly to give this map a name, because it really doesn't do anything. The idea is that one can view $\ker T$ as a separate vector space as well as a subspace of V, and the inclusion map just sends every element of the first space to the corresponding element of the second.

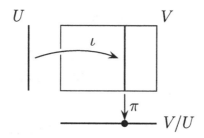

Figure 1.1 The quotient space construction.

1.4 Quotient spaces

Let V be a vector space, and let U be a subspace of V. We define a natural equivalence relation[5] on V by setting $v \sim w$ if $v - w \in U$. The set of equivalence classes
of this equivalence relation is denoted V/U and is called the **quotient space of** V
modulo U. The **(canonical) projection map** is the map $\pi : V \to V/U$ given by
$v \to [v]$, where $[v]$ denotes the equivalence class of v in V/U; v is called a **class
representative** of $[v]$. The canonical quotient space construction is illustrated in
Figure 1.1. The intuition behind the picture is that two distinct points in V that
differ by an element of U (or, if you wish, by an element of the image of U under
the inclusion map ι) get squashed (or projected) to the same point in the quotient.

Theorem 1.4 *If U is a subspace of the vector space V then V/U carries a natural
vector space structure, the projection map is linear, and* $\dim(V/U) = \dim V -
\dim U$.

Proof Let $v, w \in V$. Define $[v] + [w] = [v + w]$. We must show that this is well
defined, or independent of class representative, meaning that we would get the
same answer if we chose different class representatives on the left-hand side. But if
$v' \sim v$ and $w' \sim w$ then $v' - v \in U$ and $w' - w \in U$, so $(v' + w') - (v + w) \in U$,
which means that $[v' + w'] = [v + w]$. Next, define $a[v] = [av]$ for some scalar
a. If $v \sim w$ then $v - w \in U$ so that $a(v - w) \in U$, which means that $[av] =
[aw]$. Thus scalar multiplication is also well defined. Hence, V/U equipped with
these operations is a vector space. Observe that the zero vector of V/U is just
the zero class in V, namely the set of all vectors in U. The definitions imply that
$[av + bw] = a[v] + b[w]$, so the projection map is linear. The statement about the
dimensions follows from Theorem 1.3 because

$$0 \longrightarrow U \overset{\iota}{\longrightarrow} V \overset{\pi}{\longrightarrow} V/U \longrightarrow 0 \qquad (1.10)$$

is exact, as π is surjective with $\ker \pi = U$. $\square$

---

[5] See Appendix A.

Remark   We have followed standard usage and called the map $\pi : V \to V/U$ a projection, but technically it is not a projection in the sense of Exercise 1.9 because, as we have defined it, V/U is not a subspace of V. For this reason, although you may be tempted to do so by part (a) of Exercise 1.9, you cannot write $V = U \oplus V/U$. However, by Theorem 1.2, the exact sequence (1.10) splits, so there is a linear injection $S : V/U \to V$. If we define $W = S(V/U)$ then, by Theorem 1.3, $V = U \oplus W$. Conversely, if $V = U \oplus W$ then $W \cong V/U$ (and $U \cong V/W$).

EXERCISE 1.15   Let U be a subspace of V and let W be a subspace of X. Show that any linear map $\varphi : V \to X$ with $\varphi(U) \subseteq W$ induces a natural linear map $\widetilde{\varphi} : V/U \to X/W$. *Hint:* Set $\widetilde{\varphi}([v]) = [\varphi(v)]$. Show this is well defined.

1.5 Matrix representations

Let V be n-dimensional. Given an endomorphism $T : V \to V$ together with a basis $\{e_i\}$ of V we can construct an $n \times n$ matrix whose entries T_{ij} are given by[6]

$$T e_j = \sum_i e_i T_{ij}. \tag{1.11}$$

One writes (T_{ij}) or $\boldsymbol{T}$ to indicate the matrix whose entries are T_{ij}. The map $T \to \boldsymbol{T}$ is called a **representation** of T (in the basis $\{e_i\}$). A different choice of basis leads to a different matrix, but they both represent the same endomorphism.

Let $v = \sum_i v_i e_i \in V$. Then

$$v' := T v = \sum_j v_j T e_j = \sum_{ij} v_j e_i T_{ij}$$

$$= \sum_i \left(\sum_j T_{ij} v_j \right) e_i = \sum_i v'_i e_i,$$

so the components of v' are related to those of v according to the rule

$$v'_i = \sum_j T_{ij} v_j. \tag{1.12}$$

EXERCISE 1.16   Let S and T be two endomorphisms of V. Show that if $S \to \boldsymbol{S}$ and $T \to \boldsymbol{T}$ then $ST \to \boldsymbol{ST}$, where $ST := S \circ T$, the composition of S and T, and matrix multiplication is defined by

$$(\boldsymbol{ST})_{ij} = \sum_k S_{ik} T_{kj}. \tag{1.13}$$

(This shows why matrix multiplication is defined in the way it is.)

---

[6] Note carefully the placement of the indices.

EXERCISE 1.17 Let $T : V \to V$ be an automorphism represented by T. Show that $T^{-1} \to T^{-1}$, where T^{-1} denotes the inverse matrix.

The **row rank** (respectively, **column rank**) of a matrix T is the maximum number of linearly independent rows (respectively, columns), when they are considered as vectors in $\mathbb{R}^n$. The row rank and column rank are always equal, and they equal the **rank** rk T of T. If rk T equals n then we say that T has **maximal rank**; otherwise it is said to be **rank deficient**.

EXERCISE 1.18 Show that the rank of the endomorphism T equals the rank of the matrix T representing it in any basis.

1.6 The dual space

A **linear functional** on V is a linear map $f : V \to \mathbb{F}$. The set V^* of all linear functionals on V is called the **dual space** (of V), and is often denoted $\mathrm{Hom}(V, \mathbb{R})$. If f is a linear functional and a is a scalar, af is another linear functional, defined by $(af)(v) = af(v)$ (pointwise multiplication). Also, if f and g are two linear functionals then we can obtain a third linear functional $f + g$ by $(f + g)(v) = f(v)+g(v)$ (pointwise addition). These two operations turn V^* into a vector space, and when one speaks of the dual space one always has this vector space structure in mind.

It is customary to write $\langle v, f \rangle$ or $\langle f, v \rangle$ to denote $f(v)$.[7] When written this way it is called the **natural pairing** or **dual pairing** between V and V^*. Elements of V^* are often called **covectors**.

If $\{e_i\}$ is a basis of V, there is a canonical **dual basis** or **cobasis** $\{\theta_j\}$ of V^*, defined by $\langle e_i, \theta_j \rangle = \delta_{ij}$, where δ_{ij} is the **Kronecker delta**:

$$\delta_{ij} := \begin{cases} 1 & \text{if } i = j, \text{ and} \\ 0 & \text{otherwise.} \end{cases} \tag{1.14}$$

Any element $f \in V^*$ can be expanded in terms of the dual basis as

$$f = \sum_i f_i \theta_i, \tag{1.15}$$

where $f_i \in \mathbb{F}$. The scalars f_i are called the **components** of f with respect to the basis $\{\theta_i\}$.

EXERCISE 1.19 Show that $\{\theta_j\}$ is indeed a basis for V^*..

---

[7] Note the angular shape of the brackets.

It follows from Exercise 1.19 that dim $V^* = $ dim V. Because they have the same dimension, V and V^* are isomorphic, but not in any natural way.[8] On the other hand, V and V^{**} (the **double dual** of V) are always isomorphic via the natural map $v \mapsto (f \mapsto f(v))$.[9]

> **EXERCISE 1.20** Show that the dual pairing is **nondegenerate** in the following sense. If $\langle f, v \rangle = 0$ for all v then $f = 0$ and if $\langle f, v \rangle = 0$ for all f then $v = 0$.

> **EXERCISE 1.21** Let W be a subspace of V. The **annihilator** of W, denoted Ann W, is the set of all linear functionals that map every element of W to zero:
>
> $$\text{Ann } W := \{\theta \in V^* : \theta(w) = 0, \text{ for all } w \in W\}. \tag{1.16}$$
>
> Show that Ann W is a subspace of V^* and that every subspace of V^* is Ann W for some W. In the process verify that dim $V = $ dim $W + $ dim Ann W. *Hint:* For the second part, let U^* be a subspace of V^* and define
>
> $$W := \{v \in V : f(v) = 0, \text{ for all } f \in U^*\}.$$
>
> Use this to show that $U^* \subseteq$ Ann W. Equality then follows by a dimension argument.

> **EXERCISE 1.22** Let W be a subspace of V. Then $(V/W)^* \cong V^*/W^*$ just by dimension counting, even though there is no natural isomorphism (because there is no natural isomorphism between a vector space and its dual). But there is an interesting connection to annihilators that gives an indirect relation between the two spaces.
>
> (a) Show that $(V/W)^* \cong$ Ann W. *Hint:* Define a map $\varphi :$ Ann $W \to (V/W)^*$ by $f \mapsto \varphi(f)$, where $\varphi(f)([v]) := f(v)$, and extend by linearity. Show that φ is well defined and an isomorphism.
>
> (b) Show that $V^*/$Ann $W \cong W^*$. *Hint:* Let $\pi : V \to W$ be any projection onto W, and define $\pi^* : V^*/$Ann $W \to W^*$ by $[f] \mapsto f \circ \pi$, so $\pi^*([f])(v) = f(\pi(v))$. Again, show that π^* is well defined and an isomorphism.

1.7 Change of basis

Let $\{e_i\}$ and $\{e'_i\}$ be two bases of V. Each new (primed) basis vector can be written as a linear combination of the old (unprimed) basis vectors, and by convention we write

$$e'_j = \sum_i e_i A_{ij} \tag{1.17}$$

---

[8] The situation is different if V is an inner product space. See Section 1.10.

[9] When V is infinite dimensional, the situation is more subtle. In that case the statements $V \cong V^*$ and $V \cong V^{**}$ are both generally false. A notable exception occurs when V is a Hilbert space and V^* is defined as the continuous linear functionals on V. See e.g. [74].

for some nonsingular matrix $A = (A_{ij})$, called the **change of basis matrix**.[10] By definition, a change of basis leaves all the vectors in V untouched – it merely changes their *description*. Thus, if $v = \sum_i v_i e_i$ is the expansion of v relative to the old basis and $v = \sum_i v'_i e'_i$ is the expansion of v relative to the new basis then

$$\sum_j v'_j e'_j = \sum_{ij} v'_j e_i A_{ij} = \sum_i \left(\sum_j A_{ij} v'_j \right) e_i = \sum_i v_i e_i.$$

Hence,

$$v_i = \sum_j A_{ij} v'_j \quad \text{or} \quad v'_i = \sum_j (A^{-1})_{ij} v_j. \tag{1.18}$$

Note that the basis vectors and the components of vectors transform differently under a change of basis.

A change of basis on V induces a change of basis on V^* owing to the requirement that the natural dual pairing be preserved. Thus, if $\{\theta_i\}$ and $\{\theta'_i\}$ are the dual bases corresponding to $\{e_i\}$ and $\{e'_i\}$, respectively, then we demand that

$$\langle e'_i, \theta'_j \rangle = \langle e_i, \theta_j \rangle = \delta_{ij}. \tag{1.19}$$

Writing

$$\theta'_j = \sum_i \theta_i B_{ij} \tag{1.20}$$

for some matrix $B = (B_{ij})$ and using (1.19) gives[11]

$$\delta_{ij} = \langle e'_i, \theta'_j \rangle = \sum_{k\ell} \langle e_k A_{ki}, \theta_\ell B_{\ell j} \rangle = \sum_{k\ell} A_{ki} B_{\ell j} \langle e_k, \theta_\ell \rangle$$

$$= \sum_{k\ell} A_{ki} B_{\ell j} \delta_{k\ell} = \sum_k A_{ki} B_{kj}. \tag{1.21}$$

Writing A^T for the **transpose** matrix, whose entries are $(A^T)_{ij} := A_{ji}$, (1.21) can be written compactly as $A^T B = I$, where I is the identity matrix. Equivalently, we have $B = (A^T)^{-1} = (A^{-1})^T$, the so-called **contragredient matrix** of A.

---

[10] Equation (1.17) looks a lot like (1.11), so much so that one can define a linear map $A : V \to V$, called the **change of basis map**, given by $e_i \mapsto e'_i$ and extended by linearity. This map can be a bit confusing, though, for the following reason. In general, a map $T : V \to W$ is represented by a matrix (T_{ij}) obtained from $T e_i = \sum_j T_{ji} f_j$, where $\{e_i\}$ is a basis for V and $\{f_i\}$ is a basis for W. Viewing A as a map from V equipped with the basis $\{e_i\}$ to V equipped with the basis $\{e'_i\}$, the map A is represented by the identity matrix even though it is not the identity map. Instead, the change of basis matrix is what we obtain if we use the same basis $\{e_i\}$ for both copies of V. To avoid these sorts of confusion we shall speak only of the change of basis matrix when discussing basis changes.

[11] Sometimes the last step in (1.21) gives beginners a little trouble. We are using the substitution property of the Kronecker delta. For example, $\sum_\ell v_\ell \delta_{\ell m} = v_m$, because the Kronecker delta vanishes unless $\ell = m$. To be really explicit, if, say, $m = 2$ and the indices run from 1 to 3, then $\sum_\ell v_\ell \delta_{\ell 2} = v_1 \delta_{12} + v_2 \delta_{22} + v_3 \delta_{32} = v_2$, because $\delta_{12} = \delta_{32} = 0$ and $\delta_{22} = 1$.

If $f \in V^*$ is any covector then under a change of basis we have

$$f' = \sum_j f'_j \theta'_j = \sum_{ij} f'_j \theta_i B_{ij}$$

$$= \sum_i \left(\sum_j B_{ij} f'_j \right) \theta_i = \sum_i f_i \theta_i = f,$$

from which we conclude that

$$f_i = \sum_j B_{ij} f'_j \quad \text{or} \quad f'_i = \sum_j (B^{-1})_{ij} f_j. \tag{1.22}$$

It is instructive to rewrite (1.20) and (1.22) in terms of the matrix A instead of the matrix B. Thus, (1.20) becomes

$$\theta'_i = \sum_j \theta_j B_{ji} = \sum_j (B^T)_{ij} \theta_j = \sum_j (A^{-1})_{ij} \theta_j, \tag{1.23}$$

while (1.22) becomes

$$f'_j = \sum_i (B^{-1})_{ji} f_i = \sum_i (A^T)_{ji} f_i = \sum_i f_i A_{ij}. \tag{1.24}$$

1.8 Upstairs or downstairs?

At this point we encounter a serious ambiguity in the notation. How can you tell, just by looking, that the numbers a_i are the components of a vector or covector (or neither)? The answer is that you cannot. You must know from the context. Of course, there is no problem as long as the numbers are accompanied by the corresponding basis elements e_i or θ_i. In many applications, however, the basis elements are omitted for brevity.

This ambiguity can be partially ameliorated by introducing a new notational convention that, while convenient, requires a sharp eye for detail: one changes the position of the index i. The logic behind this is the following. If one compares (1.17) with (1.24) and (1.18) with (1.23) one sees something very interesting. Under a change of basis, the components of a covector transform like the basis vectors whereas the components of a vector transform like the basis covectors. We say that the components of a covector transform **covariantly** ("with the basis vectors"), whereas the components of a vector transform **contravariantly** ("against the basis vectors").

By convention, we continue to write e_i with a lower index for the basis vectors but we use a raised index to denote the basis covectors. Thus, we write θ^i instead of θ_i. Consistency with the aforementioned transformation properties then requires that vector components be written with upstairs (contravariant) indices and covector components be written with downstairs (covariant) indices. We therefore write

$$v = \sum_i v^i e_i \tag{1.25}$$

instead of (1.5) and

$$f = \sum_i f_i \theta^i \tag{1.26}$$

instead of (1.15).

This new notation has the advantage that the transformation properties of the object under discussion are clear by inspection.[12] For example, you notice that in (1.25) and (1.26) the indices are paired one up and one down, then summed. Whenever this occurs, we say the indices have been **contracted**. Any expression in which all the indices are contracted pairwise is necessarily invariant under a change of basis. So, for example, we see immediately that both v and f are invariant. Moreover, the positions of the indices acts as an aid in computations, for one can only equate objects with the same transformation properties. One *can* write an expression such as $a_i = b^i$ to indicate an equality of numbers, but neither side could be the components of a vector or covector because the two sides have different transformation properties under a change of basis. Consequently, such expressions are to be avoided.

Under this new convention, many of the previous formulae are modified. For instance, the natural pairing between vectors and covectors becomes

$$\langle e_i, \theta^j \rangle = \delta_i^j, \tag{1.27}$$

where δ_i^j is just another way to write the Kronecker symbol:

$$\delta_i^j := \begin{cases} 1 & \text{if } i = j, \quad \text{and} \\ 0 & \text{otherwise.} \end{cases} \tag{1.28}$$

Similarly, the change of basis formulae (1.17) and (1.18) are replaced by

$$e'_j = \sum_i e_i A^i{}_j, \tag{1.29}$$

and

$$v'^i = \sum_j (A^{-1})^i{}_j v^j, \tag{1.30}$$

---

[12] This becomes especially important when dealing with tensors, which have many more indices.

respectively, where the first (row) index on the matrix is up and the second (column) index is down.

The problem with this notation is that it is sometimes unnecessarily pedantic, and can also lead to awkward formulas. For example, with this same convention, (1.20) and (1.22) must be written

$$\theta'^j = \sum_i \theta^i B_i{}^j, \tag{1.31}$$

and

$$f_i' = \sum_j (B^{-1})_i{}^j f_j, \tag{1.32}$$

respectively. In both these formulae, the first index on the matrix is down and the second is up. Unless you are very careful about writing such a formula, you can easily get the wrong result if you place the indices in wrong locations. This can be avoided by using (1.23) instead of (1.20) and (1.24) instead of (1.22) and then raising the appropriate indices. It can also be avoided by adopting a convenient matrix notation, which we will have occasion to use later. Write $A = (A^i{}_j)$ and introduce the shorthand notation

$$e = (e_1 \ e_2 \ \cdots \ e_n), \qquad \theta = \begin{pmatrix} \theta^1 \\ \theta^2 \\ \vdots \\ \theta^n \end{pmatrix}, \tag{1.33}$$

$$v = \begin{pmatrix} v^1 \\ v^2 \\ \vdots \\ v^n \end{pmatrix}, \qquad f = (f_1 \ f_2 \ \cdots \ f_n). \tag{1.34}$$

Then (1.17), (1.18), (1.23), and (1.24) can be written

$$e' = eA, \tag{1.35}$$
$$v' = A^{-1}v, \tag{1.36}$$
$$\theta' = A^{-1}\theta, \tag{1.37}$$

and

$$f' = fA, \tag{1.38}$$

respectively. This notation makes the invariance of v and f under a change of basis easy to see. For example, $v' = e'v' = eAA^{-1}v = ev = v.$

Although there are times when it is necessary to write all the indices in their proper places, there are many times when it is not. Unless we are specifically concerned with the transformation properties of the quantities in an equation, it is not necessary to worry too much about index placement. Consequently, although we will mostly adhere to this notational convention, there will be occasions when we will not. Nonadherence will cause no problems, though, because precisely in those instances the transformation properties of the various intermediate quantities will be irrelevant to the final result.

1.9 Inner product spaces

So far everything we have said about vector spaces and linear maps works in essentially the same way regardless of the underlying field $\mathbb{F}$, but there are are a few places where the underlying field matters and here we consider one of them. For simplicity we mostly restrict ourselves to real fields in this book, but in the next three sections we discuss complex fields explicitly in order to save time later. The advantage of working over the complex numbers is that the real numbers are included in the complex numbers, so expressions valid for the complex numbers can usually be specialized to the real numbers without any additional effort.

Thus, let $\mathbb{F}$ be a subfield of $\mathbb{C}$, and let V be a vector space over $\mathbb{F}$. A **sesquilinear form** on V is a map $g : V \times V \to \mathbb{F}$ satisfying the following two properties. For all $u, v, w \in V$ and $a, b \in \mathbb{F}$, the map g is

(1) **linear on the second entry**: $g(u, av + bw) = ag(u, v) + bg(u, w)$, and
(2) **Hermitian**: $g(v, u) = \overline{g(u, v)}$,

where $\bar{a}$ is the complex conjugate of a. Note that these two properties together imply that g is **antilinear** on the first entry:

$$g(au + bv, w) = \bar{a}g(u, w) + \bar{b}g(v, w).$$

However, if $\mathbb{F}$ is a real field (a subfield of $\mathbb{R}$) then $\bar{a} = a$ and $\bar{b} = b$ and the above condition condition just says that g is linear on the first entry as well. In that case we say that g is a **bilinear form**. Moreover, the Hermiticity condition becomes the symmetry condition $g(u, v) = g(v, u)$, so a real sesquilinear form is in fact a **symmetric bilinear form**.[13] If the sesquilinear form g is

(3) **nondegenerate**, so that $g(u, v) = 0$ for all v implies $u = 0$,

then it is called an **inner product**. A vector space equipped with an inner product is called an **inner product space**.

---

[13] The prefix "sesqui" means "one and a half", which is appropriate because a sesquilinear form is in some sense halfway between linear and bilinear.

By Hermiticity $g(u, u)$ is always a real number. We may thus distinguish some important subclasses of inner products. If $g(u, u) \geq 0$ (respectively, $g(u, u) \leq 0$) then g is **nonnegative definite** (respectively, **nonpositive definite**). A nonnegative definite (respectively, nonpositive definite) inner product satisfying the condition that

$$g(u, u) = 0 \qquad \text{implies} \qquad u = 0$$

is **positive definite** (respectively, **negative definite**). A positive definite or negative definite inner product is always nondegenerate but the converse is not true, as you can verify by doing Exercise 1.23.

Example 1.2 The standard example of a complex vector space (a vector space over the complex numbers) is $\mathbb{C}^n$, the Cartesian product of $\mathbb{C}$ with itself n times: $\mathbb{C} \times \cdots \times \mathbb{C}$. A vector in $\mathbb{C}^n$ is just an n-tuple $(a_1, \ldots, a_n)$ of complex numbers, with scalar multiplication and vector addition defined componentwise as in the real case. If $u = (u_1, \ldots, u_n)$ and $v = (v_1, \ldots, v_n)$ then the standard inner product is given by

$$g(u, v) := \sum_{i=1}^{n} \overline{u_i}\, v_i.$$

Example 1.3 (The **Euclidean inner product** or **dot product** on $\mathbb{R}^n$) If $u = (u_1, u_2, \ldots, u_n)$ and $v = (v_1, v_2, \ldots, v_n)$ then we may define

$$g(u, v) := \sum_{i=1}^{n} u_i v_i. \tag{1.39}$$

The vector space $\mathbb{R}^n$ equipped with the inner product (1.39) is usually denoted $\mathbb{E}^n$ and is called **Euclidean space**. People often fail to distinguish between $\mathbb{R}^n$ and $\mathbb{E}^n$, because the Euclidean inner product is the usual default. But, as the next example shows, we ought not be quite so cavalier.

Example 1.4 (The **Lorentzian inner product** on $\mathbb{R}^n$) Let $u = (u_0, u_1, \ldots, u_{n-1})$ and $v = (v_0, v_1, \ldots, v_{n-1})$, and define

$$g(u, v) := -u_0 v_0 + \sum_{i=1}^{n-1} u_i v_i. \tag{1.40}$$

The vector space $\mathbb{R}^n$ equipped with the inner product (1.40) is usually denoted $\mathbb{M}^n$ and is called **Minkowski space** (or, more properly, **Minkowski spacetime**). The inner product space $\mathbb{M}^4$ is the domain of Einstein's special theory of relativity.

EXERCISE 1.23 Verify that the standard inner product on $\mathbb{C}^n$ and the Euclidean and Lorentzian inner products are indeed inner products, but that only the first two are positive definite. (The Lorentzian inner product is **indefinite** because $g(v, v)$

may be positive, zero, or negative.) Note that a positive definite inner product is automatically nondegenerate (just take $v = u$ in the definition); this exercise shows that the converse is false.

A set $\{v_i\}$ of vectors is **orthogonal** if

$$g(v_i, v_j) = 0 \qquad \text{whenever } i \neq j, \tag{1.41}$$

and it is **orthonormal** if

$$g(v_i, v_j) = \pm\delta_{ij}. \tag{1.42}$$

A vector v satisfying $g(v, v) = \pm 1$ is called a **unit vector**.

> **Example 1.5**   The standard basis vectors $\{e_i = (0, \ldots, 1, \ldots, 0)\}$ of $\mathbb{R}^n$ form an orthonormal basis for $\mathbb{E}^n$ as well as $\mathbb{M}^n$.

Not only do orthonormal sets of vectors exist, but we can always find enough of them to span the space, as the next result shows.

Theorem 1.5   *Every inner product space has an orthonormal basis.*

We offer two proofs of this fact, because both are instructive and interesting. The first is more "abstract", using only the tools introduced so far, while the second is more "algebraic" and uses the idea of diagonalizability.

First proof of Theorem 1.5   We use induction on $k = \dim V$. If $\dim V = 1$ then there must be at least one nonzero vector $v \in V$ such that $g(v, v) \neq 0$. Otherwise, for all $v, w \in V$,

$$0 = g(v + w, v + w) = g(v, v) + 2g(v, w) + g(w, w) = 2g(v, w), \tag{1.43}$$

which contradicts the nondegeneracy condition. So $v/\sqrt{|g(v, v)|}$ is an orthonormal basis for V.[14] Now suppose that we have found $k - 1$ orthonormal vectors $e_1, e_2, \ldots, e_{k-1}$ in V. They span a $(k - 1)$-dimensional subspace W. Let $\pi : V \to W$ be the orthogonal projection map given by

$$\pi(v) = \sum_{i=1}^{k-1} g(e_i, v)e_i. \tag{1.44}$$

It is linear by construction and surjective because $\pi(w) = w$ for all $w \in W$. Thus we get an exact sequence

$$0 \longrightarrow W^\perp \longrightarrow V \overset{\pi}{\longrightarrow} W \longrightarrow 0,$$

---

[14] The absolute value of x is denoted $|x|$.

where

$$W^\perp = \ker \pi = \{v \in V : g(v, w) = 0 \text{ for all } w \in W\} \qquad (1.45)$$

is the **orthogonal complement** of W. By Theorem 1.3, $V = W \oplus W^\perp$ (where we have identified W with its image under the inclusion map, which is a section of π), so every $v \in V$ can be written uniquely as $v = w + w^\perp$ for some $w \in W$ and $w^\perp \in W^\perp$. The inner product on V induces an inner product on $W^\perp$. We need only show that this induced inner product is nondegenerate, for then by the above argument there is a unit vector in $W^\perp$ which by construction is orthogonal to all the e_i. Let $w' \in W^\perp$ be fixed. Suppose $g(w', w^\perp) = 0$ for all $w^\perp \in W^\perp$. Then, for any $v \in V$, $g(w', v) = g(w', w + w^\perp) = g(w', w) = 0$, so we must have $w' = 0$. $\qquad\square$

Below we offer a second proof of Theorem 1.5, which is valid only for real fields, in order to define the concept of the signature of a real inner product. Recall that the **determinant** of a matrix (A_{ij}) is defined by

$$\det A := \sum_{\sigma \in \mathfrak{S}_n} (-1)^\sigma A_{1\sigma(1)} A_{2\sigma(2)} \cdots A_{n\sigma(n)}, \qquad (1.46)$$

where $\mathfrak{S}_n$ is the set of all permutations of n elements and $(-1)^\sigma$ denotes the sign of the permutation σ.[15] Given a real symmetric bilinear form g and a set $\{v_1, \ldots, v_n\}$ of vectors we can construct a real symmetric $n \times n$ matrix $G := (g(v_i, v_j))$, called the **Gram matrix** of the set. The determinant of this matrix is called a Gram determinant, or **Grammian**, of the set of vectors.

The following lemma will come in handy.

Lemma 1.6 *Let V be a real vector space with basis $\{e_i\}$, and let g be a real symmetric bilinear form. Let $g_{ij} := g(e_i, e_j)$ and set $G = (g_{ij})$. Then g is an inner product if and only if the Grammian $\det G \neq 0$.*

Proof Let $v = \sum_i a_i e_i$, and suppose that $g(v, w) = 0$ for all w. Then, for all j,

$$0 = \sum_i a_i g(e_i, e_j) = \sum_i a_i g_{ij}. \qquad (1.47)$$

If $\deg(g_{ij}) = 0$ then these equations have a nontrivial solution for the a_i, whereas if $\det(g_{ij}) \neq 0$ then all the a_i vanish, so $v = 0$. (See Exercise 1.34.) $\qquad\square$

Second proof of Theorem 1.5[16] Pick a basis $\{e_i\}_{i=1}^n$ for V. The Gram matrix $G = (g(e_i, e_j))$ is a real symmetric matrix. By the spectral theorem of linear algebra (see Appendix B), there exists an orthogonal matrix R such that

---

[15] For the reader's convenience, some basic facts about permutations are collected in Appendix A. For more about determinants, see the additional exercises at the end of this chapter as well as Section 2.8.

[16] For the terminology used herein see the additional exercises at the end of this chapter.

$\Sigma := R^{-1}GR = R^T GR$ is a diagonal matrix. (The diagonal elements $\lambda_i := \Sigma_{ii}$ of Σ are the eigenvalues of G and the columns of R are the eigenvectors of G.) By Lemma 1.6 all the eigenvalues are nonzero (because the determinant is the product of its eigenvalues). Define a new diagonal matrix B with $B_{ii} = |\lambda_i|^{-1/2}$. Then $D := B^T \Sigma B$ is a diagonal matrix with $D_{ii} = \pm 1$. Tracing back the definitions gives $D = M^T GM$, where $M := RB$. Define vectors $v_i := \sum_k M_{ki} e_k$. Then

$$g(v_i, v_j) = \sum_{k,\ell} M_{ki} M_{\ell j} g(e_k, e_\ell) = (M^T GM)_{ij} = D_{ij}.$$

It follows that $v_1, \ldots, v_n$ are the orthonormal vectors we seek. $\qquad\square$

By reordering the basis elements, if necessary, we may write the diagonal matrix appearing above as

$$D = \mathrm{diag}(\underbrace{+1, \ldots, +1}_{p}, \underbrace{-1, \ldots, -1}_{q}),$$

where there are p plus signs and q minus signs. The **signature** of the inner product g may refer to any of the following: (1) the pair (p, q), (2) the number $p - q$, or (3) the signs of the diagonal elements themselves.[17] The numbers p and q depend only on the real bilinear form g and not on the basis used to define the Gram matrix G, a result known as **Sylvester's law of inertia**.[18]

Example 1.6 Let $\{e_i\}$ be the standard basis of $\mathbb{R}^n$. Then the Gram matrix elements of the Euclidean inner product are given by

$$g_{ij} = g(e_i, e_j) = \delta_{ij}, \tag{1.48}$$

so G is already diagonal, with ones along the diagonal and zeros elsewhere. Hence this inner product has signature n or $(+ + \cdots +)$. Similarly, in that same basis, the Gram matrix elements of the Lorentzian inner product are

$$g_{ij} = g(e_i, e_j) = \eta_{ij}, \tag{1.49}$$

---

[17] Definitions (1) and (2) are sometimes preferred, because there is no material difference between, say, $(- + + \cdots +)$ and $(+ - + \cdots +)$. Indeed, only $|p - q|$ is really of interest, because, for example, inner product spaces with signature $(- + + \cdots +)$ are essentially equivalent to spaces with signature $(+ - - \cdots -)$.

[18] The proof of Sylvester's law is rather subtle. The second proof of Theorem 1.5 shows that under a change of basis effected by a matrix A, the Gram matrix G changes by a **congruence**:

$$G \quad \rightarrow \quad G' = A^T GA.$$

But a congruence does not generally preserve eigenvalues, so the spectra of G' and G are not necessarily the same. (In contradistinction, a similarity transformation $G' = A^{-1}GA$ *does* preserve the spectrum. See Exercise 1.43.) The content of Sylvester's law is that the *signs* of the eigenvalues are preserved by a congruence. For a proof see [28], Chapter X, §2, Theorem 1 or [31], Section 9.9.

where

$$
\eta_{ij} := \begin{cases} -1 & \text{if } i = j = 0, \\ +1 & \text{if } i = j \neq 0, \text{ and} \\ 0 & \text{otherwise.} \end{cases} \tag{1.50}
$$

Hence, the Lorentzian inner product has signature $n - 2$ or $(-++\cdots+)$.

1.10 The Riesz lemma

In Section 1.6 we mentioned that, in general, there is no natural isomorphism between a vector space V and its dual space V^*. Things are different if there is an inner product around. In that case there is a natural isomorphism $\psi : V \to V^*$, and this is the content of the Riesz lemma. Actually, things are more subtle than that, because there is a difference between the real and the complex case. The real case is the most important for our purposes, but we need to discuss the complex case as well in order to understand adjoint maps.

Define a map $\psi : V \to V^*$ by $v \mapsto f_v$, where

$$
f_v(\cdot) := g(v, \cdot). \tag{1.51}
$$

Equivalently, for all $w \in V$, $f_v(w) = g(v, w)$.

Lemma 1.7 (Riesz) *If $\mathbb{F} = \mathbb{R}$ then the map ψ is an isomorphism, whereas if $\mathbb{F} = \mathbb{C}$ then the map ψ is an antiisomorphism.*[19]

Proof[20] Assume the field is complex. First observe that ψ really does map to V^*, because g is always linear on the second entry. Next we have

$$
\begin{aligned}
\psi(au + bv)(w) &= g(au + bv, w) \\
&= \bar{a}g(u, w) + \bar{b}g(v, w) \\
&= (\bar{a}\psi(u) + \bar{b}\psi(v))(w),
\end{aligned}
$$

so ψ is antilinear in general. Clearly it becomes linear when the field is real. By nondegeneracy of the inner product if follows that $\ker \psi = 0$, so ψ is injective. But V and V^* have the same dimension, so by the rank–nullity theorem, ψ is also surjective. □

[19] An **antiisomorphism** between two complex vector spaces – sometimes called a **conjugate linear isomorphism** – is just an antilinear bijection.

[20] We only show this in finite dimensions, where the proof is easy. Riesz's actual lemma is valid for Hilbert spaces, which are a special class of infinite-dimensional vector spaces. For another proof of Lemma 1.7, see Exercise 1.54.

1.11 Adjoint maps, transpose maps, and duality

Let V and W be two vector spaces. If $A : V \to W$ is a linear map, there is an induced linear map of dual spaces $A^* : W^* \to V^*$, given by

$$(A^* f)(v) = f(Av) \tag{1.52}$$

or, equivalently,

$$\langle A^* f, v \rangle = \langle f, Av \rangle \tag{1.53}$$

for all $f \in W^*$ and $v \in V$. (The dual pairing on the left connects V^* and V, while that on the right connects W^* and W.) The map A^* has many names in the literature: "adjoint", "transpose", "pullback", etc. Unfortunately, all these words have other meanings as well. In this section we'll discuss the reason for the first two names (one of which is not really correct), while the third name is discussed in the context of more general maps in Section 3.14. If we need to refer to A^*, we'll simply call it the **dual map** of A.

> **EXERCISE 1.24**   Show that A^* is linear.

For simplicity let us now restrict our attention to the case $V = W$.[21] The next exercise explains why the name "transpose" is sometimes used for the dual map (and also shows that A^* is uniquely defined by A).

> **EXERCISE 1.25**   Let $\{e_i\}$ be a basis for V with corresponding dual basis $\{e_i^*\}$. Show that if A is represented by the matrix A relative to $\{e_i\}$ then A^* is represented by the transpose matrix A^T relative to $\{e_i^*\}$.

To explain the name "adjoint" we suppose that V is equipped with an inner product g. In that case, the map $A : V \to V$ induces another map $A^\dagger : V \to V$, called the **adjoint** of A, by the requirement that

$$g(A^\dagger w, v) = g(w, Av), \tag{1.54}$$

for all $v, w \in V$.

> **EXERCISE 1.26**   Given a matrix A the **adjoint matrix** $A^\dagger$ is defined to be the conjugate transpose matrix $\overline{A}^T$, namely the matrix obtained by conjugating every entry and taking the transpose. Show that if the linear map A is represented by the matrix A in some orthonormal basis $\{e_i\}$ then its adjoint map $A^\dagger$ is represented by the adjoint matrix $A^\dagger$ with respect to the same basis.

---

[21] The following results generalize easily to the case in which $V \neq W$, but to do so we would have to keep track of four bases rather than just two.

Now here is the point. When the underlying field is real, conjugation is trivial and the adjoint map is represented by the transpose matrix (assuming an orthonormal basis). It is for this reason that some authors refer to the dual map as the adjoint map. But in general they are different, so you should really avoid using the same terminology for both.

Additional exercises

1.27 Linear independence Let $\{v_1, \ldots, v_n\}$ be a linearly independent set of vectors. Show that the vectors $v_1, \ldots, v_n, w$ are linearly dependent if and only if w is a linear combination of $v_1, \ldots, v_n$.

1.28 Vector spaces of polynomials Let V be the vector space of all polynomials in one variable x of degree at most 3, where the vector space operations are defined pointwise: $(f + g)(x) = f(x) + g(x)$ and $(af)(x) = af(x)$. Show that the monomials $1, x, x^2$, and x^3 form a basis for V.

1.29 The cokernel Suppose that $T : V \to W$ is linear. The kernel of T is the set of all vectors in V that are sent to zero by T. It measures the failure of T to be injective. The failure of T to be surjective is measured by the **cokernel** of T. But the cokernel is not just the set of vectors in W that are not in the image of T. Instead, coker T is defined to be the quotient space $W/\operatorname{im} T$. Show that this is in some sense "natural", because it makes the sequence

$$0 \longrightarrow \ker T \longrightarrow V \overset{T}{\longrightarrow} W \longrightarrow \operatorname{coker} T \longrightarrow 0 \quad (1.55)$$

exact.

1.30 An alternating sum If

$$0 \longrightarrow V_0 \longrightarrow V_1 \longrightarrow \cdots \longrightarrow V_n \longrightarrow 0$$

is an exact sequence of vector spaces, show that

$$\sum_{i=0}^{n} (-1)^i \dim V_i = 0.$$

Hint: Use exactness and the rank–nullity theorem.

1.31 Determinant of the transpose Starting from the definition (1.46), show that, for any matrix A,

$$\det A = \det A^T. \quad (1.56)$$

Hint: The sign of a permutation equals that of its inverse.

1.32 The Laplace expansion of the determinant Let $A = (A_{ij})$ and let $A(i|j)$ denote the matrix obtained from A by deleting the ith row and jth column of A. The **minor** of the element A_{ij} is $\det A(i|j)$. The **signed minor** or **cofactor** of A_{ij} is $\widetilde{A}_{ij} := (-1)^{i+j} \det A(i|j)$. Prove the **Laplace expansion** formula for the determinant: that is, for any i,

$$\det A = \sum_{j=1}^{n} A_{ij} \widetilde{A}_{ij} \qquad (1.57)$$

(the Laplace expansion by the ith row) or, for any j,

$$\det A = \sum_{i=1}^{n} A_{ij} \widetilde{A}_{ij} \qquad (1.58)$$

(the Laplace expansion by the jth column). *Hint:* Starting from (1.46), show that the coefficient of A_{11} in $\det A$ is just $\det A(1|1) = \widetilde{A}_{11}$. By executing row and column flips, show that the coefficient of A_{ij} in $\det A$ is $\widetilde{A}_{ij}$. Here you need to use the fact, easily proved from the definition of $\det A$, that switching any two rows or columns of a matrix flips the sign of its determinant. Next, consider the sum (1.57) with $i = 1$, say, and show that it contains each term of $\det A$ once and only once.

1.33 The adjugate and the inverse The transpose of the matrix of cofactors is called the **adjugate** of A, written adj A. Thus, $(\text{adj } A)_{ij} = \widetilde{A}_{ji}$. Show that

$$A(\text{adj } A) = (\text{adj } A)A = (\det A)I, \qquad (1.59)$$

where I is the identity matrix. *Hint:* We have

$$[A(\text{adj } A)]_{ik} = \sum_{j=1}^{n} A_{ij}(\text{adj } A)_{jk} = \sum_{j=1}^{n} A_{ij} \widetilde{A}_{kj}. \qquad (1.60)$$

Now think about the two cases $i = k$ and $i \neq k$.

1.34 Invertibility and the determinant A matrix A is **singular** if $\det A = 0$ and **nonsingular** otherwise. Show that A is invertible if and only if it is nonsingular, and thereby show that $Av = 0$ has a nontrivial solution for v if and only if $\det A = 0$. *Big hint:* Use (1.59).

1.35 Cramer's rule Let $A = (A_{ij})$ be a nonsingular matrix, let $x = (x_1, \ldots, x_n)^T$ be a vector of indeterminates, let b be a fixed vector, and consider the simultaneous set of linear equations $Ax = b$. Prove **Cramer's rule**, namely that $x_i = \det A^{(i)} / \det A$, where $A^{(i)}$ is the matrix A with the ith column replaced by the vector b. *Hint:* Use the adjugate.

1.36 Derivative of the determinant Viewing the individual entries of the matrix A as independent variables and the determinant as a function of those variables, show that

$$\frac{\partial}{\partial A_{ij}}(\det A) = \tilde{A}_{ij}, \qquad (1.61)$$

where A_{ij} is the cofactor of A_{ij} in det A. Show that, when A is invertible,

$$\frac{\partial}{\partial A_{ij}}(\det A) = (\det A)(A^{-1})_{ji}. \qquad (1.62)$$

Hint: Specialize to the case $i = 1$, $j = 2$, say, and consider the Laplace expansion by the first row; then generalize. You may want to use (1.59).

1.37 Nonsingular matrices and rank Let $T : V \to V$ be an automorphism and T its matrix representation relative to some basis. Show that:
(a) T has maximal rank;
(b) T is nonsingular.

1.38 Orthogonal transformations and the orthogonal group In the following exercises, V is a real vector space with a positive definite inner product g. An **orthogonal transformation** is a linear map $R : V \to V$ that preserves the inner product:

$$g(Rv, Rw) = g(v, w).$$

(a) Show that R maps an orthonormal basis to an orthonormal basis and that every linear map carrying one orthonormal basis to another must be orthogonal.
(b) Let $\{e_i\}$ be an orthonormal basis for V, and let R represent the orthogonal transformation R in this basis. The matrix R is **orthogonal** if $R^T R = RR^T = I$. Show that the linear transformation R is orthogonal if and only if the matrix R is orthogonal.
(c) Let R be an orthogonal matrix. Show that det $R = \pm 1$. *Hint:* Use (2.54). *Remark:* If det $R = 1$, R is a **rotation** matrix; otherwise it is a **roto-reflection** matrix.
(d) The set of all orthogonal transformations on $\mathbb{R}^n$ that preserve the Euclidean inner product is denoted $O(n)$ and called the **orthogonal group**. Show that $O(n)$ is ($\frac{1}{2}n(n-1)$)-dimensional, in the sense that it requires this many parameters to specify an element of $O(n)$. *Hint:* How many conditions must an $n \times n$ orthogonal matrix satisfy?

1.39 The determinant of the inverse is the inverse of the determinant Show that det $A^{-1} = (\det A)^{-1}$. *Hint:* Use (2.54).

1.40 Similarity transformations Let e and e' be two bases for V, with $e' = eS$ (so that S is the change of basis matrix). Let $A : V \to V$ be a linear operator, represented by A relative to e and by A' relative to e'. Show that $A' = S^{-1}AS$. We say that A and A' are related by a **similarity transformation**.

1.41 Diagonalizability Let A be an $n \times n$ matrix and v an $n \times 1$ matrix (i.e., a column vector or column matrix). If $Av = \lambda v$ for some scalar λ then v is an **eigenvector** of A with **eigenvalue** λ. If there exists a nonsingular matrix S such that $\Sigma := S^{-1}AS$ is a diagonal matrix (so that $\Sigma_{ij} = 0$ if $i \neq j$) we say that A is **diagonalizable** (by a similarity transformation). Show that A is diagonalizable if and only if it has n linearly independent eigenvectors. *Hint:* Let the columns of S be the eigenvectors of A.

1.42 Eigenvalues and the characteristic polynomial Show that the eigenvalues of a matrix A are the roots of its **characteristic polynomial** $p_A(\lambda) := \det(A - \lambda I)$. *Remark:* Although you can calculate the eigenvalues in this way it does not mean that the matrix is diagonalizable. The story is a bit involved, but worth recalling. The characteristic polynomial $p_A(\lambda)$ is a polynomial of degree n. By the fundamental theorem of algebra, it always has n (complex) roots. If μ is a root of $p_A(\lambda)$ then $\lambda - \mu$ divides it. The **algebraic multiplicity** a_μ of the eigenvalue μ is the maximum degree of this factor in the characteristic polynomial. The **geometric multiplicity** g_μ is the dimension of the eigenspace corresponding to μ (namely the number of linearly independent eigenvectors with eigenvalue μ). One always has $g_\mu \leq a_\mu$. The matrix A is diagonalizable if and only if $a_\mu = g_\mu$ for all eigenvalues μ; then the eigenvectors of A span the vector space.

1.43 Eigenvalues are similarity invariants If $f(A) = f(S^{-1}AS)$ for some function f we say that f is a **similarity invariant**. Show that the eigenvalues of A are similarity invariants. *Hint:* Combine the result of Exercise 1.42 with (2.54).

1.44 The trace and determinant as sums and products of eigenvalues The **trace** of a matrix A is the sum of its diagonal elements: $\operatorname{tr} A := \sum_i A_{ii}$. Show that $\det A$ and $\operatorname{tr} A$ are respectively the product and the sum of the eigenvalues of A. It follows from Exercise 1.43 that the determinant and the trace are similarity invariants.

1.45 Cyclic invariance of the trace Show that the trace is invariant under cyclic permutations. That is, show that, for any matrices $\{A_1, A_2, \ldots, A_n\}$,

$$\operatorname{tr} A_1 A_2 \cdots A_n = \operatorname{tr} A_n A_1 A_2 \cdots A_{n-1}. \tag{1.63}$$

Hint: First try the case $n = 2$.

1.46 Expansion of the determinant in terms of traces The **elementary symmetric functions** are defined by

$$e_k(x_1, \ldots, x_n) := \sum_{1 \le i_1 < \cdots < i_k \le n} x_{i_1} \cdots x_{i_k}, \tag{1.64}$$

for $k \ge 1$; we also set $e_0 = 1$. For example, when $n = 3$ we have

$$e_1(x_1, x_2, x_3) = x_1 + x_2 + x_3,$$
$$e_2(x_1, x_2, x_3) = x_1 x_2 + x_1 x_3 + x_2 x_3,$$
$$e_3(x_1, x_2, x_3) = x_1 x_2 x_3.$$

The **power sum symmetric functions** are defined by

$$p_k(x_1, \ldots, x_n) := \sum_{i=1}^{n} x_i^k, \tag{1.65}$$

for $k \ge 1$; we also set $p_0 := 1$. For example, when $n = 3$ we have

$$p_1(x_1, x_2, x_3) = x_1 + x_2 + x_3,$$
$$p_2(x_1, x_2, x_3) = x_1^2 + x_2^2 + x_3^2,$$
$$p_3(x_1, x_2, x_3) = x_1^3 + x_2^3 + x_3^3.$$

It is often convenient to allow an infinite number of variables, in which case we write

$$e_k := \sum_{1 \le i_1 < \cdots < i_k} x_{i_1} \cdots x_{i_k} \tag{1.66}$$

and

$$p_k := \sum_{i \ge 1} x_i^k. \tag{1.67}$$

To recover $e_k(x_1, \ldots, x_n)$ and $p_k(x_1, \ldots, x_n)$ from e_k and p_k just set $x_{n+1} = x_{n+2} = \cdots = 0$.

(a) Define

$$E(t) := \sum_{j=1}^{\infty} e_j t^j, \tag{1.68}$$

the generating function of the elementary symmetric functions. Show that

$$E(t) = \prod_{j=1}^{\infty} (1 + x_j t). \tag{1.69}$$

Hint: Expand the product in powers of t.

(b) Define

$$P(t) := \sum_{j=1}^{\infty} p_j t^{j-1}, \tag{1.70}$$

the generating function of the power sum symmetric functions. (Note the index shift.) Show that

$$P(t) = \sum_{j=1}^{\infty} \frac{x_j}{1 - x_j t}. \tag{1.71}$$

Hint: Recall the sum of a geometric series.

(c) Derive **Newton's identities**, i.e., for all k,

$$k e_k = \sum_{i=1}^{k} (-1)^{i-1} e_{k-i} p_i. \tag{1.72}$$

Hint: Show that $P(-t) = (dE(t)/dt)/E(t)$.

(d) By repeated substitutions we can write the e's in terms of the p's, but this brute force method becomes rapidly too cumbersome. Instead, show that

$$e_n = \frac{1}{n!} \begin{vmatrix} p_1 & 1 & 0 & \cdots & \cdots \\ p_2 & p_1 & 2 & 0 & \cdots \\ \vdots & \vdots & \ddots & \ddots & \vdots \\ p_{n-1} & p_{n-2} & \cdots & p_1 & n-1 \\ p_n & p_{n-1} & \cdots & p_2 & p_1 \end{vmatrix}. \tag{1.73}$$

Hint: Write out the first k Newton identities as a matrix equation for the unknown e_i and use Cramer's rule. *Remark:* There is a better way to write this formula using partitions of integers: see ([81], Chapter 7).

(e) Write down an expression for the determinant of a 4×4 matrix A as a sum of products of the traces of powers of A. For example, for $n = 3$ we have

$$\det A = \frac{1}{3!} \left[(\operatorname{tr} A)^3 - 3(\operatorname{tr} A)(\operatorname{tr} A^2) + 2 \operatorname{tr} A^3 \right]. \tag{1.74}$$

Hint: Assume A to be diagonalizable and compute the determinant and the traces of powers of A in terms of the eigenvalues.

1.47 The trace of an endomorphism The **trace** of an endomorphism A is defined to be the trace of any matrix that represents A. Show that the trace of an endomorphism is well defined by showing that it is invariant under a change of basis.

1.48 The trace, the determinant, and the exponential Show that, for any diagonalizable matrix A,

$$e^{\operatorname{tr} A} = \det e^{A}, \qquad (1.75)$$

where the exponential of a matrix is defined by its Taylor expansion:

$$e^{A} = 1 + A + \frac{1}{2}A^{2} + \frac{1}{3!}A^{3} + \cdots .$$

Hint: Consider the eigenvalues of A. Ignore convergence issues, which all work out correctly in the end. *Remark:* A theorem due to Schur guarantees that any matrix A can be decomposed into the sum of a diagonal matrix D and a strictly upper triangular matrix N. For extra credit, use Schur's theorem to show that (1.75) holds for *any* matrix A.

1.49 The Cauchy–Schwarz inequality Assume that g is a positive definite real inner product. Prove the Cauchy–Schwarz inequality: for any two nontrivial vectors u and v,

$$g(u, v)^{2} \le g(u, u)g(v, v),$$

with equality holding if and only if $u = \lambda v$ for some scalar λ. *Hint:* Minimize $g(u + \alpha v, u + \alpha v)$ with respect to α.

1.50 The $L^{2}(\mathbb{R})$ inner product Let $L^{2}(\mathbb{R})$ be the vector space of square integrable functions on $\mathbb{R}$, where the vector space operations are defined pointwise: $(f + g)(x) = f(x) + g(x)$ and $(af)(x) = af(x)$. A function f is square integrable if $\int_{-\infty}^{\infty} f^{2}(x)\, dx < \infty$. Show that the map $(\cdot, \cdot) : L^{2}(\mathbb{R}) \times L^{2}(\mathbb{R}) \to \mathbb{R}$ defined by

$$(f, g) := \int_{-\infty}^{\infty} f(x)g(x)\, dx$$

is a positive definite inner product.

1.51 Linear independence and the Grammian Let g be an inner product on a real vector space V with $\dim V \ge n$. Show that $v_{1}, \ldots, v_{n}$ are linearly independent if and only if the Grammian $\det(g(v_{i}, v_{j}))$ is nonvanishing. *Hint:* You may find it easier to prove the negation instead, that is, to show that $v_{1}, \ldots, v_{n}$ are linearly dependent if and only if the Grammian vanishes. To that end, start with $\sum_{j} c_{j} v_{j} = 0$ and analyze the system of equations you get by taking the inner product of this equation with all the v_{i}.

1.52 Gram–Schmidt orthonormalization   Following the ideas given in the first proof of Theorem 1.5 we can deduce an algorithm, called **Gram–Schmidt orthonormalization**, for constructing an orthonormal basis from a linearly independent set of vectors. The algorithm is easiest to explain if we assume $\mathbb{F} = \mathbb{R}$ and a positive definite inner product, but can be made to work in general with suitable modifications.

Let $\{v_1, \ldots, v_n\}$ be linearly independent, where $n = \dim V$. The first step is to set

$$e_1 := \frac{v_1}{\sqrt{g(v_1, v_1)}}.$$

Next, project v_2 onto the orthogonal complement of e_1 and then normalize:

$$e_2' := v_2 - e_1 g(e_1, v_2), \qquad e_2 := \frac{e_2'}{\sqrt{g(e_2', e_2')}}.$$

Now, project v_3 onto the orthogonal complement of e_1 and e_2 and normalize:

$$e_3' := v_3 - e_1 g(e_1, v_3) - e_2 g(e_2, v_3), \qquad e_3 := \frac{e_3'}{\sqrt{g(e_3', e_3')}}.$$

Continuing in this way, we end up with an orthonormal basis $\{e_1, \ldots, e_n\}$.

Let V be the four-dimensional vector space consisting of the polynomials of degree at most 3 equipped with the inner product

$$g(p, q) = \int_{-1}^{1} p(x)q(x)\, dx.$$

Using Gram–Schmidt orthonormalization starting from the linearly independent set $\{1, x, x^2, x^3\}$ (cf. Exercise 1.28), construct an orthonormal basis of polynomials for V. *Remark:* The resulting basis functions are the (normalized) **Legendre polynomials**.

1.53 Dual maps and annihilators   Let $T : V \to W$ be a linear map with dual map T^*. Show that $\ker T^* = \operatorname{Ann} \operatorname{im} T$ and $\operatorname{im} T^* = \operatorname{Ann} \ker T$.

Warning: $\ker T^* \neq (\ker T)^*$ and $\operatorname{im} T^* \neq (\operatorname{im} T)^*$. *Hint:* The first assertion is easy, but the second is not. For the second, let $f \in \operatorname{Ann} \ker T$ and consider the function

$$g(w) := \begin{cases} f(S(w)) & \text{if } w \in \operatorname{im} T, \quad \text{and} \\ 0 & \text{otherwise,} \end{cases}$$

where S is a section of T. Recall Theorem 1.3.

1.54 Riesz's lemma revisited   Riesz's lemma admits a more direct proof. For
simplicity let us assume that the field is real. As previously observed the map
$v \rightarrow f_v$ is then linear, so we need only show it is invertible. Let $\{e_i\}$ be an
orthonormal basis of V. Let $f \in V^*$, and suppose that $f(e_i) = f_i$. Define

$$v_f := \sum_i g(e_i, e_i) f_i e_i.$$

Show that the map $f \mapsto v_f$ is the desired inverse.

2

Multilinear algebra

Come, let us hasten to a higher plane
Where dyads tread the fairy fields of Venn,
Their indices bedecked from one to n
Commingled in an endless Markov chain!
Stanislaw Lem[†]

Multilinear algebra is basically just linear algebra with many vector spaces at the same time. The fundamental objects are generalizations of vectors called tensors. Tensors can be viewed from many different perspectives. Mathematicians introduce tensors formally as a quotient of a certain module, while physicists introduce tensors using objects with many indices that transform in a specific way under a change of basis. Each definition is viewed disparagingly by the other camp, even though they are equivalent and both have their uses. We follow a middle approach here.

2.1 The tensor product

To start, we define a new kind of vector product, called the **tensor product**, usually denoted by the symbol $\otimes$. Given two vectors v and w, we can form their tensor product $v \otimes w$. The product $v \otimes w$ is called a **tensor of order 2** or a **second-order tensor** or a **2-tensor**. The order of the factors matters, as $w \otimes v$ is generally different from $v \otimes w$. In fancy language, the tensor product is *noncommutative*. We can form higher-order tensors by repeating this procedure. For example, given another vector u we can construct $u \otimes v \otimes w$, a third-order tensor. (The tensor product is associative, so we need not worry about parentheses.) Order-0 tensors are just scalars, while order-1 tensors are vectors.

[†] Excerpt from "Love and tensor algebra", from *The Cyberiad: Fables For The Cybernetic Age* by Stanislaw Lem, translated by Michael Kandel. English translation copyright ©1974 by Houghton Mifflin Harcourt Publishing Company. Reprinted by permission of Houghton Mifflin Harcourt Publishing Company. All rights reserved.

In older books the tensor $v \otimes w$ is sometimes called a *dyadic* product (of the vectors v and w) and is written vw. That is, the tensor product symbol $\otimes$ is simply dropped. Even in modern works the tensor product symbol is often omitted. This generally leads to no confusion, as the only way to understand the proximate juxtaposition of two vectors is as a tensor product.

The set T^r of tensors of order r forms a vector space in a natural way; if S and T are both tensors of order r then $aT + bS$ is again a tensor of order r. One writes

$$T^r = \underbrace{V \otimes V \otimes \cdots \otimes V}_{r \text{ times}} = V^{\otimes r}. \tag{2.1}$$

The set $T = \bigcup_r T^r$ of all tensors forms a mathematical object called an **algebra**, which basically means that addition and the tensor product are compatible in the usual way. Specifically, if R is a tensor of order r and S is a tensor of order s then $R \otimes S$ is a tensor of order $r + s$.[1] In addition, scalars "pull through" tensor products,

$$T \otimes (aS) = (aT) \otimes S = a(T \otimes S), \tag{2.2}$$

and tensor products are distributive over addition:

$$R \otimes (S + T) = R \otimes S + R \otimes T, \tag{2.3}$$
$$(S + T) \otimes R = S \otimes R + T \otimes R. \tag{2.4}$$

Just as a vector has components in some basis, so does a tensor. Let e_1, e_2, e_3 be the canonical basis of $\mathbb{R}^3$. Then the canonical basis for the vector space $\mathbb{R}^3 \otimes \mathbb{R}^3$ of order-2 tensors on $\mathbb{R}^3$ is given by the set $e_i \otimes e_j$ as i and j run from 1 to 3. Written out in full, these basis elements are

$$\begin{array}{ccc} e_1 \otimes e_1, & e_1 \otimes e_2, & e_1 \otimes e_3, \\ e_2 \otimes e_1, & e_2 \otimes e_2, & e_2 \otimes e_3, \\ e_3 \otimes e_1, & e_3 \otimes e_2, & e_3 \otimes e_3. \end{array} \tag{2.5}$$

The most general second-order tensor on $\mathbb{R}^3$ is a linear combination of these basis tensors:

$$T = \sum_{ij} T^{ij} e_i \otimes e_j. \tag{2.6}$$

Almost always the basis is understood and fixed throughout. For this reason, physicists and engineers often identify tensors with their components. So, for example, T^{ij} is often called a tensor, when really it just gives the components of some tensor in some basis. Under certain circumstances this transgression can be convenient.

---

[1] Technically, this makes T into a **graded algebra**, just like the algebra of polynomials.

But it can also be dangerous because, to be the components of a tensor, the quantities T^{ij} must behave in a very specific way under a change of basis. We shall return to this issue in Section 2.3.

As a simple illustration of tensor products and components, let us find the components of the tensor $v \otimes w$.[2] We have

$$
\begin{aligned}
v \otimes w &= \left(\sum_i v^i e_i \right) \otimes \left(\sum_j w^j e_j \right) \\
&= \sum_{ij} v^i w^j (e_i \otimes e_j),
\end{aligned}
\tag{2.7}
$$

so the components of $v \otimes w$ are just $v^i w^j$. The same idea works in general. For example, the components of $u \otimes v \otimes w$ are just $u^i v^j w^k$ and so on.

It is perhaps worth observing that a tensor of the form $v \otimes w$ for some vectors v and w is not the most general order-2 tensor. The reason is that the most general order-2 tensor has nine algebraically independent components, whereas $v_i w_j$ has only six algebraically independent components (three from each vector). For example, $v_2 w_1 = (v_1 w_1)(v_2 w_2)/v_1 w_2$.

> **Example 2.1** Some examples of tensors you may have already encountered include the inertia tensor and the electromagnetic field strength tensor. Given a rigid body consisting of a bunch of point masses m_α at positions $\boldsymbol{r}_\alpha = (x_{\alpha,1}, x_{\alpha,2}, x_{\alpha,3})$, its **inertia tensor** is
>
> $$ I_{ij} = \sum_\alpha m_\alpha (r_\alpha^2 \delta_{ij} - x_{\alpha,i} x_{\alpha,j}), $$
>
> where $r_\alpha^2 = \boldsymbol{r}_\alpha \cdot \boldsymbol{r}_\alpha$, and $\cdot$ is the usual Euclidean dot product. Of course, this is typical physics sloppiness: I_{ij} is not a tensor, it is the component of a tensor in the standard basis $\{e_1, e_2, e_3\}$ of $\mathbb{R}^3$. Moreover, the indices are in the wrong place. If we were being really pedantic we would write
>
> $$ I^{ij} = \sum_\alpha m_\alpha (r_\alpha^2 \delta^{ij} - x_\alpha^i x_\alpha^j), $$
>
> in which case the tensor itself would be $I = \sum_{ij} I^{ij} e_i \otimes e_j$. The components of the inertia tensor of a continuous body are given by
>
> $$ I^{ij} = \int_V \rho(\boldsymbol{r}) \left(r^2 \delta^{ij} - x^i x^j \right) d^3 x, $$

---

[2] In some books $v \otimes w$ is called a *tensor of rank 2*. The problem with this terminology is that it conflicts with standard usage in matrix theory. If we consider the components $v_i w_j$ of the tensor $v \otimes w$ in some basis to be the components of a matrix, then this matrix only has rank 1. (The rows are all multiples of each other.) To avoid this problem, one usually says that a tensor of the form $v \otimes w \otimes \cdots$ has rank 1. Any tensor is a sum of rank-1 tensors, and we say that the **rank** of the tensor is the minimum number of rank-1 tensors needed to write it as such a sum. Finding methods to compute the rank of a tensor efficiently is a difficult open problem.

where $\rho(\boldsymbol{r})$ is the density at the point $\boldsymbol{r}$. (Here the integrand is something called a "tensor density", which we discuss in Exercise 8.52.) Similarly, if A_μ denotes the electromagnetic vector potential, the field strength tensor is

$$F_{\mu\nu} = \partial_\mu A_\nu - \partial_\nu A_\mu,$$

where $\partial_\mu := \partial/\partial x^\mu$. Once again this is sloppy nomenclature because the $F_{\mu\nu}$ are the components of the field strength tensor F in some basis. Exactly what basis is being used here will have to wait until we reach Example 3.14.

2.2 General tensors

It turns out that the previous definition of a tensor is a bit too restrictive for many purposes and must be expanded to include dual objects. Let V be a vector space and V^* its dual space. Then **a tensor T of type** (r, s) is an element of the tensor product space

$$T_s^r = \underbrace{V \otimes V \otimes \cdots \otimes V}_{r \text{ times}} \otimes \underbrace{V^* \otimes V^* \otimes \cdots \otimes V^*}_{s \text{ times}} = V^{\otimes r} \otimes (V^*)^{\otimes s}. \tag{2.8}$$

What we previously called a tensor of order r is just a tensor of type $(r, 0)$. The properties of the tensor product ensure that the space of all tensors forms a **multigraded algebra**, as before.[3]

Pick a basis $\{e_i\}$ for V and a dual basis $\{\theta^i\}$ for V^*. Then a basis for the space T_s^r comprises objects of the form

$$e_{i_1} \otimes e_{i_2} \otimes \cdots \otimes e_{i_r} \otimes \theta^{j_1} \otimes \theta^{j_2} \otimes \cdots \otimes \theta^{j_s}, \tag{2.9}$$

where all indices run from 1 to $\dim V$. A general tensor T of type (r, s) is a linear combination of the basis elements:

$$T = \sum_{\substack{i_1,\dots,i_r \\ j_1,\dots,j_s}} T^{i_1 i_2 \dots i_r}{}_{j_1 j_2 \dots j_s} e_{i_1} \otimes e_{i_2} \otimes \cdots \otimes e_{i_r} \otimes \theta^{j_1} \otimes \theta^{j_2} \otimes \cdots \otimes \theta^{j_s}. \tag{2.10}$$

As before the elements $T^{i_1 i_2 \dots i_r}{}_{j_1 j_2 \dots j_s}$ are called the components of T. In order to avoid having to write out the summation symbols, Einstein introduced his **summation convention**: whenever two indices with the same labels appear on the same side of an equation, one assumes that there is an *implicit sum* over those indices. This allows us to rewrite (2.10) as

$$T = T^{i_1 i_2 \dots i_r}{}_{j_1 j_2 \dots j_s} e_{i_1} \otimes e_{i_2} \otimes \cdots \otimes e_{i_r} \otimes \theta^{j_1} \otimes \theta^{j_2} \otimes \cdots \otimes \theta^{j_s}. \tag{2.11}$$

---

[3] The term "multigraded" in this context means that if S is a tensor of type (r, s) and T is a tensor of type (p, q) then $S \otimes T$ is a tensor of type $(r + p, s + q)$.

This often saves much writing, but it does require that one should keep a careful track of all the indices. We shall adopt this convention later in the book, but for now we leave the summation symbols in place.

2.3 Change of basis

It follows immediately from our discussion and the results of Section 1.7 that the components of a tensor transform under a change of basis as one would expect from the placement of their indices. Upstairs indices transform contravariantly, whereas downstairs indices transform covariantly. Before we can write down the actual transformation law, though, we need to introduce another notational convention.

In Section 1.7 we wrote $\{v'^i\}$ to denote the components of the vector v relative to the new basis $\{e'_i\}$, but this notation can be problematic in the context of tensors, because tensors have more than one index. Each index goes with a different basis element, so it is convenient to put the primes on the indices rather than on the tensor label itself. Thus, for example, (1.18) would be written $v^{i'} = \sum_j (A^{-1})^{i'}{}_j v^j$. According to this convention, the transformation law for tensor components under a change of basis is written

$$T^{i'_1 \dots i'_r}{}_{j'_1 \dots j'_s} = \sum_{\substack{i_1, \dots, i_r \\ j_1, \dots, j_s}} T^{i_1 \dots i_r}{}_{j_1 \dots j_s} (A^{-1})^{i'_1}{}_{i_1} \cdots (A^{-1})^{i'_r}{}_{i_r} A^{j_1}{}_{j'_1} \cdots A^{j_s}{}_{j'_s}. \qquad (2.12)$$

This notation has advantages and disadvantages. The primary advantage is that, by putting the primes on the indices rather than on the tensor symbol itself, it suggests that we are dealing with the same tensor but in different coordinate systems. The primary disadvantage is that it is not clear to which components we are referring, new or old, when writing something like $T^{312}{}_{134}$, say. So, in this notation, one must really write $T^{3'1'2'}{}_{1'3'4'}$ to refer to a particular component of T in the new coordinate system.

2.4 Tensors as multilinear maps

There is yet another way of looking at tensors that is very handy in practice. It exploits the isomorphism between a vector space and its dual. The map

$$T : \underbrace{V^* \times \cdots \times V^*}_{r \text{ times}} \times \underbrace{V \times \cdots \times V}_{s \text{ times}} \to \mathbb{R} \qquad (2.13)$$

is said to be **multilinear** if it is linear in each entry:

$$T(v_1, \dots, au + bw, \dots, v_{r+s})$$
$$= aT(v_1, \dots, u, \dots, v_{r+s}) + bT(v_1, \dots, w, \dots, v_{r+s}), \qquad (2.14)$$

where $v_1, \ldots, v_{r+s}$, u, and w are elements of V or V^* as appropriate, and a and b are scalars. The space of all such maps is linear under pointwise addition and scalar multiplication. Let us denote it by $\widetilde{T}_s^r$.

By employing the natural pairing $\langle,\rangle$ (see Section 1.6) between V and V^*, we may view the tensor

$$e_{i_1} \otimes \cdots \otimes e_{i_r} \otimes \theta^{j_1} \otimes \cdots \otimes \theta^{j_s} \tag{2.15}$$

as a multilinear map on the Cartesian product space $(V^*)^{\times r} \times V^{\times s}$ that acts according to the rule

$$(e_{i_1} \otimes \cdots \otimes e_{i_r} \otimes \theta^{j_1} \otimes \cdots \otimes \theta^{j_s})(\theta^{k_1}, \ldots, \theta^{k_r}, e_{\ell_1}, \ldots, e_{\ell_s})$$
$$= \langle e_{i_1}, \theta^{k_1} \rangle \cdots \langle e_{i_r}, \theta^{k_r} \rangle \langle \theta^{j_1}, e_{\ell_1} \rangle \cdots \langle \theta^{j_s}, e_{\ell_s} \rangle$$
$$= \delta_{i_1}^{k_1} \cdots \delta_{i_r}^{k_r} \delta_{\ell_1}^{j_1} \cdots \delta_{\ell_s}^{j_s}. \tag{2.16}$$

In other words, the tensor product symbol acts like a placeholder that just tells you which covectors or vectors should be paired. Viewing the tensor product in this way as a multilinear map, we have

$$T(\theta^{k_1}, \ldots, \theta^{k_r}, e_{\ell_1}, \ldots, e_{\ell_s})$$
$$= \sum_{\substack{i_1,\ldots,i_r \\ j_1,\ldots,j_s}} T^{i_1 i_2 \ldots i_r}{}_{j_1 j_2 \ldots j_s}$$
$$\times (e_{i_1} \otimes e_{i_2} \otimes \cdots \otimes e_{i_r} \otimes \theta^{j_1} \otimes \theta^{j_2} \otimes \cdots \otimes \theta^{j_s})(\theta^{k_1}, \ldots, \theta^{k_r}, e_{\ell_1}, \ldots, e_{\ell_s})$$
$$= \sum_{\substack{i_1,\ldots,i_r \\ j_1,\ldots,j_s}} T^{i_1 i_2 \ldots i_r}{}_{j_1 j_2 \ldots j_s} \delta_{i_1}^{k_1} \cdots \delta_{i_r}^{k_r} \delta_{\ell_1}^{j_1} \cdots \delta_{\ell_s}^{j_s}$$
$$= T^{k_1 k_2 \ldots k_r}{}_{\ell_1 \ell_2 \ldots \ell_s}. \tag{2.17}$$

This establishes an isomorphism from $\widetilde{T}_s^r$ to T_s^r. Essentially, one can view a tensor either *passively* as an element of a certain vector space (the tensor product space) or *actively* as a multilinear functional on the dual of that vector space. They are two sides of the same coin. Therefore we may interchange multilinear maps and tensors as we please.

2.5 Symmetry types of tensors

Let T be a tensor of type $(0, 2)$ with components T_{ij} in some basis. If $T_{ij} = T_{ji}$ we say T is **symmetric**, while if $T_{ij} = -T_{ji}$ we say T is **antisymmetric**. Of course, a general $(0, 2)$ tensor has no such property. But if a $(0, 2)$ tensor has either property, we say it has **definite symmetry**.

EXERCISE 2.1 As we have defined it, the condition of having definite symmetry would appear to be basis dependent. Using the tensor component transformation law (2.12) show that this is not the case. That is, if T is symmetric (respectively, antisymmetric) for one choice of basis then it is symmetric (respectively, antisymmetric) for all choices of basis.

EXERCISE 2.2 Let A_{ij} be symmetric and B^{ij} be antisymmetric. Show that $\sum_{ij} A_{ij} B^{ij} = 0$.

The **symmetric part** T_{sym} of T is the symmetric tensor with components

$$(T_{\text{sym}})_{ij} := \frac{1}{2} \left(T_{ij} + T_{ji} \right) \tag{2.18}$$

while the **antisymmetric part** T_{asym} of T is the antisymmetric tensor with components

$$(T_{\text{asym}})_{ij} := \frac{1}{2} \left(T_{ij} - T_{ji} \right). \tag{2.19}$$

Evidently, for a $(0, 2)$ tensor, T is the sum of its symmetric and antisymmetric parts.

These ideas can be generalized to higher-order tensors, but the results are not as simple. Defining what is meant by "definite symmetry" for higher-order tensors requires some understanding of the representation theory of the symmetric group, which is beyond the scope of this work. Fortunately, for most purposes we need only consider the higher-order analogues of symmetric and antisymmetric tensors, and these are easy enough to describe using basic ideas of permutation theory, as follows.

Let T be a tensor of type $(0, p)$ with components $T_{i_1 i_2 \ldots i_p}$. Define the symmetric and antisymmetric parts of T by

$$(T_{\text{sym}})_{i_1 i_2 \ldots i_p} := \frac{1}{p!} \sum_{\sigma \in \mathfrak{S}_p} T_{i_{\sigma(1)} i_{\sigma(2)} \ldots i_{\sigma(p)}} =: T_{(i_1 i_2 \ldots i_p)} \tag{2.20}$$

and

$$(T_{\text{asym}})_{i_1 i_2 \ldots i_p} := \frac{1}{p!} \sum_{\sigma \in \mathfrak{S}_p} (-1)^{\sigma} T_{i_{\sigma(1)} i_{\sigma(2)} \ldots i_{\sigma(p)}} =: T_{[i_1 i_2 \ldots i_p]}, \tag{2.21}$$

where $\mathfrak{S}_p$ is the set of all permutations of p elements and $(-1)^{\sigma}$ denotes the sign of the permutation σ.[4] The rightmost expressions give a convenient shorthand notation. Observe that if $p > 2$ then it is no longer true that T is the sum of its symmetric and antisymmetric parts.

---

[4] See Appendix A.

A $(0, p)$ tensor is **(totally) symmetric** (or simply, **symmetric**), if

$$T_{i_{\sigma(1)} i_{\sigma(2)} \dots i_{\sigma(p)}} = T_{i_1 i_2 \dots i_p} \qquad \text{for every } \sigma \in \mathfrak{S}_p. \tag{2.22}$$

It is **(totally) antisymmetric** (or **skew symmetric**, or **alternating**) if

$$T_{i_{\sigma(1)} i_{\sigma(2)} \dots i_{\sigma(p)}} = (-1)^\sigma T_{i_1 i_2 \dots i_p} \qquad \text{for every } \sigma \in \mathfrak{S}_p. \tag{2.23}$$

The notions of symmetric and antisymmetric tensors extend *mutatis mutandis* to tensors of type $(p, 0)$ on just raising all the indices. Moreover, symmetry type is preserved under the taking of linear combinations of tensors of the same symmetry type; thus, the set of all symmetric and the set of all antisymmetric tensors are both subspaces of the space of all tensors. Specifically, we write $\mathrm{Sym}^p V$ (respectively, $\mathrm{Alt}^p V$) for the subspace of symmetric (respectively, antisymmetric) $(0, p)$ tensors or $(p, 0)$ tensors.

EXERCISE 2.3 Show that, as the names imply, if T is a general $(0, p)$ tensor then T_{sym} is totally symmetric and T_{asym} is totally antisymmetric. *Extended hint:* To get you started, we prove the antisymmetric case here and leave the symmetric case to you. We have the following chain of equalities:

$$(T_{\mathrm{asym}})_{i_{\tau(1)} \dots i_{\tau(p)}} = \frac{1}{p!} \sum_{\sigma \in \mathfrak{S}_p} (-1)^\sigma T_{i_{\sigma(\tau(1))} i_{\sigma(\tau(2))} \dots i_{\sigma(\tau(p))}} \tag{2.24}$$

$$= \frac{1}{p!} \sum_{\sigma \in \mathfrak{S}_p} (-1)^\sigma T_{i_{\sigma\tau(1)} i_{\sigma\tau(2)} \dots i_{\sigma\tau(p)}} \tag{2.25}$$

$$= \frac{1}{p!} \sum_{\pi\tau^{-1} \in \mathfrak{S}_p} (-1)^{\pi\tau^{-1}} T_{i_{\pi(1)} i_{\pi(2)} \dots i_{\pi(p)}} \tag{2.26}$$

$$= (-1)^\tau \frac{1}{p!} \sum_{\pi\tau^{-1} \in \mathfrak{S}_p} (-1)^\pi T_{i_{\pi(1)} i_{\pi(2)} \dots i_{\pi(p)}} \tag{2.27}$$

$$= (-1)^\tau \frac{1}{p!} \sum_{\pi \in \mathfrak{S}_p} (-1)^\pi T_{i_{\pi(1)} i_{\pi(2)} \dots i_{\pi(p)}} \tag{2.28}$$

$$= (-1)^\tau (T_{\mathrm{asym}})_{i_1 i_2 \dots i_p}. \tag{2.29}$$

These equalities hold for the following reasons: (2.24) and (2.25) hold by definition; in (2.26) we have set $\pi = \sigma\tau$; (2.27) holds because the sign function is a group homomorphism from $\mathfrak{S}_p$ to ± 1, so that $(-1)^{\pi\tau^{-1}} = (-1)^\pi (-1)^{\tau^{-1}}$, and because a permutation has the same sign as its inverse, since $1 = (-1)^{id} = (-1)^{\tau\tau^{-1}} = (-1)^\tau (-1)^{\tau^{-1}}$; (2.28) holds because $\pi\tau^{-1}$ runs over all permutations as π runs over all permutations; lastly, (2.29) is true by definition.

Remark We have introduced the idea of symmetry from the point of view of tensor components, as this seems to be the most elementary way to do it. But, as discussed in Section 2.4, we can view all this from the perspective of multilinear maps. Viewed as a multilinear map, the condition that T be symmetric is just

$$T(v_{\sigma(1)}, v_{\sigma(2)}, \ldots, v_{\sigma(p)}) = T(v_1, v_2, \ldots, v_p), \tag{2.30}$$

while the condition that T be antisymmetric is

$$T(v_{\sigma(1)}, v_{\sigma(2)}, \ldots, v_{\sigma(p)}) = (-1)^{\sigma} T(v_1, v_2, \ldots, v_p), \tag{2.31}$$

for any collection of p vectors $(v_1, v_2, \ldots, v_p)$ and any permutation $\sigma \in \mathfrak{S}_p$. From this perspective, the basis independence of the symmetry conditions is obvious.

2.6 Alternating tensors and the space $\bigwedge^p V$ of p-vectors

In this section we restrict our attention to alternating tensors, for which one can construct a very powerful and elegant stand-alone theory that leads eventually to the concept of a differential form. The theory is made possible by the use of a convenient shorthand notation for an alternating tensor product.

Given two vectors $v, w \in V$, define their **wedge** or **exterior product** by[5]

$$v \wedge w = v \otimes w - w \otimes v. \tag{2.32}$$

The wedge product of two vectors is called a **2-vector**, and the vector space generated by the set of all 2-vectors is denoted $\bigwedge^2 V$. Observe that, for any vector v, $v \wedge v = 0$.

We claim that $\bigwedge^2 V$ is naturally isomorphic to the vector space $\mathrm{Alt}^2 V$ of alternating $(2, 0)$ tensors. First note that $\bigwedge^2 V$ is spanned by all 2-vectors of the form $e_i \wedge e_j$ where $i < j$. To see this, choose a basis $\{e_1, e_2, \ldots, e_n\}$ for V. If

$$v = \sum_i v^i e_i \qquad \text{and} \qquad w = \sum_j w^j e_j,$$

then, by the linearity of the tensor product,

$$v \wedge w = \sum_{ij} \left(v^i e_i \otimes w^j e_j - w^j e_j \otimes v^i e_i \right) = \sum_{ij} v^i w^j \, e_i \wedge e_j.$$

By a swap of dummy indices and by the antisymmetric nature of the wedge product, we have

$$\sum_{ij} v^i w^j \, e_i \wedge e_j = \sum_{ji} v^j w^i \, e_j \wedge e_i = -\sum_{ij} v^j w^i \, e_i \wedge e_j.$$

---

[5] Unfortunately, there are competing conventions here. Many authors define $v \wedge w$ as $(v \otimes w - w \otimes v)/2$. See the discussion in footnote 8.

Therefore we can write (again using the antisymmetry property of the wedge product)

$$v \wedge w = \frac{1}{2} \sum_{ij} (v^i w^j - v^j w^i) e_i \wedge e_j = \sum_{i<j} (v^i w^j - v^j w^i) e_i \wedge e_j.$$

It follows that any linear combination of 2-vectors can be written as a linear combination of the elements $e_i \wedge e_j$ for $i < j$. Moreover, these 2-vectors are all linearly independent (check!), so they form a basis for the space of 2-vectors. There are $\binom{n}{2}$ such vectors, where $\binom{n}{k} = \frac{n!}{k!(n-k)!}$ is the binomial coefficient "n choose k", so this is the dimension of $\bigwedge^2 V$. For example, if V is four dimensional, the $\binom{4}{2} = 6$ vectors $e_1 \wedge e_2$, $e_1 \wedge e_3$, $e_1 \wedge e_4$, $e_2 \wedge e_3$, $e_2 \wedge e_4$, and $e_3 \wedge e_4$ form a basis for the space $\bigwedge^2 V$.

However, $\binom{n}{2}$ is also the dimension of $\mathrm{Alt}^2 V$, because the components of an alternating $(2, 0)$ tensor satisfy $T^{ij} = -T^{ji}$, which means that the only independent components of T are those for which $i < j$. The linear map $\psi : \mathrm{Alt}^2 V \to \bigwedge^2 V$ given by $T \mapsto \sum_{i<j} T^{ij} e_i \wedge e_j$ furnishes the natural isomorphism.

Although we could now go on to define wedge products of more than two vectors in terms of tensor products, it is much simpler and more elegant to proceed axiomatically. Thus, let us begin again and define the wedge product $v_1 \wedge v_2 \wedge \cdots \wedge v_p$ of p vectors $v_1, v_2, \ldots, v_p$ so that it satisfies the following two axioms.

(1) (multilinearity) For every i and all scalars a and b,

$$v_1 \wedge \cdots \wedge \overbrace{(au + bv)}^{i th \ position} \wedge \cdots \wedge v_p = a(v_1 \wedge \cdots \wedge u \wedge \cdots \wedge v_p)$$
$$+ b(v_1 \wedge \cdots \wedge v \wedge \cdots \wedge v_p). \quad (2.33)$$

(2) (antisymmetry) For every i, j,

$$v_1 \wedge \cdots \wedge v_i \wedge \cdots \wedge v_j \wedge \cdots \wedge v_p = -v_1 \wedge \cdots \wedge v_j \wedge \cdots \wedge v_i \wedge \cdots \wedge v_p. \quad (2.34)$$

A wedge product of p vectors is called a p-**vector**, and the vector space generated by all the p-vectors is denoted $\bigwedge^p V$. Obviously $\bigwedge^1 V = V$, and by convention $\bigwedge^0 V = \mathbb{F}$, where $\mathbb{F}$ is the underlying field. A basis for $\bigwedge^p V$ is given by all p-vectors of the form

$$e_I := e_{i_1} \wedge \cdots \wedge e_{i_p}, \quad (2.35)$$

where the **multi-index** I denotes the *ordered* index collection[6] $(i_1, \ldots, i_p)$ with

$$1 \leq i_1 < \cdots < i_p \leq \dim V. \tag{2.36}$$

We write $|I| = p$ to indicate that I contains p indices. The dimension of $\bigwedge^p V$ is $\binom{n}{p}$, because there are this many basis vectors of the form (2.35). In ordinary and multi-index notation, a general p-vector in $\bigwedge^p V$ is written as

$$\omega = \frac{1}{p!} \sum_{i_1, \cdots, i_p} a^{i_1 \cdots i_p} e_{i_1} \wedge \cdots \wedge e_{i_p} = \sum_I a^I e_I, \tag{2.37}$$

where $a^{i_1 \cdots i_p}$ is totally antisymmetric and $a^I = a^{(i_1, \ldots, i_p)} \in \mathbb{F}$. A single term in this sum is called a **monomial**.[7]

Example 2.2   In (2.37) we have stipulated that the components $a^{i_1 \cdots i_p}$ be totally antisymmetric. Suppose for the moment that we had not insisted on this. Consider what would happen in the case $p = 2$, $n = 3$. Writing out all the components explicitly gives

$$\frac{1}{2} \left(a^{11} e_1 \wedge e_1 + a^{12} e_1 \wedge e_2 + a^{21} e_2 \wedge e_1 + a^{22} e_2 \wedge e_2 \right.$$

$$\left. + a^{13} e_1 \wedge e_3 + a^{31} e_3 \wedge e_1 + a^{23} e_2 \wedge e_3 + a^{32} e_3 \wedge e_2 + a^{33} e_3 \wedge e_3 \right)$$

$$= \frac{1}{2} \left(a^{12} - a^{21} \right) e_1 \wedge e_2 + \frac{1}{2} \left(a^{13} - a^{31} \right) e_1 \wedge e_3 + \frac{1}{2} \left(a^{23} - a^{32} \right) e_2 \wedge e_3.$$

In other words, the components of the basis vectors $e_i \wedge e_j$ with $i < j$ are the antisymmetric parts of a. If a had a symmetric part, it would disappear from this expression. This is why we may as well demand that a be antisymmetric from the start. Moreover, when a is antisymmetric, the above expression reduces to

$$a^{12} e_1 \wedge e_2 + a^{13} e_1 \wedge e_3 + a^{23} e_2 \wedge e_3,$$

which is precisely $\sum_I a^I e_I$ in this case.

EXERCISE 2.4   Generalize the previous example by showing that

$$\sum_{i_1, \cdots, i_p} a^{i_1 \cdots i_p} e_{i_1} \wedge \cdots \wedge e_{i_p} = \sum_{i_1, \cdots, i_p} a^{[i_1 \cdots i_p]} e_{i_1} \wedge \cdots \wedge e_{i_p}. \tag{2.38}$$

---

[6] We adopt the standard convention that parentheses denote an ordered set whereas curly braces denote an unordered set.

[7] Warning: Be careful to distinguish $a^{(i_1 \cdots i_p)}$ (without the commas) from $a^{(i_1, \ldots, i_p)}$ (with the commas). The former is the totally symmetric part of a as defined in (2.20), whereas the latter *vanishes* unless the indices are in increasing order. For example, if $i_1 = 3$, $i_2 = 1$, and $i_3 = 2$, then

$$a^{(312)} = \frac{1}{3!}(a^{123} + a^{132} + a^{213} + a^{231} + a^{312} + a^{321}),$$

whereas $a^{(3,1,2)} = 0$ (and $a^{(1,2,3)} = a^{123}$).

In other words, a general p-vector is determined uniquely by the antisymmetric components of a. It follows that $\bigwedge^p V$ is isomorphic to $\mathrm{Alt}^p V$. *Hint:* Feel free to use the fact easily proved from axiom (2.34) that, for any permutation $\sigma \in \mathfrak{S}_p$,

$$v_{i_{\sigma(1)}} \wedge v_{i_{\sigma(2)}} \wedge \cdots \wedge v_{i_{\sigma(p)}} = (-1)^\sigma v_{i_1} \wedge v_{i_2} \wedge \cdots \wedge v_{i_p}. \tag{2.39}$$

You will also need to use dummy indices cleverly.

Axioms (2.33) and (2.34) imply that

$$v_1 \wedge v_2 \wedge \cdots \wedge v_p = c_p \sum_{\sigma \in \mathfrak{S}_p} (-1)^\sigma v_{\sigma(1)} \otimes v_{\sigma(2)} \otimes \cdots \otimes v_{\sigma(p)}, \tag{2.40}$$

where c_p is some constant. We have chosen $c_2 = 1$ in (2.32), so to be consistent we should probably take $c_p = 1$; other authors take $c_p = 1/p!$. Fortunately, the choice is mostly irrelevant. As long as we understand the wedge product via the axioms (2.33) and (2.34), we need never go back and forth between wedge products and tensor products.[8] It is much easier to deal with wedge products than with alternating sums such as those on the right-hand side of (2.40), and for the most part this is what everyone does.

2.7 The exterior algebra

The wedge product turns the collection $\bigwedge V$ of all $\bigwedge^p V$ for $p = 0, 1, 2, \ldots$ into a graded algebra, called the **exterior algebra** of V. Given $\lambda \in \bigwedge^p V$ and $\mu \in \bigwedge^q V$, we define $\lambda \wedge \mu \in \bigwedge^{p+q} V$ in the obvious way, namely we set

$$(v_1 \wedge \cdots \wedge v_p) \wedge (w_1 \wedge \cdots \wedge w_q) = v_1 \wedge \cdots \wedge v_p \wedge w_1 \wedge \cdots \wedge w_q \tag{2.41}$$

and extend by linearity. What could be simpler? It follows that the exterior product satisfies the usual properties of distributivity and associativity, together with what is sometimes called **graded antisymmetry**:

$$\lambda \wedge (a\mu + bv) = a\lambda \wedge \mu + b\lambda \wedge v \quad \text{(distributivity)}, \tag{2.42}$$

$$\lambda \wedge (\mu \wedge v) = (\lambda \wedge \mu) \wedge v \quad \text{(associativity)}, \tag{2.43}$$

$$\mu \wedge \lambda = (-1)^{pq} \lambda \wedge \mu \quad \text{(graded antisymmetry)}. \tag{2.44}$$

EXERCISE 2.5 Prove property (2.44).

EXERCISE 2.6 The only purpose of this very tedious exercise is to emphasize that when dealing with alternating tensors it is far preferable to stick with the

---

[8] The one exception to this rule is when one views tensors as multilinear maps, for then those annoying prefactors again rear their ugly heads. But here again axiomatics save the day for us, for we will define the natural pairing of p-vectors and p-tuples of vectors via the interior product; this allows us to avoid any mention of combinatorial prefactors. See Section 3.13.

axiomatically defined wedge product. Suppose that, in spite of our suggestion, you insisted on defining p-vectors in terms of tensor products. You would then define a linear map alt : $T_p^0 \to$ Alt$^p V$ by

$$\text{alt}(e_{i_1} \otimes \cdots \otimes e_{i_p}) = \frac{1}{p!} \sum_{\sigma \in \mathfrak{S}_p} (-1)^\sigma e_{i_{\sigma(1)}} \otimes \cdots \otimes e_{i_{\sigma(p)}}, \qquad (2.45)$$

extended by linearity. You might go on to define the wedge product of $S \in T_p^0$ and $T \in T_q^0$ by

$$S \wedge T = \text{alt}(S \otimes T). \qquad (2.46)$$

(Note that this differs from our previous convention by a numerical factor.) Use (2.46) to prove that the wedge product so defined is associative. *Remark:* The wedge product defined this way also satisfies properties (2.42) and (2.44). Property (2.42) is immediate from the multilinearity of alt, while (2.44) follows from the antisymmetry of the wedge product for vectors, as before. *Hint:* Proceed in steps. Let S and T be tensors. (i) Show that alt(alt(T)) = alt(T). (ii) Show that if alt(S) = 0 then $S \wedge T = T \wedge S = 0$. (iii) Use the linearity of alt and properties (i) and (ii) to show that $(R \wedge S) \wedge T - \text{alt}(R \otimes S \otimes T) = 0$, where, say, $R \in T_r^0$. Argue that a similar result holds for $R \wedge (S \wedge T)$. If you're really stuck, look in [33] or [79].

2.8 The induced linear transformation $\bigwedge T$

Let $T : V \to V$ be a linear map. We define the *pth exterior power of T* to be the linear map

$$\bigwedge{}^p T : \bigwedge{}^p V \to \bigwedge{}^p V \qquad (2.47)$$

given by

$$\left(\bigwedge{}^p T\right)(v_1 \wedge \cdots \wedge v_p) = T v_1 \wedge \cdots \wedge T v_p. \qquad (2.48)$$

This map is natural, in the sense that

$$\bigwedge{}^p (ST)(v_1 \wedge \cdots \wedge v_p) = (ST)v_1 \wedge \cdots \wedge (ST)v_p$$
$$= \left(\bigwedge{}^p S\right)(T v_1 \wedge \cdots \wedge T v_p)$$
$$= \left(\bigwedge{}^p S\right)\left(\bigwedge{}^p T\right)(v_1 \wedge \cdots \wedge v_p), \qquad (2.49)$$
$$(2.50)$$

so that

$$\bigwedge{}^p (ST) = \left(\bigwedge{}^p S\right)\left(\bigwedge{}^p T\right). \qquad (2.51)$$

Similarly, if $\lambda \in \bigwedge^p V$ and $\mu \in \bigwedge^q V$ then

$$(\textstyle\bigwedge^{p+q} T)(\lambda \wedge \mu) = (\bigwedge^p T)(\lambda) \wedge (\bigwedge^q T)(\mu). \tag{2.52}$$

Of particular interest is the special case in which $p = n = \dim V$. In that case, because $\dim \bigwedge^n V = 1$ the map $\bigwedge^n T$ is a linear map between one-dimensional vector spaces and hence is multiplication by a scalar. If $v_1, \ldots, v_n$ are linearly independent vectors then

$$(\textstyle\bigwedge^n T)(v_1 \wedge \cdots \wedge v_n) = T v_1 \wedge \cdots \wedge T v_n = (\det T)(v_1 \wedge \cdots \wedge v_n). \tag{2.53}$$

The scalar $\det T$ is called the **determinant** of the map T, because if $\mathbf{T}$ represents T in some basis then $\det T$ is the usual matrix determinant $\det \mathbf{T}$.

EXERCISE 2.7 Let $T : \mathbb{R}^3 \to \mathbb{R}^3$ be given by $T e_1 = e_1 + 2e_2$, $T e_2 = 3e_2 + 2e_3$, and $T e_3 = e_1 + e_3$, where $\{e_1, e_2, e_3\}$ is the standard basis. Compute $\bigwedge^3 T(e_1 \wedge e_2 \wedge e_3)$ directly from (2.48) and the properties of the wedge product and compare it to $\det \mathbf{T}$.

EXERCISE 2.8 Prove that $\det T = \det \mathbf{T}$ in any basis, where $\det \mathbf{T}$ is defined by (1.46). (Actually, you need to use (1.46) together with (1.56).)

Pick a basis $\{e_i\}$ of V, and recall the multi-index notation e_I introduced in (2.35). Then we have[9]

$$
\begin{aligned}
(\textstyle\bigwedge^p T) e_I &= T e_{i_1} \wedge \cdots \wedge T e_{i_p} \\
&= \sum_{j_1 \cdots j_n} (T_{j_1 i_1} \cdots T_{j_p i_p}) e_{j_1} \wedge \cdots \wedge e_{j_p} \\
&= \sum_J e_J T_{JI},
\end{aligned}
$$

which says that the map $\bigwedge^p T$ is represented by the matrix (T_{JI}) in the natural basis. This matrix is called the *pth* **compound matrix** of (T_{ij}) and is the matrix of minors of (T_{ij}) indexed by rows J and columns I.[10]

EXERCISE 2.9 Let V be a three-dimensional vector space with basis $\{e_1, e_2, e_3\}$, and choose the basis $\{e_1 \wedge e_2, e_1 \wedge e_3, e_2 \wedge e_3\}$ for $\bigwedge^2 V$. Show by explicit computation that the second compound matrix of (T_{ij}) is given by

$$
\begin{pmatrix}
T_{11} T_{22} - T_{21} T_{12} & T_{11} T_{23} - T_{21} T_{13} & T_{12} T_{23} - T_{22} T_{13} \\
T_{11} T_{32} - T_{31} T_{12} & T_{11} T_{33} - T_{31} T_{13} & T_{12} T_{33} - T_{32} T_{13} \\
T_{21} T_{32} - T_{31} T_{22} & T_{21} T_{33} - T_{31} T_{23} & T_{22} T_{33} - T_{32} T_{23}
\end{pmatrix}.
$$

[9] As discussed previously, this is one of those cases where the upstairs/downstairs placement of indices is entirely irrelevant and would just clutter up the formulae.

[10] The minor of a matrix (T_{ij}) indexed by rows J and columns I is the determinant of the matrix obtained from (T_{ij}) by crossing out all rows except those indexed by J and all columns except those indexed by I.

By (2.51), the *p*th compound of the product of two matrices is the product of the *p*th compounds of each matrix, a useful fact which is very difficult to prove directly. In particular, when $p = n$ this is just the statement that the determinant of a product of two matrices is the product of their determinants:

$$\det(\boldsymbol{ST}) = (\det \boldsymbol{S})(\det \boldsymbol{T}). \tag{2.54}$$

EXERCISE 2.10   Using (2.53) and the result of Exercise 2.8, prove the following statements for any matrix. (1) The determinant changes sign whenever any two rows or columns are interchanged. (2) The determinant vanishes if any two rows or columns are equal. (3) The determinant is unchanged if we add a multiple of any row to another row or a multiple of any column to another column. (4) Multiplying a column or row by a constant multiplies the determinant by the same constant. *Hint:* Let $\{e_i\}$ be a basis for V, and consider a linear map $T : V \to V$ satisfying $T e_i = v_i$, so that v_i is the *i*th column of the matrix $\boldsymbol{T}$ that represents T in the basis $\{e_i\}$. You may want to invoke (1.56).

2.9 The Hodge dual

An inner product g on V induces a natural inner product (also denoted g) on $\bigwedge^p V$, as follows. If $\lambda = v_1 \wedge \cdots \wedge v_p$ and $\mu = w_1 \wedge \cdots \wedge w_p$, we define

$$g(\lambda, \mu) := \det(g(v_i, w_j)), \tag{2.55}$$

where the right-hand side is the determinant of the matrix whose (i, j)th entry is $g(v_i, w_j)$, and then extend by bilinearity. By (1.56) g is symmetric, so we need only verify its nondegeneracy. For this, let $\{e_i\}$ be an orthonormal basis for V. Then the claim is that the set $\{e_I = e_{i_1} \wedge \cdots \wedge e_{i_p}\}$, as I varies over all ordered index sets of size p, is an orthonormal basis for $\bigwedge^p V$ relative to g. Set $e_J = e_{j_1} \wedge \cdots \wedge e_{j_p}$. Define $(g_{IJ})_{k\ell} := g(e_{i_k}, e_{j_\ell})$. By orthonormality, $(g_{IJ})_{k\ell} = \pm\delta_{i_k j_\ell}$, so we have

$$g(e_I, e_J) = \det((g_{IJ})_{k\ell}) = \det(\pm\delta_{i_k j_\ell}).$$

Suppose $j_p \notin I$, say, so that $I \neq J$. Then the rightmost determinant vanishes, because the *p*th column of $(\delta_{i_k j_\ell})$ vanishes. Similar reasoning shows that $g(e_I, e_J) = 0$ whenever $I \neq J$. However, if $I = J$ then $g(e_I, e_J) = \pm 1$. Hence the claim is proved. But if an orthonormal basis exists, the metric must be nondegenerate by Lemma 1.6.

Example 2.3   Let $\{e_1, e_2, \ldots, e_n\}$ be a basis for V, and define $g_{ij} = g(e_i, e_j)$. Choose $\sigma = e_1 \wedge \cdots \wedge e_n$. Then $g(\sigma, \sigma) = \det(g_{ij})$. If the basis is orthonormal then $g(\sigma, \sigma) = (-1)^d$, where $n - 2d$ is the signature of g.

In preparation for the main definition of this section, fix $\lambda \in \bigwedge^p V$. Then we have a natural linear map from $\bigwedge^{n-p} V$ to $\bigwedge^n V$ given by

$$\mu \to \lambda \wedge \mu. \tag{2.56}$$

But $\bigwedge^n V$ is a one-dimensional vector space spanned by some element σ, so

$$\lambda \wedge \mu = f_\lambda(\mu)\sigma \tag{2.57}$$

for some linear functional f_λ on $\bigwedge^{n-p} V$. Given an inner product g, the Riesz lemma guarantees the existence of a unique $(n-p)$-vector $\star\lambda$ such that

$$g(\star\lambda, \mu) = f_\lambda(\mu). \tag{2.58}$$

The element $\star\lambda \in \bigwedge^{n-p} V$ is called the **Hodge dual** or **Hodge star** of λ. Combining (2.57) and (2.58) we may write

$$\lambda \wedge \mu = g(\star\lambda, \mu)\sigma. \tag{2.59}$$

Written this way it is clear that the Hodge dual depends on both the inner product g and the choice of basis element σ for $\bigwedge^n V$.

At first sight the definition of the Hodge dual appears to be worthless, because it asserts the existence of an object but seems to give you no means of computing it. It is true that the definition is indirect but, as we shall soon see, one can actually use (2.59) to compute things. Of course, you might be wondering what it's good for. The answer is basically that Hodge duality provides an elegant means of writing various important formulae (see e.g. Example 3.14).

Arguably the most important properties of the Hodge dual are given in the following theorem.

Theorem 2.1 *Let V be n-dimensional with inner product g. Let $\eta, \lambda \in \bigwedge^p V$, and choose $\sigma \in \bigwedge^n V$ to satisfy $g(\sigma, \sigma) = (-1)^d$. Then*

$$\star\star\lambda = (-1)^{p(n-p)+d}\lambda, \tag{2.60}$$

and

$$\eta \wedge \star\lambda = \lambda \wedge \star\eta = (-1)^d g(\eta, \lambda)\sigma. \tag{2.61}$$

Proof Choose an orthonormal basis $\{e_i\}$ for V. We must have

$$\sigma = a\, e_1 \wedge \cdots \wedge e_n$$

for some constant a, and the hypothesis requires that $a = \pm 1$. For definiteness we choose $a = 1$. (The proof for $a = -1$ is identical.)

We prove (2.60) and leave the proof of (2.61) as an exercise. By linearity it suffices to verify (2.60) for $\lambda = e_I$ with $|I| = p$. Without loss of generality we may assume that $I = (1, 2, \ldots, p)$, so that

$$\lambda = e_1 \wedge e_2 \wedge \cdots \wedge e_p.$$

If J is any multi-index other than I, (2.59) gives

$$\lambda \wedge e_J = g(\star\lambda, e_J)\sigma. \tag{2.62}$$

The left-hand side vanishes unless $J = \{p+1, p+2, \ldots, n\}$, so we must have

$$\star\lambda = c\, e_{p+1} \wedge \cdots \wedge e_n$$

for some constant c that can only be ± 1. The constant is determined by choosing $J = (p+1, p+2, \ldots, n)$, so that (2.62) becomes

$$\sigma = cg(e_J, e_J)\sigma.$$

It follows that

$$c = g(e_J, e_J),$$

so we may write

$$\star\, e_I = g(e_J, e_J)e_J. \tag{2.63}$$

A similar argument shows that

$$\star e_J = bg(e_I, e_I)e_I$$

for some constant b, determined as follows. We have

$$(-1)^{p(n-p)}\sigma = (-1)^{p(n-p)}e_I \wedge e_J = e_J \wedge e_I = g(\star e_J, e_I)\sigma = g(e_I, \star e_J)\sigma,$$

so that

$$(-1)^{p(n-p)} = g(e_I, \star e_J) = b[g(e_I, e_I)]^2 = b$$

and thus

$$\star\, e_J = (-1)^{p(n-p)}g(e_I, e_I)e_I. \tag{2.64}$$

We conclude that

$$\begin{aligned}\star(\star e_I) &= (-1)^{p(n-p)}g(e_J, e_J)g(e_I, e_I)e_I \\ &= (-1)^{p(n-p)}g(\sigma, \sigma)e_I \\ &= (-1)^{p(n-p)+d}e_I,\end{aligned}$$

where the equality used in the penultimate step,

$$g(e_I, e_I)g(e_J, e_J) = g(\sigma, \sigma), \tag{2.65}$$

follows from (2.55). $\qquad\square$

EXERCISE 2.11 Prove (2.61).

Example 2.4 Let $V = \mathbb{M}^4$ and let $e_0 = (1, 0, 0, 0)$, $e_1 = (0, 1, 0, 0)$, $e_2 = (0, 0, 1, 0)$, and $e_3 = (0, 0, 0, 1)$. Thus

$$(g_{ij}) = \begin{pmatrix} -1 & 0 & 0 & 0 \\ 0 & 1 & 0 & 0 \\ 0 & 0 & 1 & 0 \\ 0 & 0 & 0 & 1 \end{pmatrix}. \tag{2.66}$$

Choose $\sigma = e_0 \wedge e_1 \wedge e_2 \wedge e_3$, so that $g(\sigma, \sigma) = \det(g_{ij}) = -1$. Let us compute $\star e_0$, for which $p = 1$. As $n = 4$, $\star e_0$ will be a 3-vector, so it can be written as

$$\star e_0 = \sum a^{i_1 i_2 i_3} e_{i_1} \wedge e_{i_2} \wedge e_{i_3}$$

for some constants $a^{i_1 i_2 i_3}$. Equation (2.61) gives

$$e_0 \wedge \star e_0 = -g(e_0, e_0)\sigma = \sigma,$$
$$e_1 \wedge \star e_0 = -g(e_1, e_0)\sigma = 0,$$
$$e_2 \wedge \star e_0 = -g(e_2, e_0)\sigma = 0,$$
$$e_3 \wedge \star e_0 = -g(e_3, e_0)\sigma = 0,$$

from which we conclude that

$$\star e_0 = e_1 \wedge e_2 \wedge e_3. \tag{2.67}$$

Similar calculations yield

$$\star e_1 = e_0 \wedge e_2 \wedge e_3, \tag{2.68}$$
$$\star e_2 = e_0 \wedge e_3 \wedge e_1, \tag{2.69}$$
$$\star e_3 = e_0 \wedge e_1 \wedge e_2. \tag{2.70}$$

Applying (2.60) to (2.67) gives

$$\star (e_1 \wedge e_2 \wedge e_3) = \star \star e_0 = (-1)^{1(4-1)+1} e_0 = e_0. \tag{2.71}$$

Similarly, we get

$$\star(e_0 \wedge e_2 \wedge e_3) = e_1, \tag{2.72}$$
$$\star(e_0 \wedge e_3 \wedge e_1) = e_2, \tag{2.73}$$
$$\star(e_0 \wedge e_1 \wedge e_2) = e_3. \tag{2.74}$$

Next, let's find $\star(e_0 \wedge e_1)$, which is a 2-vector. Equation (2.61) gives

$$(e_0 \wedge e_1) \wedge \star(e_0 \wedge e_1) = -g(e_0 \wedge e_1, e_0 \wedge e_1)\sigma = \sigma,$$

because

$$g(e_0 \wedge e_1, e_0 \wedge e_1) = \begin{vmatrix} g(e_0, e_0) & g(e_0, e_1) \\ g(e_1, e_0) & g(e_1, e_1) \end{vmatrix} = \begin{vmatrix} -1 & 0 \\ 0 & 1 \end{vmatrix} = -1.$$

When $(i, j) \notin \{(0, 1), (1, 0)\}$ we have

$$(e_i \wedge e_j) \wedge \star(e_0 \wedge e_1) = -g(e_i \wedge e_j, e_0 \wedge e_1)\sigma = 0.$$

Therefore

$$\star (e_0 \wedge e_1) = e_2 \wedge e_3. \tag{2.75}$$

By analogy we have

$$\star(e_0 \wedge e_2) = e_3 \wedge e_1, \tag{2.76}$$

$$\star(e_0 \wedge e_3) = e_1 \wedge e_2. \tag{2.77}$$

But the analogy stops here, because e_0 is different from the other basis vectors. For example,

$$\star(e_1 \wedge e_2) = -e_0 \wedge e_3, \tag{2.78}$$

$$\star(e_2 \wedge e_3) = -e_0 \wedge e_1, \tag{2.79}$$

$$\star(e_3 \wedge e_1) = -e_0 \wedge e_2. \tag{2.80}$$

These equations can be obtained in two ways. For example, to obtain (2.78) we can proceed as before to get

$$(e_1 \wedge e_2) \wedge \star(e_1 \wedge e_2) = -g(e_1 \wedge e_2, e_1 \wedge e_2)\sigma = -\sigma,$$

because

$$g(e_1 \wedge e_2, e_1 \wedge e_2) = \begin{vmatrix} g(e_1, e_1) & g(e_1, e_2) \\ g(e_2, e_1) & g(e_2, e_2) \end{vmatrix} = \begin{vmatrix} 1 & 0 \\ 0 & 1 \end{vmatrix} = 1.$$

Similarly, we compute

$$(e_i \wedge e_j) \wedge \star(e_1 \wedge e_2) = -g(e_i \wedge e_j, e_1 \wedge e_2)\sigma = 0,$$

whenever $(i, j) \notin \{(1, 2), (2, 1)\}$, from which we conclude that we must have (2.78). Alternatively, we could simply notice from (2.77) that

$$\star(e_1 \wedge e_2) = \star^2(e_0 \wedge e_3),$$

and then use the fact that, on 2-vectors,

$$\star^2 = (-1)^{p(n-p)+d} = (-1)^{2(4-2)+1} = -1.$$

Lastly, observe that

$$\sigma \wedge \star\sigma = -g(\sigma, \sigma)\sigma = \sigma,$$

giving

$$\star \sigma = 1 \quad \text{and} \quad \star 1 = \sigma, \tag{2.81}$$

a result valid more generally for any signature inner product on $\mathbb{R}^n$.

Additional exercises

2.12 Linear independence and tensor products Let $\{v_1, \ldots, v_k\}$ be a set of linearly independent vectors in V. Show that, for any vectors $w_1, \ldots, w_k$,

$$\sum_{i=1}^{k} v_i \otimes w_i = 0 \qquad \Rightarrow \qquad w_1 = \cdots = w_k = 0.$$

Hint: Complete $\{v_1, \ldots, v_k\}$ to a basis.

2.13 The Kronecker product of two matrices Let (A_{ij}) be an $m \times m$ matrix and let (B_{ij}) be a $n \times n$ matrix. The **Kronecker product** $A \otimes B$ is the $mn \times mn$ block matrix

$$A \otimes B = \begin{pmatrix} A_{11} B & \cdots & A_{1m} B \\ \vdots & \ddots & \vdots \\ A_{m1} B & \cdots & A_{mm} B \end{pmatrix}.$$

More explicitly, $A \otimes B$ is

$$\begin{pmatrix}
A_{11}B_{11} & A_{11}B_{12} & \cdots & A_{11}B_{1n} & \cdots & \cdots & A_{1m}B_{11} & A_{1m}B_{12} & \cdots & A_{1m}B_{1n} \\
A_{11}B_{21} & A_{11}B_{22} & \cdots & A_{11}B_{2n} & \cdots & \cdots & A_{1m}B_{21} & A_{1m}B_{22} & \cdots & A_{1m}B_{2n} \\
\vdots & \vdots & \ddots & \vdots & & & \vdots & \vdots & \ddots & \vdots \\
A_{11}B_{n1} & A_{11}B_{n2} & \cdots & A_{11}B_{nn} & \cdots & \cdots & A_{1m}B_{n1} & A_{1m}B_{n2} & \cdots & A_{1m}B_{nn} \\
\vdots & \vdots & & \vdots & \ddots & & \vdots & \vdots & & \vdots \\
\vdots & \vdots & & \vdots & & \ddots & \vdots & \vdots & & \vdots \\
A_{m1}B_{11} & A_{m1}B_{12} & \cdots & A_{m1}B_{1n} & \cdots & \cdots & A_{mm}B_{11} & A_{mm}B_{12} & \cdots & A_{mm}B_{1n} \\
A_{m1}B_{21} & A_{m1}B_{22} & \cdots & A_{m1}B_{2n} & \cdots & \cdots & A_{mm}B_{21} & A_{mm}B_{22} & \cdots & A_{mm}B_{2n} \\
\vdots & \vdots & \ddots & \vdots & & & \vdots & \vdots & \ddots & \vdots \\
A_{m1}B_{n1} & A_{m1}B_{n2} & \cdots & A_{m1}B_{nn} & \cdots & \cdots & A_{mm}B_{n1} & A_{mm}B_{n2} & \cdots & A_{mm}B_{nn}
\end{pmatrix}.$$

Now let $A : V \to V$ and $B : W \to W$ be linear maps and let $v \in V$ and $w \in W$. The **tensor product map** $A \otimes B : V \otimes W \to V \otimes W$ is naturally defined by

$$(A \otimes B)(v \otimes w) = Av \otimes Bw,$$

extended by linearity. Pick a basis $\{e_1, \ldots, e_m\}$ for V and $\{f_1, \ldots, f_n\}$ for W. If A is represented by the matrix A and B is represented by the matrix B relative to the chosen basis, show that $A \otimes B$ is represented by the Kronecker product matrix $A \otimes B$ relative to the basis $\{e_i \otimes f_j\}$, $1 \le i \le m, 1 \le j \le n$, of $V \otimes W$, provided we order the basis elements of $V \otimes W$ in lexicographic order, namely as $e_1 \otimes f_1, e_1 \otimes f_2, \ldots, e_2 \otimes f_1, e_2 \otimes f_2, \ldots, e_m \otimes f_n$.

2.14 Linear independence and wedge products   Show that $v_1 \wedge \cdots \wedge v_p = 0$ if and only if the set $\{v_1, v_2, \ldots, v_p\}$ is linearly dependent.

2.15 Cartan's lemma   Let $\{v_1, \ldots, v_k\}$ be a set of linearly independent vectors in V, and suppose that

$$v_1 \wedge w_1 + v_2 \wedge w_2 + \cdots + v_k \wedge w_k = 0$$

for some vectors $w_1, \ldots, w_k$. Prove that

$$w_i = \sum_{j=1}^{k} A_{ij} v_j$$

for some symmetric matrix A (i.e., $A_{ij} = A_{ji}$). *Hint:* Complete $v_1, \ldots, v_k$ to a basis of V.

2.16 The Sylvester–Franke theorem   Let V be an n-dimensional vector space and $A : V \rightarrow V$ a linear map. Show that

$$\det \bigwedge^p A = (\det A)^{\binom{n-1}{p-1}}.$$

Hint: Assume A to be diagonalizable. The general case then follows by a continuity argument.

2.17 The Hodge dual of a vector   Let V be Euclidean n-space, namely $\mathbb{R}^n$ equipped with the standard basis $e_1, \ldots, e_n$ and the Euclidean inner product $g(e_i, e_j) = \delta_{ij}$. Choose $\sigma = e_1 \wedge \cdots \wedge e_n$. Let $\mu = \sum_i a_i e_i$. Compute $\star\mu$.

2.18 The Hodge dual as pseudo-isometry   Show that the Hodge star operator preserves the natural inner product on p-forms up to sign. That is, for any p-forms λ and μ,

$$g(\star\lambda, \star\mu) = (-1)^d g(\lambda, \mu),$$

where $d = (n - t)/2$.

2.19 The wedge product of a direct sum of vector spaces   Given two vector spaces V and W, define a linear map

$$\bigwedge^k V \otimes \bigwedge^\ell W \rightarrow \bigwedge^{k+\ell}(V \oplus W)$$

by

$$(v_1 \wedge \cdots \wedge v_k) \otimes (w_1 \wedge \cdots \wedge w_\ell) \mapsto (v_1, 0) \wedge \cdots \wedge (v_k, 0) \wedge (0, w_1) \wedge \cdots \wedge (0, w_\ell),$$

where for clarity we represent v_k as an element of $V \oplus W$ by $(v_k, 0)$, etc. Use this map to show that

$$\bigoplus_{k=0}^{n} \left(\bigwedge^k V \otimes \bigwedge^{n-k} W \right) \cong \bigwedge^n (V \oplus W). \tag{2.82}$$

2.20 The characteristic polynomial and the exterior algebra Let V be an n-dimensional vector space and $A : V \to V$ a linear map. Then

$$\det(I + zA) = \sum_{k=0}^{n} (\operatorname{tr} \textstyle\bigwedge^k A) z^k, \tag{2.83}$$

where I is the identity matrix and tr means "trace". We will show this in two ways.

(a) First assume A to be diagonalizable. Prove (2.83) by relating the left-hand side to the characteristic polynomial of A. Recall that

$$\prod_{k=1}^{n} (1 + a_k z) = \sum_{k=1}^{n} e_k(a_1, \dots, a_n) z^k,$$

where $e_k(a_1, \dots, a_n)$ is the kth elementary symmetric function.

(b) Now prove (2.83) without assuming diagonalizability, by letting $B := I + zA$ and using the distributive property of the wedge product and the definition of the trace. For simplicity (as the general proof works similarly) you may wish to restrict to the case $n = 3$.

2.21 The symmetric algebra Given p vectors $v_1, \dots, v_p \in V$, their **symmetric tensor product**

$$v_1 \odot v_2 \odot \cdots \odot v_p$$

is defined by the following two properties.

(1) (multilinearity) For every i and all scalars a and b,

$$v_1 \odot \cdots \odot \overbrace{(au + bv)}^{i\text{th position}} \odot \cdots \odot v_p = a(v_1 \odot \cdots \odot u \odot \cdots \odot v_p)$$
$$+ b(v_1 \odot \cdots \odot v \odot \cdots \odot v_p).$$

(2) (symmetry) For every i, j,

$$v_1 \odot \cdots \odot v_i \odot \cdots \odot v_j \odot \cdots \odot v_p$$
$$= v_1 \odot \cdots \odot v_j \odot \cdots \odot v_i \odot \cdots \odot v_p.$$

The symmetric tensor product of p vectors is clearly a symmetric tensor, and the vector space of all such tensors is commonly denoted $\operatorname{Sym}^p V$. Obviously $\operatorname{Sym}^1 V = V$ and by convention $\operatorname{Sym}^0 V = \mathbb{F}$. If $\{e_i\}$ is a basis of V, a basis for $\operatorname{Sym}^p V$ is given by all elements of the form

$$e_{i_1} \odot \cdots \odot e_{i_p} \qquad (1 \le i_1 \le \cdots \le i_p \le \dim V),$$

so the most general element of $\operatorname{Sym}^p V$ can be written

$$\frac{1}{p!} \sum_{i_1,\dots,i_p} a^{i_1\dots i_p} e_{i_1} \odot \cdots \odot e_{i_p},$$

where $a^{i_1,\dots,i_p}$ are the components of a symmetric p-tensor.

(a) Let dim $V = n$. Show that the dimension of $\mathrm{Sym}^p V$ is $\binom{n+p-1}{n-1}$.

(b) The symmetric tensor product turns the collection $\mathrm{Sym}\, V$ of all $\mathrm{Sym}^p V$ for $p = 0, 1, 2, \dots$ into a graded algebra, called the **symmetric algebra** of V, in the obvious way:

$$(v_1 \odot \cdots \odot v_p) \odot (w_1 \odot \cdots \odot w_q) := v_1 \odot \cdots \odot v_p \odot w_1 \odot \cdots \odot w_q.$$

Show that $\mathrm{Sym}\, V$ is isomorphic to the polynomial algebra $\mathbb{F}[e_1, e_2, \dots, e_n]$, namely, the set of all polynomials with the e_i acting as indeterminates.

(c) Show that

$$\mathrm{Sym}^p(V \oplus W) \cong \bigoplus_{j=0}^{p} \mathrm{Sym}^j V \otimes \mathrm{Sym}^{p-j} W.$$

(d) A linear map $A : V \to V$ induces a linear map $\mathrm{Sym}^k A : \mathrm{Sym}^k V \to \mathrm{Sym}^k V$ in the natural way, namely

$$\mathrm{Sym}^k A(v_1 \odot \cdots \odot v_k) = Av_1 \odot \cdots \odot Av_k.$$

Assuming A to be diagonalizable, verify that

$$\frac{1}{\det(I - zA)} = \sum_{k=0}^{\infty} (\mathrm{Sym}^k A) z^k.$$

Hint: Look up "homogeneous symmetric functions."

2.22 Fock space   In quantum mechanics the space of possible states of a single particle is a special kind of (often infinite-dimensional) vector space, called a **Hilbert space**, denoted here by $\mathcal{H}$. A state containing p particles is an element of $\mathcal{H}^{\otimes p}$, the tensor product of $\mathcal{H}$ with itself p times. But in relativistic quantum mechanics (or quantum field theory) the number of particles is not necessarily constant, so the state of the system is now an element of the direct sum

$$\mathcal{F} := \bigoplus_{p=0}^{\infty} \mathcal{H}^{\otimes p},$$

called **Fock space**. The symmetric and antisymmetric subspaces

$$\mathcal{F}^+ := \bigoplus_{p=0}^{\infty} \mathrm{Sym}^p \mathcal{H},$$

$$\mathcal{F}^- := \bigoplus_{p=0}^{\infty} \bigwedge^p \mathcal{H},$$

are of particular importance as they represent the spaces of **bosonic** and **fermionic** particle states, respectively.

Given a vector $v \in \mathcal{H}$, the bosonic creation and annihilation operators $\widehat{a}^\dagger(v)$ and $\widehat{a}(v)$ and fermionic creation and annihilation operators $\widehat{b}^\dagger(v)$ and $\widehat{b}(v)$ act linearly on their respective Fock spaces as follows:

$$\widehat{a}^\dagger(v)(v_1 \odot \cdots \odot v_p) = v \odot v_1 \odot \cdots \odot v_p,$$

$$\widehat{a}(v)(v_1 \odot \cdots \odot v_p) = \sum_{i=1}^{p} (v, v_i)\, v_1 \odot \cdots \odot \widehat{v}_i \odot \cdots \odot v_p,$$

$$\widehat{b}^\dagger(v)(v_1 \wedge \cdots \wedge v_p) = v \wedge v_1 \wedge \cdots \wedge v_p,$$

$$\widehat{b}(v)(v_1 \wedge \cdots \wedge v_p) = \sum_{i=1}^{p} (-1)^i (v, v_i)\, v_1 \wedge \cdots \wedge \widehat{v}_i \wedge \cdots \wedge v_p,$$

where (u, v) is the standard Hermitian inner product on $\mathcal{H}$ (see Appendix B), and the caret on v_i means that the vector is missing.

Given two operators $\widehat{A}$ and $\widehat{B}$, their **commutator** (respectively, **anticommutator**) is $[\widehat{A}, \widehat{B}]_- := \widehat{A}\widehat{B} - \widehat{B}\widehat{A}$ (respectively, $[\widehat{A}, \widehat{B}]_+ := \widehat{A}\widehat{B} + \widehat{B}\widehat{A}$). Show that the creation and annihilation operators obey the **canonical commutation relations**

$$[\widehat{a}(u), \widehat{a}(v)]_- = 0,$$
$$[\widehat{a}^\dagger(u), \widehat{a}^\dagger(v)]_- = 0,$$
$$[\widehat{a}(u), \widehat{a}^\dagger(v)]_- = (u, v)\widehat{1}$$

on $\mathcal{F}^+$ and the **canonical anticommutation relations**

$$[\widehat{b}(u), \widehat{b}(v)]_+ = 0,$$
$$[\widehat{b}^\dagger(u), \widehat{b}^\dagger(v)]_+ = 0,$$
$$[\widehat{b}(u), \widehat{b}^\dagger(v)]_+ = (u, v)\widehat{1}$$

on $\mathcal{F}^-$, where $\widehat{1}$ is the identity operator. *Hint:* To show that an operator is zero on $\mathcal{F}^+$ (respectively, $\mathcal{F}^-$) it suffices to show that it is zero on every basis element and, to do this, it is enough to show that the operator vanishes on a basis element of the form $v_1 \odot \cdots \odot v_p$ (respectively, $v_1 \wedge \cdots \wedge v_p$).

3

Differentiation on manifolds

> When in doubt, differentiate.
> *Shing-Shen Chern (1979).*
> *Raoul Bott (1982).*

The idea of a differentiable manifold had its genesis in the nineteenth century with the work of Carl Friedrich Gauss and of Georg Friedrich Bernhard Riemann. Gauss was interested in surveying and cartography, which led him to develop the tools of calculus on curved surfaces. His famous *theorema egregium*, or remarkable theorem,[1] revealed that one could consider the intrinsic properties of a surface independently of the way in which it was embedded in three-dimensional space, and this led him, Riemann, and others, to abstract these concepts even further. Their ideas have had far reaching applications in many areas of mathematics and the natural sciences.

Roughly, an n-dimensional manifold (or n-manifold) can be thought of as a kind of patchwork quilt built from pieces of $\mathbb{R}^n$. Classic examples of 2-manifolds are the 2-sphere S^2 and the 2-torus T^2 (see Figure 3.1). Usually one pictures these as living in $\mathbb{R}^3$, but one can consider them in their own right just as bits of $\mathbb{R}^2$ sewn together in certain ways. The technical definition of a manifold requires considerable background, which we will try to keep to a minimum. First, we need the idea of a topology.

3.1 Basic topology*

Consider a basketball. When it is inflated, its surface is a sphere.[2] But when it is deflated *its surface is still a topological sphere*. In fact, we could deform the sphere

---

[1] For the younger generation, *theorema egregium* might be better translated as "egregious theorem".

[2] Ignore the dimples and the hole for the needle.

 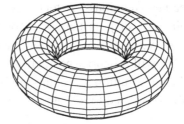

Figure 3.1 The 2-sphere and the 2-torus.

 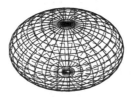 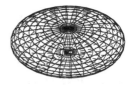

Figure 3.2 Topological 2-spheres.

in any way we like and, as long as we do not tear it anywhere, it is still topologically a sphere.[3] We say that all these shapes have the same *topology* but, since the distance between the points on the surface has changed, they have different *geometries* (see Figure 3.2).

At first sight the actual definition of a topology appears to have nothing to do with these notions. Only after much study does one begin to see why the following definition is reasonable.[4] A **topology** τ on a set X is a family of subsets of X, called *open* sets, satisfying the following.

(1) Arbitrary unions of open sets are open.
(2) Finite intersections of open sets are open.
(3) The empty set $\emptyset$ and X are both open.

A **topological space** (or, simply, a **space**) is a set X endowed with a topology.

> **Example 3.1** Let X be a finite set, and let τ be the set of all subsets of X. This is called the **discrete topology** on X.

A **neighborhood** of $p \in X$ is any open set containing p. If q lies in a neighborhood of p we say that q is **near** p. Topology is therefore sometimes called the

---

[3] In this book, the phrase "2-sphere" will always mean the standard unit 2-sphere, namely $\{(x, y, z) \in \mathbb{R}^3 : x^2 + y^2 + z^2 = 1\}$.
[4] Readers may wish to skip the rest of this section, and only refer back to it as needed.

study of nearness relations.[5] A topology on X is called **Hausdorff** if the points in every pair lie in disjoint neighborhoods or, more technically, if for every two points $p, q \in X$, there exist two disjoint open sets U and V such that $p \in U$ and $q \in V$. We will primarily be interested in Hausdorff topologies, because these coincide with our intuition that points are isolated objects; other sorts of topologies are generally considered to be pathological (at least by non-topologists).

> **EXERCISE 3.1** Let Y be a subset of a topological space X. Suppose that for every $y \in Y$ there is an open set U with $y \in U \subset Y$. Show that Y is open in X.

A subset $Y \subseteq X$ is **closed** if its complement $\overline{Y} := X - Y$ is open.

> **EXERCISE 3.2** Let X be a topological space. Show that the following properties hold.
>
> (1) Arbitrary intersections of closed sets are closed.
> (2) Finite unions of closed sets are closed.
> (3) The empty set $\emptyset$ and X are both closed.
>
> (This exercise demonstrates that we could equally well take these axioms to be the definition of a topology, then define open sets to be the complements of closed sets.)

The **interior** Y^0 of a set Y is the union of all the open sets contained in Y, and the **closure** $\operatorname{cl} Y$ of Y is the intersection of all the closed sets containing Y. A subset $A \subset X$ is **dense** in X if $\operatorname{cl} A = X$. A point $x \in X$ is an **accumulation point** of Y if $x \in \operatorname{cl}(Y - \{x\})$.

> **EXERCISE 3.3** Let Y be a subset of a topological space X. Show that $y \in \operatorname{cl} Y$ if and only if every neighborhood of y meets Y.

> **EXERCISE 3.4** Show that a set Y is closed if and only if it contains all its accumulation points.

> **EXERCISE 3.5** Show that a single point is a closed set in a Hausdorff topology.

An **open cover** of X is a collection $\{U_\alpha\}$ of open sets whose union is X. An open cover is **locally finite** if, for every $p \in X$, there is an open neighborhood U of p such that $|\{\alpha : U \cap U_\alpha \neq \emptyset\}|$ is finite. (Note that a locally finite open cover is not necessarily finite.) An open cover $\{V_\beta\}$ is a **refinement** of $\{U_\alpha\}$ if for all β there is an α such that $U_\alpha \supset V_\beta$.

A topological space X is **compact** if every open cover has a finite subcollection that also covers X ("every open cover has a finite subcover"). Intuitively, compact spaces can be thought of as being finite in extent. For example, a subset

---

[5] The word "topology" derives from the Greek τόποϛ, meaning place, and λόγοϛ, meaning a speaking or discourse.

of Euclidean space (see Example 3.4) is compact if and only if it is closed and bounded (the Heine–Borel theorem).

EXERCISE 3.6 Show that a closed subset of a compact set is compact.

A **basis** for a topology on X is a collection $\mathcal{B}$ of subsets of X such that

(1) every x belongs to some element of $\mathcal{B}$;
(2) given any two basis elements B_1 and B_2 in $\mathcal{B}$, if $x \in B_1 \cap B_2$ then there exists a B_3 such that $x \in B_3 \subset B_1 \cap B_2$.

The topology **generated by** $\mathcal{B}$ is just the collection of open subsets of X, where U is declared open if for every $x \in U$ there is an element $B \in \mathcal{B}$ such that $x \in B \subset U$.

Example 3.2 The **standard topology** on $\mathbb{R}$ is the topology generated by all the intervals $(a, b) := \{x : a < x < b\}$.

Given two sets X and Y, the **product topology** on $X \times Y$ is the topology generated by all sets of the form $U \times V$, where U is open in X and V is open in Y.

Example 3.3 The **standard topology** on $\mathbb{R}^n$ is just the product topology on $\underbrace{\mathbb{R} \times \cdots \times \mathbb{R}}_{n \text{ times}}$.

Example 3.4 Recall that Euclidean n-space $\mathbb{E}^n$ is $\mathbb{R}^n$ equipped with the Euclidean inner product g. The **Euclidean distance** $d(x, y)$ between two points x and y is defined to be $d(x, y) = \|x - y\|$.[6] The **open ball of radius r about the point p** is the set $B_r(p) = \{x \in \mathbb{R}^n : d(x, p) < r\}$. The set of all these open balls generates the **Euclidean topology** on $\mathbb{E}^n$. It can be shown that the Euclidean topology on $\mathbb{E}^n$ coincides with the standard topology on $\mathbb{R}^n$. For this reason, most people refer to $\mathbb{R}^n$ as Euclidean space! Unfortunately, this convention is so ingrained in many people (including this author) that it would be difficult to ignore. *Henceforth, whenever $\mathbb{R}^n$ is treated as a topological space it is understood that we are using the (standard) Euclidean topology.*

Let $f : X \to Y$ be a map between topological spaces. The map f is **continuous** if the inverse image of an open set in Y is open in X.

EXERCISE 3.7 Show that f is continuous if the inverse image of a closed set is closed.

EXERCISE 3.8 In calculus you learn that $f : \mathbb{R} \to \mathbb{R}$ is continuous at a point $p \in \mathbb{R}$ if for every $\varepsilon > 0$ there exists a $\delta > 0$ such that, for all $x \in \mathbb{R}$, $|f(x) - f(p)| < \epsilon$ whenever $|x - p| < \delta$. The map f is said to be continuous if it is continuous

---

[6] We use the standard notation $\|x\| := [g(x, x)]^{1/2}$, where the positive root is understood.

at every point $p \in \mathbb{R}$. Show that the epsilon-delta definition of continuity and the inverse-image definition of continuity coincide for maps from $\mathbb{R}$ to $\mathbb{R}$.

EXERCISE 3.9 The topological space Y is a **continuous image** of X if there is a continuous surjection $f : X \to Y$. Show that the continuous image of a compact set is compact.

EXERCISE 3.10 Let τ be a topology on X, and let $Y \subset X$. The collection

$$\tau_Y = \{Y \cap U : U \in \tau\} \tag{3.1}$$

is a topology on Y, called the **induced topology** or **subspace topology** on Y. The space Y equipped with this topology is a **subspace** of X. Show that τ_Y is the coarsest topology for which the inclusion map $\iota : Y \hookrightarrow X$ is continuous. (A topology σ is **coarser** than a topology τ if $\sigma \subseteq \tau$; τ is **finer** than σ because it has more open sets.)

EXERCISE 3.11 Show that a subspace of a Hausdorff space is Hausdorff.

EXERCISE 3.12 Let $f : X \to Y$ be a continuous map of Hausdorff topological spaces, and suppose that X is compact. Show that $f^{-1}(y)$ is compact for $y \in Y$.

EXERCISE 3.13 If X is a topological space (with topology τ) and $\sim$ is an equivalence relation on X, the quotient $Y := X/\sim$ is naturally a topological space under the **quotient topology** σ, defined as

$$\sigma = \{U \subseteq Y : \pi^{-1}(U) \in \tau\},$$

where $\pi : X \to Y$ is the natural projection map $x \mapsto [x]$. In other words, the quotient topology is the finest topology for which π is continuous. (Note that the quotient topology σ does not generally preserve the properties of τ. For example, σ need not be Hausdorff even if τ is.) Suppose that $g : X \to Z$ is a continuous map of topological spaces that respects the equivalence relation $\sim$, so that $x_1 \sim x_2$ implies $g(x_1) = g(x_2)$. Show there is a unique continuous map $f : Y \to Z$ such that $g = f \circ \pi$. (One says that g "descends to the quotient".)

A map $f : X \to Y$ between topological spaces is a **homeomorphism** if it is continuous with continuous inverse, meaning that there is a continuous map $g : Y \to X$ such that $g \circ f = f \circ g = 1$. If such a pair of maps exist, we write $X \approx Y$ and say that X and Y are **homeomorphic** (or **topologically equivalent**). A property $P(X)$ is a **topological invariant** of X if $X \approx Y$ implies $P(X) = P(Y)$. That is, $P(X)$ depends only on the topology of X. The property $P(X)$ is a **complete topological invariant** of a space X provided that $P(X) = P(Y)$ if and only if $X \approx Y$. Topology can be loosely characterized as the study of the topological invariants of spaces.

3.2 Multivariable calculus facts

Let $U \subset \mathbb{R}^n$ be an open set and suppose that $f : U \to \mathbb{R}$ is a function. Label the points of $\mathbb{R}^n$ by the n-tuples $x = (x^1, \ldots, x^n)$. Then the **partial derivative** $\partial f / \partial x^i$ is defined by

$$\frac{\partial f}{\partial x^i} = \lim_{t \to 0} \frac{f(x + te_i) - f(x)}{t}, \tag{3.2}$$

where $e_i = (0, \ldots, 1, \ldots, 0)$ has a "1" in the ith slot. It is sometimes convenient to use multi-index notation to denote higher-order partial derivatives. Let $\alpha = (i_1, \ldots, i_k)$. Then

$$\frac{\partial f}{\partial x^\alpha}(x) := \frac{\partial^k f}{\partial x^{i_1} \cdots \partial x^{i_k}}. \tag{3.3}$$

A function $f : \mathbb{R}^n \to \mathbb{R}$ is C^∞, or **smooth**, if $\partial^k f / \partial x^\alpha$ exists and is continuous for all α. The composition of smooth functions is smooth.

Again let $U \subset \mathbb{R}^n$ be an open set but now suppose that $f : U \to \mathbb{R}^m$ is a map, given by $x \mapsto (f^1(x), \ldots, f^m(x))$. The map f is **smooth** if each **component function** f^i is smooth. The **derivative** $Df(x)$ of f at x is just the matrix of partial derivatives $Df(x) = (\partial f^i / \partial x^j)$; this matrix is called the **Jacobian matrix**.[7] When $n = m$, its determinant $\det(\partial f^i / \partial x^j)$ is called the **Jacobian determinant** or more simply the **Jacobian** of the map f.

> **EXERCISE 3.14** Show that if $f : U \subset \mathbb{R}^n \to \mathbb{R}^m$ is *linear* then $Df = f$. ("The derivative of a linear map equals the map itself.")

Now let $U, V \subset \mathbb{R}^n$ be two open sets, and let $f : U \to V$ be a homeomorphism. If f and f^{-1} are both smooth then f is called a **diffeomorphism**. This brings us to the all-important **inverse function theorem**, which says that if the matrix representing the derivative of a function is invertible at some point then the function itself is a local diffeomorphism in the neighborhood of that point.[8]

Theorem 3.1 *Let $W \subset \mathbb{R}^n$ and suppose that $f : W \to \mathbb{R}^n$ is a smooth map. If $a \in W$ and $Df(a)$ is nonsingular then there exists an open neighborhood U of a*

---

[7] Technically, the derivative $Df(x)$ is the unique linear map $T : \mathbb{R}^n \to \mathbb{R}^m$ satisfying

$$\lim_{\|h\| \to 0} \frac{\|f(x+h) - f(x) - T(h)\|}{\|h\|} = 0,$$

where $h \in \mathbb{R}^n$ and $\|x\|$ is the usual Euclidean length of x. The Jacobian matrix is just the representation of this linear map relative to the standard bases of $\mathbb{R}^n$ and $\mathbb{R}^m$. You can prove this for yourself or else look at [79], Chapter 2. The distinction is not really critical in Euclidean space, where the standard basis is almost always understood, but it will be more important later when we discuss general manifolds. By that time, though, we will have a fancier way of understanding the derivative map.

[8] For a proof, see *e.g.* [11], pp. 42ff or [79], pp. 35ff.

*in W such that $V = f(U)$ is open and $f : U \to V$ is a diffeomorphism. If $x \in U$
and $y = f(x)$ then $(Df^{-1})(y) = (Df(x))^{-1}$.*

The importance of the inverse function theorem for us lies in what it tells us about
coordinate systems, as we shall see in the next section.

> **EXERCISE 3.15** Prove the converse of the inverse function theorem: if $f : U \to V$
> is a diffeomorphism of open sets then $Df(x)$ is invertible at all points $x \in U$. *Hint:*
> Consider $f \circ f^{-1}$ and use the chain rule.

3.3 Coordinates

3.3.1 Coordinate functions

In the usual presentation of $\mathbb{R}^n$ we write the points as n-tuples $(x^1, \ldots, x^n)$. The
individual numbers x^i are called the **coordinates** of the point. We are so accus-
tomed to this presentation that we do not usually think too much about it. But there
is a hidden conceptual confusion because the presentation conflates two different
ideas, namely the points and their coordinates. The difficulty is resolved by thinking of the coordinates as **coordinate functions** that assign the coordinates to the
point. According to this point of view, x^i is a *function* from $\mathbb{R}^n$ to $\mathbb{R}$ that sends the
point $a = (a_1, \ldots, a_n)$ to the number a_i. Equivalently, we write

$$x^i(a) = a_i. \tag{3.4}$$

This seems perfectly fine. The problem comes when we want to express a point of
$\mathbb{R}^n$ as $x = (x^1, \ldots, x^n)$, for then we have the somewhat strange looking equation

$$x^i(x) = x^i. \tag{3.5}$$

In this equation, the x^i on the left is a function from $\mathbb{R}^n$ to $\mathbb{R}$, while the x^i on the
right is a simple number, namely the ith coordinate of the point $x = (x^1, \ldots, x^n)$.
Unfortunately, there is no way to avoid this notational crisis, because bad habits
have become so ingrained. You must decide from the context which meaning of x^i
is intended. Henceforth we will make every effort to avoid writing equations such
as (3.5).

Part of the reason why this problem is so pernicious is that we generally describe
points in terms of their coordinates. For example, the canonical presentation of the
unit 2-sphere S^2 defines it to be the subset of $\mathbb{R}^3$ given by $\{x \in \mathbb{R}^3 : (x^1)^2 +
(x^2)^2 + (x^3)^2 = 1\}$. How else would you define it? Almost every way you can
think of uses coordinates. Even if it were possible to avoid coordinates in order to
describe a sphere (and it is), it is essentially impossible to describe any moderately
complicated set of points without using coordinates. After all, this is the reason why
Descartes first introduced coordinates and why analytic geometry is so powerful.

Indeed, we will often refer to a point using its coordinates in some coordinate system. Nevertheless, we must remember in our own minds that *coordinates are just arbitrary labels* for the points.

3.3.2 Coordinate systems

This gives rise to a related problem, one which leads eventually to the idea of a manifold. The description of the 2-sphere given above uses too many coordinates. The 2-sphere S^2 is two dimensional, and our intuition says that a two-dimensional object ought to be describable in terms of two coordinates, not three. But it turns out that there is no single coordinate system of two coordinates that will do the job. Instead, we must use at least two different coordinate systems (each comprising two coordinates) to label the points of S^2.

To get a feel for the difficulty, let's try to describe S^2 using only a single pair of coordinates. The obvious choice is to use polar coordinates (θ, ϕ), so that points on the sphere are given by

$$x^1 = \sin\theta \cos\phi,$$
$$x^2 = \sin\theta \sin\phi,$$
$$x^3 = \cos\theta.$$

The problem lies in the *range* of the coordinates. For example, if we were to choose $0 \leq \theta < 2\pi$ and $0 \leq \phi < 2\pi$, some points would end up getting assigned more than one pair of coordinates, such as $(\pi/2, 0)$ and $(3\pi/2, \pi)$, both of which label the point $(1, 0, 0)$. Not only would it be confusing to have two distinct labels for the same object, it also makes taking derivatives impossible because a function must be single valued in order to be continuous and hence differentiable. To ameliorate this problem one usually takes the range to be $0 \leq \theta \leq \pi$ and $0 \leq \phi < 2\pi$. But even this is no good, because the north and south poles are each labeled by an infinity of pairs, namely $(0, \phi)$ and (π, ϕ), respectively. To avoid the problem, one can take $0 < \theta < \pi$ and $0 \leq \phi < 2\pi$, but then the poles are not labeled at all.

In fact, no matter how hard you try, you will never find a single set of coordinates to describe the 2-sphere. The reason is that the sphere and a part of the plane are topologically distinct. But, as we shall see, we can label all the points of S^2 with two sets of coordinates because we can cut the sphere into two pieces, each of which can be flattened to look like a piece of the plane.

3.3.3 Change of coordinates

Consider the transformation in $\mathbb{R}^2$ between Cartesian and polar coordinates:

$$x = r \cos\theta,$$
$$y = r \sin\theta.$$

(3.6)

Every point in the plane can be described uniquely in the (x, y) coordinate system, but the origin is a problem for the polar coordinate system because it is described by the infinity of pairs $(r, \theta) = (0, \text{anything})$. Another way to see that something strange happens at the origin is to compute the Jacobian of the transformation. Doing so gives

$$\begin{vmatrix} \partial x/\partial r & \partial x/\partial \theta \\ \partial y/\partial r & \partial y/\partial \theta \end{vmatrix} = \begin{vmatrix} \cos \theta & -r \sin \theta \\ \sin \theta & r \cos \theta. \end{vmatrix} = r, \tag{3.7}$$

which vanishes at the origin.

The inverse function theorem provides the link between these two ways of determining the validity of a given coordinate transformation. A coordinate transformation is a good one if there is a one-to-one correspondence between the two sets of coordinates and if the transformation is differentiable. In other words, a set of functions $\{f^i(x)\}$ on $\mathbb{R}^n$ constitutes a good coordinate system in the neighborhood of a point x if the transformation $(x^1, \ldots, x^n) \mapsto (f^1(x^1, \ldots, x^n), \ldots, f^n(x^1, \ldots, x^n))$ is a diffeomorphism.[9] Ascertaining whether a map is a diffeomorphism can be difficult. But checking whether a Jacobian vanishes is usually easy. The inverse function theorem assures us that, as long as the Jacobian of the transformation is nonsingular, we can coordinatize the neighborhood of x with the functions f^i.

3.4 Differentiable manifolds

This leads us at last to the definition of a smooth manifold. An n-**dimensional smooth manifold** M consists of a Hausdorff topological space together with a countable collection of open sets $\{U_i\}$, called **coordinate neighborhoods** or **coordinate patches**, that cover M and a collection of maps $\{\varphi_i\}$, called **coordinate maps**, satisfying two conditions.

(1) Each $\varphi_i : U \to \mathbb{R}^n$ is a homeomorphism onto an open subset of $\mathbb{R}^n$. (We say that M is **locally Euclidean**.)

(2) If U_i and U_j are two overlapping coordinate neighborhoods with coordinate maps φ_i and φ_j then $\varphi_j \circ \varphi_i^{-1} : \varphi_i(U_i \cap U_j) \to \varphi_j(U_i \cap U_j)$ is a diffeomorphism.[10] (We say that the coordinate maps are **compatible on overlaps**.)

---

[9] To be fair, if we didn't want to differentiate anything we would only need it to be a homeomorphism. But we do want to differentiate things.

[10] If we replace the word "diffeomorphism" with the word "homeomorphism" we simply get a manifold, sometimes called a **topological manifold**.

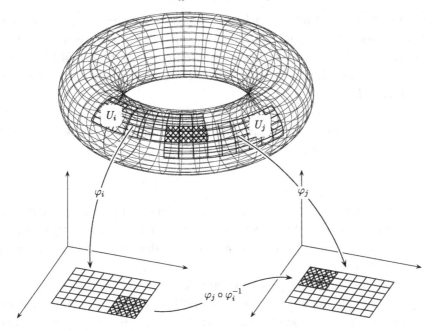

Figure 3.3 Two coordinate charts and a transition function.

Each pair (U_i, φ_i) is called a **coordinate chart**, and the collection of all coordinate charts is called an **atlas**. The maps $\varphi_j \circ \varphi_i^{-1}$ are called **transition functions** of the atlas. See Figure 3.3.

The condition that M be Hausdorff is there basically to express our intuition of space as "infinitely divisible", so that we can separate points with open sets. The condition that the cover be countable is there for a technical reason having to do with extending locally defined quantities to globally defined ones. The condition that M be locally Euclidean serves at least two purposes. First, it tells us that, in the neighborhood of a point, all n-dimensional manifolds look like a (mildly deformed) bit of Euclidean n-space. Second, it allows us to define local coordinates so that we can compute things. The compatibility condition ensures that we can patch together the coordinate systems consistently, so that we always end up with valid coordinates.

Let (U, φ) be a coordinate chart with $p \in U$ and suppose that $\varphi(p) = q$. If $x^1, \ldots, x^n$ are the standard coordinate functions on $\mathbb{R}^n$ then q has coordinates $(x^1(q), \ldots, x^n(q))$. Thus we can write

$$\varphi(p) = (x^1(q), \ldots, x^n(q)). \tag{3.8}$$

It is a convenient abuse of notation to view the x^i as coordinate functions on U instead of $\varphi(U)$ and to write

$$\varphi(p) = (x^1(p), \ldots, x^n(p)). \tag{3.9}$$

As U generally does not live in Euclidean space, (3.9) makes no sense unless it is interpreted to mean (3.8). (Of course, if U is a subset of Euclidean space then φ is just the identity map, so (3.9) makes perfect sense; this is the justification for the convention.) The functions $x^1, \ldots, x^n$, viewed as functions on U, are called **local coordinates on** U.

Let V be another coordinate neighborhood, with local coordinates $y^1, \ldots, y^n$. If $U \cap V \neq \emptyset$ then on $\varphi(U \cap V)$ the action of the transition function $\psi \circ \varphi^{-1}$ can be written in local coordinates as follows:

$$(x^1, \ldots, x^n) \mapsto (y^1(x^1, \ldots, x^n), \ldots, y^n(x^1, \ldots, x^n)). \tag{3.10}$$

The compatibility condition (2) in the definition of a manifold (see the start of this section) is just the statement that (3.10) is a diffeomorphism, which, by the inverse function theorem, is equivalent to the requirement that the Jacobian determinant $\det(\partial y^i / \partial x^j)$ be nonzero.

The sign of this determinant is important. A manifold is said to be **orientable** if it is possible to choose an ordering of the local coordinates so that the Jacobian determinants of the transition functions have the same sign on every pair of overlapping neighborhoods.[11] If this is possible then the manifold has two opposite **orientations**, according to the choice of sign. For the most part we will concern ourselves only with orientable manifolds, because they are the ones that arise most often in applications, although we will have occasion to consider non-orientable manifolds as well. As orientable manifolds enjoy many properties not shared by nonorientable manifolds, we are often obliged to distinguish the two possibilities explicitly.

Example 3.5 The Cartesian space $\mathbb{R}^n$ is an orientable n-dimensional manifold. (Surprise.)

Example 3.6 The unit 2-sphere S^2 is an orientable two dimensional manifold. To see this, we must cover it with a set of compatible coordinate charts, then check the signs of the Jacobians.[12] A traditional way to accomplish this is to cover the sphere with six patches, but here we employ **stereographic projection** in order to use only two. The first chart, (U, φ), is defined as follows. The open set U in $\mathbb{R}^3$ covers the entire sphere except for the north pole N. Pick a point $P \in U$, and let $\varphi(P)$ be the unique point Q on the $z = 0$ plane such that the line $\overline{NQ}$ contains P.

---

[11] In Appendix C we discuss an easier way to verify the orientability of smooth manifolds using differential forms. Orientations of two dimensional surfaces are the easiest to understand, especially if they are viewed as subsets of $\mathbb{R}^3$. In that case, one can show that a surface is orientable if it is possible to define on it a nowhere-vanishing normal vector field.

[12] Actually, we should also show that it is Hausdorff with a countable cover, but this follows automatically by equipping S^2 with the subspace topology.

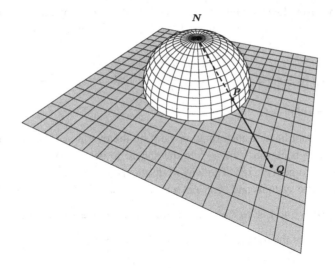

Figure 3.4 Stereographic projection.

(See Figure 3.4.) Observe that the upper hemisphere projects onto points on the plane outside the sphere while the lower hemisphere projects to points on the plane inside the sphere. This is called *stereographic projection* from N. Let S be the south pole, and define (V, ψ) similarly via stereographic projection from S.[13]

Define coordinates (u^1, u^2) such that $Q = \varphi(P) = \varphi(x_P, y_P, z_P) = (u^1, u^2, 0)$. The line from N to Q is described parametrically by $N + t(Q - N)$, so to find the value of t corresponding to the point P we write

$$(x_P, y_P, z_P) = (0, 0, 1) + t(u^1, u^2, -1) = (tu^1, tu^2, 1 - t)$$

and solve for t. The point P is on the sphere, so the sum of the squares of the coordinates is unity, which leads to

$$t = \frac{2}{1 + \eta}$$

where $\eta := (u^1)^2 + (u^2)^2$. We thus obtain

$$x_P = \frac{2u^1}{1 + \eta}, \qquad y_P = \frac{2u^2}{1 + \eta}, \qquad \text{and} \qquad z_P = -\frac{1 - \eta}{1 + \eta}.$$

The patch U is defined by the range $0 \le u^i < \infty$, $i = 1, 2$. The lower hemisphere is mapped to the disk $\eta \le 1$ while the upper hemisphere is mapped to the points with $\eta \ge 1$.

---

[13] Stereographic projection is often defined from the sphere onto the tangent plane at the south pole but, as we wish to consider projections from the south and north pole simultaneously, the median plane is more convenient here.

Projecting from the south pole instead, we coordinatize V by pairs (v^1, v^2), defined so that $Q = \psi(P) = \psi(x_P, y_P, z_P) = (v^1, v^2, 0)$. Repeating the argument above gives

$$x_P = \frac{2v^1}{1 + \xi}, \qquad y_P = \frac{2v^2}{1 + \xi}, \qquad \text{and} \qquad z_P = \frac{1 - \xi}{1 + \xi},$$

where $\xi := (v^1)^2 + (v^2)^2$.

To find the coordinate transformation valid on the overlap $\varphi(U \cap V)$ we start from (u^1, u^2), map backwards via φ^{-1} to the sphere to get (x_P, y_P, z_P), and then map down via ψ to get (v^1, v^2). The way to do this is just to equate (x_P, y_P, z_P) in the two coordinate systems and carry out a bit of tedious algebra until you discover that

$$(v^1(u^1, u^2), v^2(u^1, u^2)) = (\psi \circ \varphi^{-1})(u^1, u^2) = (u^1, u^2)/\eta.$$

The Jacobian of the transformation $\psi \circ \varphi^{-1}$ is therefore

$$\frac{\partial(v^1, v^2)}{\partial(u^1, u^2)} = -\frac{1}{\eta^2}$$

on the overlap $\varphi(U \cap V)$. Evidently this is nowhere vanishing. (Also, it does not blow up, because the two patches do not overlap at the south pole where $\eta = 0$.) Hence, by the inverse function theorem, $\psi \circ \varphi^{-1} : \varphi(U \cap V) \to \psi(U \cap V)$ is a diffeomorphism, so S^2 is a smooth manifold. As the Jacobian is everywhere of one sign, S^2 is orientable.[14]

Example 3.7 If you take a rectangular strip of paper and bend it around until the left and right edges meet then glue the edges together, you get a cylinder. But if you hold the left side and twist the right side through 180 degrees, then bend it around and glue the ends together, you get a **Möbius strip** (Figure 3.5). This process of gluing together edges is formally known as **identifying** the edges. Mathematically, the Möbius strip can be viewed as the quotient space $(\mathbb{R} \times [0, 1])/\sim$, where we have $(x, y) \sim (x + 1, -y)$. If you substitute $\mathbb{R}$ for the closed interval $[0, 1]$ in the definition, you get the (infinite) **Möbius band** B. We claim that B is a two-dimensional nonorientable smooth manifold.[15]

Let $[x, y]$ denote the equivalence class of (x, y) in B. It's a bit less confusing if we cover B with three patches rather than the traditional two. Let's choose

$$U_1 = \{[x, y] : 0 < x < 2/3, y \in \mathbb{R}\},$$
$$U_2 = \{[x, y] : 1/3 < x < 1, y \in \mathbb{R}\},$$
$$U_3 = \{[x, y] : 2/3 < x < 4/3, y \in \mathbb{R}\}.$$

---

[14] If we were to list the coordinates v^1 and v^2 in the opposite order we could force the Jacobian to be positive on the overlap. All that we require is the same sign on every overlap.

[15] So far all our manifolds have been without boundary. The Möbius strip is an example of a (two-dimensional smooth nonorientable) manifold with boundary. For more information about those sorts of things you'll need to read Chapter 6.

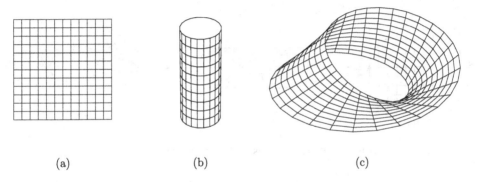

(a) (b) (c)

Figure 3.5 (a) A strip of paper. (b) A cylinder. (c) A Möbius strip.

Define coordinate maps $\varphi_i : U_i \to \mathbb{R}^2$ by $\varphi_i([x, y]) = (x, y)$, where (x, y) is the unique representative of $[x, y]$ in the parameter range corresponding to U_i. For example, $\varphi_1([1/3, y]) = (1/3, y)$ but $\varphi_1([4/3, y]) = \varphi_1([1/3, -y]) = (1/3, -y)$.

Now consider the transition maps. On $U_1 \cap U_2$ we have $(\varphi_1 \circ \varphi_2^{-1})(x, y) = (x, y)$, because $\varphi_1([x, y]) = \varphi_2([x, y])$. For example, $\varphi_1([1/2, y]) = (1/2, y) = \varphi_2([1/2, y])$. The same thing happens on $U_2 \cap U_3$. In particular, both these transition maps have Jacobians equal to 1. But on $U_1 \cap U_3$ something unexpected happens. For example, $\varphi_1^{-1}(1/6, y) = [1/6, y] = [7/6, -y]$ so $(\varphi_3 \circ \varphi_1^{-1})(1/6, y) = (7/6, -y)$. In general, $(\varphi_3 \circ \varphi_1^{-1})(x, y) = (x + 1, -y)$. Observe that the Jacobian of this transformation is

$$\begin{vmatrix} \partial(x+1)/\partial x & \partial(-y)/\partial x \\ \partial(x+1)/\partial y & \partial(-y)/\partial y \end{vmatrix} = \begin{vmatrix} 1 & 0 \\ 0 & -1 \end{vmatrix} = -1.$$

All the Jacobians are nonvanishing, so all the transition maps are diffeomorphisms. Hence B is a smooth manifold. But the Jacobians have different signs, so we cannot conclude anything about orientability. By thinking about this example a bit harder, though, you should be able to convince yourself that you will always get a sign mismatch for any set of coordinate charts. (Basically the reason is that any covering has to include an open set that contains points of the form $[0, y]$, and on these the Jacobian always flips sign.) Hence the Möbius strip is not orientable. (For a slightly more rigorous proof, see [27], Section 2.8.)

EXERCISE 3.16 Let $\mathbb{Z}^n$ be the integer lattice in $\mathbb{R}^n$, namely the set of all points whose coordinates are integers. The n-torus T^n is defined to be the quotient space $T^n = \mathbb{R}^n/\mathbb{Z}^n$. For example, the 2-torus T^2 is the set of all points $(x, y) \in \mathbb{R}^2$ such that $(x+1, y) \sim (x, y) \sim (x, y+1)$. Equivalently, it is the set of all points in the unit square with outside boundary points identified as follows. First identify the top and bottom boundaries of the unit square along the arrows to get a cylinder, then identify the left and right boundaries along the arrows to get the 2-torus (a doughnut). (See Figure 3.6.)

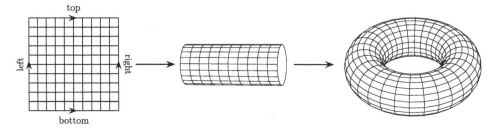

Figure 3.6 The 2-torus obtained via identifications.

Show that T^2 is an orientable manifold. *Hint:* You may assume that T^2 is Hausdorff with a countable cover. (If, however, you are itching to prove this, you may want to revisit the notion of a quotient topology.) By analogy with Example 3.7, cover T^2 with nine patches. For instance, take

$$U_1 = \{[x, y] : 0 < x < 2/3, 0 < y < 2/3\},$$
$$U_2 = \{[x, y] : 0 < x < 2/3, 1/3 < y < 1\},$$
$$U_3 = \{[x, y] : 0 < x < 2/3, 2/3 < y < 4/3\},$$
$$U_4 = \{[x, y] : 1/3 < x < 1, 0 < y < 2/3\},$$

etc., together with the analogous coordinate maps.

EXERCISE 3.17 Real projective n-space $\mathbb{RP}^n$ consists of the set of all lines through the origin in $\mathbb{R}^{n+1}$. If x is a nonzero vector in $\mathbb{R}^{n+1}$, let $[x]$ denote the line determined by x. Clearly, $[x] = [y]$ if and only if $x = \lambda y$ for some $\lambda \neq 0$. Let $U_i = \{[x^0, x^1, \dots, x^n] : x^i \neq 0\} \subset \mathbb{RP}^n$ be an open subset of $\mathbb{RP}^n$ and, for $0 \leq i \leq n$, define coordinate maps $\varphi_i : U_i \to \mathbb{R}^n$ by

$$\varphi_i([x^0, x^1, \dots, x^n]) = \left(\frac{x^0}{x^i}, \frac{x^1}{x^i}, \dots, \frac{x^{i-1}}{x^i}, \frac{x^{i+1}}{x^i} \dots, \frac{x^n}{x^i} \right).$$

Show that the pairs (U_i, φ_i) define a set of compatible coordinate neighborhoods, thereby making $\mathbb{RP}^n$ a (smooth) n-dimensional manifold. What can you say about the orientability of $\mathbb{RP}^n$? *Remark:* The tricky part about this problem is finding the correct inverse maps. Technically, you should really show that $\mathbb{RP}^n$ is Hausdorff with a countable cover in the natural (quotient) topology, but you can skip this part.

3.5 Smooth maps on manifolds

The definition of the derivative given in Section 3.2 uses the linear structure of Euclidean space in a crucial way, and there is no way to define something similar for a general curved space. Instead, we use the existence of the coordinate maps on a smooth manifold to define what we mean by differentiability.

Specifically, a function $f : M \to \mathbb{R}$ is **smooth** at $p \in M$ if, for any chart (U, φ) with $p \in U$, the map

$$\widetilde{f} := f \circ \varphi^{-1} : \varphi(U) \to \mathbb{R} \qquad (3.11)$$

is a smooth function in the usual Euclidean sense near $\varphi(p)$. This definition is independent of coordinate chart by the very definition of a manifold for, if (V, ψ) is some other chart with $p \in V$ then, as $\psi \circ \varphi^{-1}$ is smooth, $f \circ \varphi^{-1} = f \circ \psi^{-1} \circ \psi \circ \varphi^{-1}$ is smooth if and only if $f \circ \psi^{-1}$ is smooth. The space of smooth functions on M is denoted $\Omega^0(M)$. Once again, if $x^1, \ldots, x^n$ are local coordinates near p, we abuse notation and write $f = f(x^1, \ldots, x^n)$.

The same idea applies to more general maps. Let M and N be two smooth manifolds of dimensions m and n, respectively. A map $f : M \to N$ is **smooth** if for every $p \in M$ there exist charts (U, φ) on M and (V, ψ) on N, with $p \in U$ and $f(U) \subset V$ such that

$$\widetilde{f} := \psi \circ f \circ \varphi^{-1} : \varphi(U) \to \psi(V) \qquad (3.12)$$

is a smooth map of Euclidean spaces. A smooth map $f : M \to N$ is a **diffeomorphism** if f^{-1} exists and is smooth, in which case we say that M and N are **diffeomorphic**.

If $x^1, \ldots, x^n$ are local coordinates near U and $y^1, \ldots, y^m$ are local coordinates near V then we should write

$$f(x^1, \ldots, x^m) = (f^1(x^1, \ldots, x^m), \ldots, f^n(x^1, \ldots, x^m)), \qquad (3.13)$$

where $f^i = y^i \circ f$, but once again we adopt the conventional but sloppy notation and write instead[16]

$$f(x^1, \ldots, x^m) = (y^1(x^1, \ldots, x^m), \ldots, y^n(x^1, \ldots, x^m)). \qquad (3.14)$$

Writing it in this way makes it appear that $f : M \to N$ is an ordinary multivariable function on Euclidean space and in a sense it is, via the coordinate maps in a neighborhood of a point. For this reason we will use the notation (3.14) in all

---

[16] Although (3.10) and (3.14) look similar, they have different meanings. The former is a change of coordinates between two coordinate neighborhoods of the same point on M, whereas the latter is a map between a neighborhood of M and a neighborhood of N. But there is a sense in which they can be considered to be equivalent, which is related to the notion of **active** and **passive** transformations. The idea is illustrated in Figure 3.7. Let M be a disk in the plane, equipped with the usual Cartesian coordinates. According to the active point of view, we leave the coordinate system alone but deform the disk until it looks like N. A point $p \in M$ with coordinates (a, b) is mapped to a point $q \in N$ with coordinates (c, d). According to the passive point of view, we leave the points of M alone, but deform the coordinate system so that the coordinates of p change from (a, b) to (c, d). As with optical illusions either viewpoint is legitimate, but keeping both in your mind at once will drive you crazy. The problem stems from the use of coordinates, so it is best to avoid their use whenever possible. Unfortunately, one does not always have this luxury. To avoid confusion from the start, you are generally better off distinguishing the interpretations of (3.10) and (3.14).

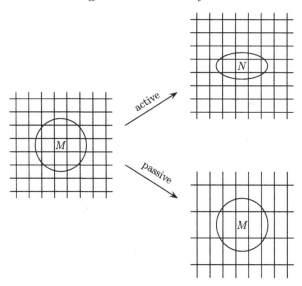

Figure 3.7 Active and passive interpretations of a diffeomorphism.

that follows, remembering that it is only true locally, and that it employs various notational liberties.

Remark Because smooth maps of manifolds reduce to smooth maps of Euclidean space, the inverse function theorem works for manifolds in the same way as it does in Euclidean space: if the Jacobian matrix of f is nonsingular then f is a local diffeomorphism. In particular, M and N can only be diffeomorphic if $m = n$.

Remark In multivariable calculus one defines curves and surfaces in $\mathbb{R}^3$ by means of a parameterization. For example, a curve is a map $\gamma : I \to \mathbb{R}^3$ where I is an open subset of $\mathbb{R}$ given by $\gamma(t) = (x^1(t), x^2(t), x^3(t))$. A surface is a map $\sigma : U \to \mathbb{R}^3$, where U is an open subset of $\mathbb{R}^2$ given by $\sigma(u, v) = (x^1(u, v), x^2(u, v), x^3(u, v))$. A coordinate map of a manifold can be thought of as a kind of inverse parameterization. More precisely, if U is an open subset of $\mathbb{R}^n$ and $\psi : U \to M$ is a diffeomorphism onto an open set $V \subset M$ then (U, ψ) is a **parameterization** of V and (V, ψ^{-1}) is a coordinate chart.

3.6 Immersions and embeddings

Let M and N be smooth manifolds of dimensions m and n, respectively, and let $f : M \to N$ be a smooth map, represented in local coordinates by (3.14). If $m \leq n$ and the Jacobian matrix $(\partial y^i / \partial x^j)$ of the transformation (3.14) has maximal rank (namely m) at $p \in M$ then f is called an **immersion at** p, and if f is an immersion for all $p \in M$ then M is an (immersed) **submanifold** of N. The definition

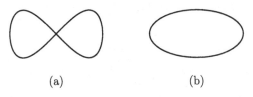

(a) (b)

Figure 3.8 (a) An immersed circle. (b) An embedded circle.

makes sense, because if the Jacobian has maximal rank then $f(M)$ is locally coor-
dinatizable according to the inverse function theorem. If $m \geq n$ and the Jacobian
of the transformation has maximal rank (namely n) at $p \in M$ then f is called a
submersion at p.

An injective immersion f is called an **embedding** (or **imbedding**), provided
that f maps M homeomorphically onto its image $f(M)$ (in the induced topology). The basic difference between immersions and embeddings is that the image
of an immersion can have self-intersections whereas the image of an embedding
cannot. (See Figure 3.8.) The celebrated **Whitney embedding theorem** says that
any n-dimensional *topological* manifold can be embedded in $\mathbb{R}^{2n+1}$, and any n-
dimensional *smooth* manifold can be embedded in $\mathbb{R}^{2n}$. For example, the Klein
bottle is a smooth 2-manifold obtained in the same manner as the torus in Example 3.16 except that the arrows on the right and at the top of the square are reversed
prior to making the identifications. If you try to construct the Klein bottle in $\mathbb{R}^3$ you
will observe that you cannot do it without creating self-intersections somewhere
(Figure 3.9). But, thanks to Whitney's theorem, we know that it can be embedded
in $\mathbb{R}^4$. Of course, the dimension $2n$ is only an upper bound. For example, we know
that the 2-sphere can be embedded into $\mathbb{R}^3$.

Whitney's result means that, without loss of generality, we could simply treat
manifolds as living in a large Euclidean space. The reason we do not do this is
primarily aesthetic. Mathematicians strive for parsimony when making definitions,
and it is inelegant to define an object with reference to something outside itself.
The intrinsic description of a manifold also has physical utility. Einstein modeled
the universe as a smooth manifold of a certain type. By definition, the universe
is everything. What could it possibly mean to embed the universe in something
outside itself?

We can use the ideas of this section to write down a very handy criterion for
obtaining manifolds from maps. Suppose that $f : M \to N$ with $m \geq n$. If f
is a submersion at $p \in M$ then p is called a **regular point** of f, otherwise it is
a **critical point** of f. A point $q \in N$ is a **regular value** of f if every point in
$f^{-1}(q)$ is regular.[17] By famous theorems of Brown and Sard, the regular values of

---

[17] This means that q is automatically a regular value if $f^{-1}(q)$ is empty.

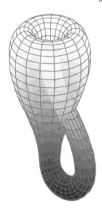

Figure 3.9 A Klein bottle immersed in $\mathbb{R}^3$. (Source: Wikipedia)

any smooth function f are dense in N, so in this sense almost all the points of N are regular values.

Theorem 3.2 (Regular value theorem) *Let $f : M \to N$ be a smooth map of manifolds, with $\dim M = m$ and $\dim N = n$. Let $m \geq n$ and let $q \in N$ be a regular value of f. Then $f^{-1}(q)$ is a smooth embedded submanifold of M of dimension $m - n$.*

Proof Let $y^1, \ldots, y^n$ be local coordinates in a neighborhood V of q, with $y^i(q) = 0$. In $U = f^{-1}(V)$, consider the functions $f^i := y^i \circ f$ for $1 \leq i \leq n$. The first claim is that $f^1, \ldots, f^n$ can be completed to a set of good coordinates on an open subset of U. To do this, let $x_1, \ldots, x_m$ be good coordinates in a neighborhood of $p \in U$, and consider the entire set $\{f^1, \ldots, f^n, x^1, \ldots, x^m\}$. By hypothesis the Jacobian $[\partial f^i / \partial x^j]_{1 \leq i \leq n, 1 \leq j \leq m}$ has rank n so, by linear algebra, some $n \times n$ minor of the Jacobian matrix is nonzero. Without loss of generality, suppose it is the principal minor indexed by the first n columns. Discarding $x^1, \ldots, x^n$ from our list we are left with the set $\{f^1, \ldots, f^n, x^{n+1}, \ldots, x^m\}$. The Jacobian of the coordinate transformation $\{x^1, \ldots, x^m\} \mapsto \{f^1, \ldots, f^n, x^{n+1}, \ldots, x^m\}$, which we may write as $[\partial f^i / \partial x^j \; \partial x^k / \partial x^j]_{1 \leq i \leq n, n+1 \leq k \leq m, 1 \leq j \leq m}$, is then nonzero, so by the inverse function theorem the coordinate transformation is a diffeomorphism (possibly on some smaller neighborhood U' of p), and the first claim is proved.

The second claim is that a subset of these coordinates provides good coordinates for $f^{-1}(q)$. Specifically, note that for all $p \in U'$ we have $f^i(p) = 0$. The map $p \mapsto (0, \ldots, 0, x^{n+1}(p), \ldots, x^m(p))$ is manifestly a homeomorphism from U' onto its image in $\mathbb{R}^{m-n}$. It remains only to show that the transition functions for charts of this form are diffeomorphisms. So, let U'' be some other open set in M containing p, with coordinates $(g^1, \ldots, g^n, z^{n+1}, \ldots z^m)$ constructed as above, and consider the coordinate transformation represented symbolically by $(f, x) \mapsto (g, z)$. The Jacobian determinant of this map can also be written symbolically, as

$$\begin{vmatrix} \partial g / \partial f & \partial g / \partial x \\ \partial z / \partial f & \partial z / \partial x \end{vmatrix},$$

and it is nonzero by the first claim. But the g^i are different coordinate functions representing the map f, so everywhere on $f^{-1}(q)$ we have $g^i(0, x) = 0$. In particular, $\partial g^i / \partial x^j = 0$ on $f^{-1}(q)$. It follows that $\det(\partial z / \partial x) \neq 0$, so the second claim is proved. □

Remark As points are closed in a Hausdorff space, $f^{-1}(q)$ is a closed subset of M.

    EXERCISE 3.18 Using Theorem 3.2, show that the n-dimensional sphere S^n is a manifold.

3.7 The tangent space

To motivate what follows, let us return to Euclidean space for a moment. Let ψ be a scalar field on $\mathbb{R}^3$ and let $\gamma : I \to \mathbb{R}^3$ be a parameterized curve given by $t \mapsto (\gamma^1(t), \gamma^2(t), \gamma^3(t)) = (x^1(t), x^2(t), x^3(t))$. Then

$$\frac{d}{dt} \psi(\gamma(t)) = \frac{\partial \psi}{\partial x} \frac{dx}{dt} + \frac{\partial \psi}{\partial y} \frac{dy}{dt} + \frac{\partial \psi}{\partial z} \frac{dz}{dt} = \frac{d\gamma}{dt} \cdot \nabla \psi, \qquad (3.15)$$

where ∇ is the gradient operator $(\partial / \partial x, \partial / \partial y, \partial / \partial z)$ and $d\gamma / dt$ is the tangent vector to the curve. Equation (3.15) is the directional derivative of ψ in the direction of $d\gamma / dt$. It tells us how ψ changes as we move along the curve. But ψ is arbitrary; if we were to strip away ψ from both sides of (3.15) we would get

$$\frac{d}{dt} = \frac{d\gamma}{dt} \cdot \nabla. \qquad (3.16)$$

    Equation (3.16) gives us a clue how the idea of a tangent vector can be generalized from Euclidean space to a general manifold. We cannot situate arrows or straight lines inside an arbitrary manifold (because a general manifold is not a vector space), but we can retain the essence of a tangent vector as something that tells you how a function changes as you move in that direction. So instead of $d\gamma / dt$, which lives in Euclidean space, we now view d/dt itself as the tangent vector to the curve. This perspective can be a bit confusing at first, because γ seems to have disappeared altogether. Nevertheless, γ is still there because d/dt makes no sense as a tangent vector in the absence of a curve.[18]

---

[18] For a detailed explanation of the correct way to understand d/dt, see the discussion of integral curves in Section 3.16.

Example 3.8 As a simple illustration of the utility of this perspective, consider again the transformation (3.6) between Cartesian and polar coordinates in the plane. By considering the effect of the gradient operator we see that the vector $\partial/\partial x$ (respectively, $\partial/\partial y$) points in the direction of increase of the x (respectively, y) coordinate. We set the length of each vector to unity and therefore write

$$e_x = \frac{\partial}{\partial x} \quad \text{and} \quad e_y = \frac{\partial}{\partial y}$$

for the usual orthonormal basis of $\mathbb{R}^2$.

From (3.6) and the chain rule we compute

$$\frac{\partial}{\partial r} = \frac{\partial x}{\partial r}\frac{\partial}{\partial x} + \frac{\partial y}{\partial r}\frac{\partial}{\partial y} = \cos\theta\frac{\partial}{\partial x} + \sin\theta\frac{\partial}{\partial y},$$

$$\frac{\partial}{\partial \theta} = \frac{\partial x}{\partial \theta}\frac{\partial}{\partial x} + \frac{\partial y}{\partial \theta}\frac{\partial}{\partial y} = -r\sin\theta\frac{\partial}{\partial x} + r\cos\theta\frac{\partial}{\partial y}.$$

The vector $\partial/\partial r$ evidently has unit length, so it must be the unit vector e_r in the r direction. The vector $\partial/\partial\theta$ points in the θ direction (i.e., the direction in which θ increases), but it must be scaled by r to yield a unit vector. Thus we have

$$e_r = \cos\theta\, e_x + \sin\theta\, e_y$$

$$e_\theta = -\sin\theta\, e_x + \cos\theta\, e_y.$$

The reader should verify that these are indeed the correct unit vectors. (See Figure 3.10.)

EXERCISE 3.19 The coordinate transformation from spherical polar to Cartesian coordinates is given by

$$x = r\sin\theta\cos\phi,$$
$$y = r\sin\theta\sin\phi,$$
$$z = r\cos\theta.$$

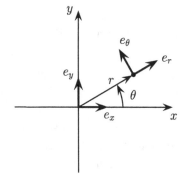

Figure 3.10 Cartesian and polar unit vectors.

By analogy with the previous example, show that the unit vectors in spherical polar coordinates are given by

$$e_r = \sin\theta\cos\phi\,e_x + \sin\theta\sin\phi\,e_y + \cos\theta\,e_z,$$
$$e_\theta = \cos\theta\cos\phi\,e_x + \cos\theta\sin\phi\,e_y - \sin\theta\,e_z,$$
$$e_\phi = -\sin\phi\,e_x + \cos\phi\,e_y.$$

To extend these ideas to a general manifold M, we define a **tangent vector** X_p at a point $p \in M$ to be a **linear derivation** at p. This means that, for $a, b \in \mathbb{F}$ and $f, g \in \Omega^0(M)$, $X_p : \Omega^0(M) \to \mathbb{R}$ satisfies

(1) linearity: $X_p(af + bg) = aX_p(f) + bX_p(g)$, and
(2) the Leibniz property: $X_p(fg) = g(p)X_p(f) + f(p)X_p(g)$.

The vector space T_pM generated by all the X_p is called the **tangent space to M at p**. Although the tangent space T_pM is an abstract linear space attached to M at p, we often imagine it to look like the picture in Figure 3.11, because this agrees with our intuitions for a manifold embedded in Euclidean space. Still, the picture is somewhat misleading (because manifolds are defined independently of any embedding); a better way to think of the tangent space is in terms of something called the "tangent bundle", which we will describe in Example 7.3.

Theorem 3.3 *Let $x^1, \ldots, x^n$ be local coordinates in a neighborhood U of a point $p \in M$. Then T_pM is spanned by the tangent vectors $\partial/\partial x^1, \ldots, \partial/\partial x^n$. In particular, $\dim T_pM = n$.*

Proof Without loss of generality we may assume that $(x^1(p), \ldots, x^n(p)) = (0, \ldots, 0) = 0$. Let X_p be a tangent vector at p, and define $X^i := X_p(x^i)$. As discussed in Section 3.5, in U the smooth map $f : M \to \mathbb{R}$ can be written as

$$(x^1, \ldots, x^n) \mapsto f(x^1, \ldots, x^n),$$

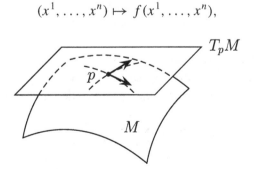

Figure 3.11 The idea of the tangent space at a point.

which means that we can apply Taylor's theorem with remainder. Writing $x = (x^1, \ldots, x^n)$ we have, in the neighborhood of p,

$$f(x) = f(0) + \sum_i x^i g_i(x) \quad \text{with} \quad g_i(0) = \left.\frac{\partial f}{\partial x^i}\right|_p. \tag{3.17}$$

As usual, we identify the constant $f(0)$ with the constant function $f(0) \cdot 1$, where 1 is the unit function. The tangent vector axioms imply that X_p acting on a constant function vanishes,[19] so, applying X_p to the equation on the left in (3.17) yields (remembering that $x^i(p) = 0$)

$$X_p(f) = X_p(f(0)) + \sum_i X_p(x^i) g_i(0) + \sum_i (0) X_p(g_i)$$

$$= 0 + \sum_i X^i \left.\frac{\partial f}{\partial x^i}\right|_p + 0,$$

from which we conclude that

$$X_p = \sum_i X^i \left.\frac{\partial}{\partial x^i}\right|_p. \tag{3.18}$$

It follows that the tangent vectors $\partial/\partial x^i$ form a basis for $T_p M$, which is therefore n-dimensional. The scalars X^i are the components of X_p in the coordinate system $x_1, \ldots, x_n$ (or, in the basis $\partial/\partial x^i$). $\qquad\square$

Remark Applying Theorem 3.3 to $M = \mathbb{R}^n$ allows us to identify $T_p \mathbb{R}^n$ with $\mathbb{R}^n$ itself, because points and vectors are naturally identified in $\mathbb{R}^n$ (cf. Figure 3.10).

If $\{y^1, \ldots, y^n\}$ is another set of local coordinates in a neighborhood of p, Theorem 3.3 says that the $\partial/\partial y^j$ constitute an equally valid basis of $T_p M$. According to the chain rule,

$$\frac{\partial}{\partial y^j} = \sum_i \frac{\partial x^i}{\partial y^j} \frac{\partial}{\partial x^i}. \tag{3.19}$$

The key fact to remember from this result is that the change of basis matrix taking us from one set of coordinate basis vectors to another is precisely the (inverse of the) Jacobian matrix. The vector X_p may be expanded in the $\partial/\partial y^j$ basis as

$$X_p = \sum_j Y^j \left.\frac{\partial}{\partial y^j}\right|_p, \tag{3.20}$$

---

[19] Proof: By the Leibniz property applied to the unit functions $f = g = 1$, $X_p(1) = X_p(1 \cdot 1) = 2X_p(1) \Rightarrow X_p(1) = 0$, so by linearity, $X_p(c \cdot 1) = cX_p(1) = 0$ for any constant c.

for some components Y^j. Substituting (3.19) into (3.18) and comparing with (3.20) gives

$$Y^j = \sum_i X^i \left.\frac{\partial y^j}{\partial x^i}\right|_p. \tag{3.21}$$

A **vector field** X on M is a smooth map $p \mapsto X_p$. In terms of the local coordinates $x_1, \ldots, x_n$,

$$X = \sum_i X^i(x)\frac{\partial}{\partial x^i}, \tag{3.22}$$

where the X^i are now smooth functions, as opposed to constants. If $\{y_1, \ldots, y_n\}$ is another set of local coordinates, and if we write

$$X = \sum_j Y^j(y)\frac{\partial}{\partial y^j}, \tag{3.23}$$

then the same argument as above gives

$$Y^j(y(x)) = \sum_i X^i(x)\frac{\partial y^j}{\partial x^i}. \tag{3.24}$$

As the indices are starting to proliferate, we will henceforth employ the Einstein summation convention whenever convenient. In particular, we write

$$X = X^i \frac{\partial}{\partial x^i}$$

in lieu of (3.22) and

$$Y^j = X^i \frac{\partial y^j}{\partial x^i}$$

instead of (3.24).

EXERCISE 3.20 Let X and Y be two vector fields. The *Lie bracket* $[X, Y]$ of the two vector fields is defined to be the differential operator

$$[X, Y] := XY - YX. \tag{3.25}$$

(a) Show that the Lie bracket of two vector fields is a vector field. *Hint:* Just show that $[X, Y]$ is a linear derivation.

(b) If, in local coordinates,[20]

$$X = X^i \frac{\partial}{\partial x^i} \qquad \text{and} \qquad Y = Y^j \frac{\partial}{\partial x^j},$$

show that

---

[20] Warning: In (3.23) we wrote Y^j for the components of the vector field X in the $\partial/\partial y^j$ basis. But in this exercise Y^j denotes the components of the vector field Y in the $\partial/\partial x^j$ basis. The two Y^j's mean very different things, so don't confuse them.

$$[X, Y] = \left(X^i \frac{\partial Y^j}{\partial x^i} - Y^i \frac{\partial X^j}{\partial x^i} \right) \frac{\partial}{\partial x^j}. \qquad (3.26)$$

(c) Show that the Lie bracket of vector fields X, Y, and Z obeys the **Jacobi identity**:

$$[X, [Y, Z]] + [Y, [Z, X]] + [Z, [X, Y]] = 0. \qquad (3.27)$$

Hint: Do not use local coordinates. Instead, expand each term as a sum of products of the form XYZ and cancel terms.

Remark: A **Lie algebra** is a vector space $\mathfrak{a}$ equipped with a binary operation $[\cdot, \cdot]$: $\mathfrak{a} \times \mathfrak{a} \to \mathfrak{a}$ that (i) is bilinear, (ii) is antisymmetric, and (iii) satsifies the Jacobi identity. The above exercise shows that the set of all vector fields on a manifold forms a Lie algebra.

It is important to note that the tangent space $T_p M$ is defined independently of any coordinate system. Thus, at each point we are free to pick any basis $\{e_i\}$, not just a coordinate basis $\{\partial/\partial x^i\}$. Any smoothly varying basis on M is called a **frame field**.[21] Generally, frame fields do not exist everywhere on M; they clearly exist locally, because $e_i = \partial/\partial x^i$ is an instance. In terms of a frame field $\{e_i\}$, we could write a vector field X as

$$X = X^i e_i. \qquad (3.28)$$

Note that, if $e_i \neq \partial/\partial x^i$ then the X^i in (3.28) is not the same as $X(x^i)$.

Two important questions arise. First, when does a frame field exist globally on M? Second, given a frame field $\{e_i\}$, can we find a local system of coordinates $\{x^i\}$ such that $e_i = \partial/\partial x^i$? We will return to these questions below.[22]

If $\{e_i'\}$ is any other frame field then there is a nonsingular matrix A (actually, a matrix of smooth functions) relating the two frame fields. Right away we have a notational problem, stemming from two competing but equally entrenched conventions. We observed in Section 1.7 that if $A : V \to V$ is a change of basis map then

$$e_j' = e_i A^i{}_j. \qquad (1.17)$$

The problem is that, if $\{e_i\}$ is a frame field then each e_i is a tangent vector and therefore a linear derivation. The matrix entries $A^i{}_j$ are functions, so the unwary reader might read (1.17) as saying that e acts on A as a derivation, which is not

---

[21] It is also called a **moving frame** (in French, *repère mobile*), especially when restricted to a curve. This terminology is due to Élie Cartan.

[22] See Sections 7.1 and 8.1.1, respectively.

true. For example, if $e_i = \partial/\partial x^i$ and $e_i' = \partial/\partial y^i$ then, to maintain consistency with (1.17), (3.19) must be written

$$\frac{\partial}{\partial y^j} = \frac{\partial}{\partial x^i} \frac{\partial x^i}{\partial y^j} \tag{3.29}$$

so that

$$A^i{}_j = \frac{\partial x^i}{\partial y^j}. \tag{3.30}$$

However, from calculus we know that the x derivatives do not act on the Jacobian matrix in (3.29), so we just have to remember that the same convention holds in (1.17).

We could avoid this problem by writing all the derivations on the right. For example, we could leave (3.19) alone and instead write (1.17) as

$$e_j' = A^i{}_j e_i \tag{3.31}$$

but then this messes up our shorthand notation (1.35), because it would conflict with the standard convention (1.13) for matrix multiplication. One way out of this latter difficulty would be two write (1.17) as

$$e_j' = A_j{}^i e_i, \tag{3.32}$$

and some writers do this, but this conflicts with the standard convention that the first index represents a row of the matrix so we shall avoid it. Instead, we will stick with the standard linear algebra conventions (1.17) and (1.35) and just remember that although e_i is a derivation it does not act on the matrix elements $A^i{}_j$.

3.8 The cotangent space $T_p^* M$

The dual space to $T_p M$ is called the **cotangent space** and is denoted $T_p^* M$. In local coordinates $\{x^i\}$ around p, $T_p M$ is spanned by the n basis vectors $\partial/\partial x^i$. The corresponding dual basis vectors are denoted dx^i. By definition we have

$$\left\langle \frac{\partial}{\partial x^i}, dx^j \right\rangle = \delta_i^j. \tag{3.33}$$

Clearly, dim $T_p^* M = n$. A general element α_p of $T_p^* M$ is a linear combination of the basis elements:

$$\alpha_p = a_i \, dx^i, \tag{3.34}$$

where the a_i are constants. A **(differential) 1-form** or **(smooth) covector field** α on M is a smooth assignment $p \to \alpha_p$. In local coordinates around p we have

$$\alpha = a_i(x) \, dx^i, \tag{3.35}$$

where the $a_i(x)$ are smooth functions on M. If $\{y^j\}$ is another system of local coordinates near p then, as you can demonstrate in the next exercise,

$$\alpha = b_j(y)\, dy^j \tag{3.36}$$

where

$$b_j(y) = a_i(x(y)) \frac{\partial x^i}{\partial y^j}. \tag{3.37}$$

Equivalently,

$$a_i(x) = b_j(y(x)) \frac{\partial y^j}{\partial x^i}. \tag{3.38}$$

In particular, we can write

$$\alpha = b_j(y) \frac{\partial y^j}{\partial x^i}\, dx^i, \tag{3.39}$$

which shows that

$$dy^j = \frac{\partial y^j}{\partial x^i}\, dx^i, \tag{3.40}$$

a reassuring result.

EXERCISE 3.21 Prove that (3.37) is equivalent to (3.38) and that both follow from (3.24). *Hint:* Consider $\langle X, \alpha \rangle$.

Just as the tangent space does not depend on any particular coordinate system for its definition, so neither does the cotangent space. If $\{e_i\}$ is a generic basis for $T_p M$ then the corresponding dual basis $\{\theta^i\}$ is called a **coframe field**. Thus

$$\langle e_i, \theta^j \rangle = \delta_i^j. \tag{3.41}$$

In terms of the coframe field, an arbitrary 1-form can be written as

$$\alpha = a_i\, \theta^i. \tag{3.42}$$

If $\{\theta'^i\}$ is another coframe field, related to the first by

$$\theta'^i = (A^{-1})^i{}_j \theta^j, \tag{3.43}$$

then

$$\alpha = a'_i \theta'^i \tag{3.44}$$

where

$$a'_i = a_j A^j{}_i. \tag{3.45}$$

3.9 The cotangent space as jet space*

In the previous section we defined a differential 1-form as a smoothly varying map
on M that assigns to every point p of M an element of the cotangent space at p.
Although this is correct, it is perhaps rather unsatisfying because the cotangent
space is defined in terms of the tangent space, which is itself defined as a space of
derivations on functions. There is, however, an elegant way to define the cotangent
space directly, using just the notion of a smooth function, which is perhaps a bit
more natural from the point of view of manifold theory. It uses the idea of a "jet".

Let $f : M \to \mathbb{R}$ be a smooth function, $p \in M$, and $\{x^i\}$ local coordinates around
p. We say that f **vanishes to first order at** p if $\partial f / \partial x^i$ vanishes at p for all i. This
is a well-defined notion (that is, independent of the coordinate system) because if
$\{y^j\}$ is any other coordinate system about p then

$$\left. \frac{\partial}{\partial x^i} \right|_p = \left. \frac{\partial y^j}{\partial x^i} \right|_p \left. \frac{\partial}{\partial y^j} \right|_p .$$

The Jacobian matrix is nonsingular and thus invertible, so $\partial f / \partial x^i$ vanishes for all
i if and only if $\partial f / \partial y^j$ vanishes for all j. Inductively, for $k \geq 1$ we say that f
vanishes to kth order at p if, for every i, $\partial f / \partial x^i$ vanishes to $(k-1)$th order at p,
where f vanishes to zeroth order at p if $f(p) = 0$. Put another way, f vanishes to
kth order at p if the first k terms in its Taylor expansion vanish at p. For example,
x vanishes to zeroth order at 0, x^2 vanishes to first order at 0, x^3 vanishes to second
order at 0, etc.

For $k > 0$ let M_p^k denote the set of all smooth functions on M vanishing to
$(k-1)$th order at p, and set $M_p^0 := \Omega^0(M)$ and $M_p := M_p^1$. Each M_p^k is a vector
space under the usual pointwise operations, and we have the series of inclusions[23]

$$M_p^0 \supset M_p^1 \supset M_p^2 \cdots .$$

We now define $T_p^* M$, the **cotangent space to M at** p, to be the quotient space[24]

$$T_p^* M = M_p / M_p^2. \tag{3.46}$$

An element of $T_p^* M$ is called a **differential 1-form** at p.

The reason that this definition is so elegant is that it uses only minimal ingre-
dients, namely the notions of a smooth function on a manifold and its order of
vanishing at a point (which is a smooth, coordinate independent, notion). It offers
a way of understanding 1-forms intrinsically, without reference to derivations or

---

[23] This series is called a **graded filtration**. The word "filtration" suggests a kind of sieving as we proceed down
the chain, while the word "graded" means that the filtration carries a natural product structure that respects
the indexing. Specifically, $M_p^r \cdot M_p^s \subset M_p^{r+s}$.

[24] It is a specific example of what is called a **jet space**.

other additional structures. Moreover, all the usual properties of 1-forms can be shown to follow easily.

For instance, let $f \in M_p^0$ be a smooth function on M. Then **the differential of** f **at** p, denoted df_p, is the element of M_p/M_p^2 given by the equivalence class of $f - f(p) \cdot 1$, where 1 is the unit function. Symbolically,

$$df_p = [f - f(p) \cdot 1]. \tag{3.47}$$

From this we obtain the next result.

Theorem 3.4 *Let* $x^1, \ldots, x^n$ *be local coordinates in a neighborhood* U *of a point* $p \in M$. *Then* $T_p^* M$ *is spanned by the 1-forms* $dx_p^1, \ldots, dx_p^n$. *In particular,* $\dim T_p^* M = n$.

Proof [25] Let f be a smooth function on M. Choose local coordinates $x^1, \ldots, x^n$ in a neighborhood of a point $p \in M$ and set $x = (x^1, \ldots, x^n)$ as before. Without loss of generality we can choose our coordinates so that $x(p) = 0$. Then, near p, Taylor's theorem with remainder gives

$$(f - f(p) \cdot 1)(x) = \sum_i a_i x^i + \sum_{ij} a_{ij}(x) x^i x^j,$$

where $a_i = \partial f / \partial x^i |_p$. But the quadratic term vanishes to first order at p, so

$$[f - f(p) \cdot 1] = [a_i x^i] = [a_i (x^i - x^i(p) \cdot 1)],$$

which implies that

$$df_p = a_i \, dx_p^i.$$

Hence the dx_p^i span $T_p^* M$. To see that they are also linearly independent, suppose that $\lambda_i \, dx_p^i = 0$. Then $\lambda_i x^i$ vanishes to first order at p (i.e., $\lambda_i x^i \in M_p^2$), which is only possible if all the λ_i vanish at p. $\qquad\square$

The proof shows that

$$df_p = \frac{\partial f}{\partial x^i} \, dx^i \bigg|_p, \tag{3.48}$$

which agrees with the familiar differential from calculus. In particular, the Leibniz rule

$$d(fg)_p = g_p \cdot df_p + f_p \cdot dg_p$$

---

[25] The reader will notice some similarities between this proof and that of Theorem 3.3 (see (3.17)). This is not accidental, of course, as both constructions essentially reduce to Taylor's theorem.

follows immediately from the ordinary rule. Moreover, if $y^1, \ldots, y^n$ is another coordinate system near p then, as the definition of df_p is coordinate independent, we have

$$\frac{\partial f}{\partial x^i} dx^i \bigg|_p = \frac{\partial f}{\partial y^j} dy^j \bigg|_p.$$

Applying this to $f = y^j$ gives

$$dy^j_p = \frac{\partial y^j}{\partial x^i} dx^i \bigg|_p,$$

namely the usual change of variables formula.

Remark One can now define the tangent space to be the dual of the cotangent space. Not surprisingly, the tangent space so defined coincides with the previous definition in terms of derivations.

3.10 Tensor fields

There are many different ways to define a tensor field on a manifold, but the essential idea is simple enough: a tensor field is just a smooth assignment of a tensor to every point of M. The problem lies in the means by which one expresses this fact. As discussed in Section 2.4, we can view a tensor either as an element of a certain tensor product space or as a multilinear map. A similar duality holds for a tensor field. Both perspectives are used throughout the literature, and each is worth knowing.[26]

3.10.1 Tensor fields as elements of a tensor product space

We begin by expressing a tensor field in terms of certain tensor product spaces. Choose local coordinates $x^1, \ldots, x^n$ on a patch U of a manifold M. Then, for any $p \in U$, the vector fields $\partial/\partial x^i$ and the 1-form fields dx^i constitute dual bases for $T_p M$ and $T_p^* M$, respectively. These combine to provide a basis

$$\frac{\partial}{\partial x^{i_1}} \otimes \cdots \otimes \frac{\partial}{\partial x^{i_r}} \otimes dx^{j_1} \otimes \cdots \otimes dx^{j_s} \tag{3.49}$$

for the tensor product space $(T_p M)^{\otimes r} \otimes (T_p^* M)^{\otimes s}$, and any element Ψ of this space can be expanded in terms of this basis:

$$\Psi = \Psi^{i_1 \cdots i_r}{}_{j_1 \cdots j_s} \frac{\partial}{\partial x^{i_1}} \otimes \cdots \otimes \frac{\partial}{\partial x^{i_r}} \otimes dx^{j_1} \otimes \cdots \otimes dx^{j_s}. \tag{3.50}$$

---

[26] Actually, the fanciest way to view a tensor field is in terms of something called a tensor product bundle. For this, see Example 7.5.

A **tensor field** Ψ of type (r, s) is a map from M to T_s^r such that, on any coordinate patch U, the components $\Psi^{i_1 i_2 \cdots i_r}{}_{j_1 j_2 \cdots j_s}$ are smoothly varying functions.[27]

If $\{y^1, \ldots, y^n\}$ is a set of local coordinates on a patch V that overlaps with U then

$$\frac{\partial}{\partial x^i} = \frac{\partial y^k}{\partial x^i} \frac{\partial}{\partial y^k} \qquad \text{and} \qquad dx^i = \frac{\partial x^i}{\partial y^k} dy^k. \tag{3.51}$$

So, writing $\Psi^{i_1' i_2' \cdots i_r'}{}_{j_1' j_2' \cdots j_s'}$ for the components of Ψ in the y coordinates, we must have

$$\begin{aligned}
&\Psi^{i_1' \cdots i_r'}{}_{j_1' \cdots j_s'} \frac{\partial}{\partial y^{i_1'}} \otimes \cdots \otimes \frac{\partial}{\partial y^{i_r'}} \otimes dy^{j_1'} \otimes \cdots \otimes dy^{j_s'} \\
&= \Psi^{i_1 \cdots i_r}{}_{j_1 \cdots j_s} \frac{\partial}{\partial x^{i_1}} \otimes \cdots \otimes \frac{\partial}{\partial x^{i_r}} \otimes dx^{j_1} \otimes \cdots \otimes dx^{j_s} \\
&= \Psi^{i_1 \cdots i_r}{}_{j_1 \cdots j_s} \frac{\partial y^{i_1'}}{\partial x^{i_1}} \frac{\partial}{\partial y^{i_1'}} \otimes \cdots \otimes \frac{\partial y^{i_r'}}{\partial x^{i_r}} \frac{\partial}{\partial y^{i_r'}} \otimes \frac{\partial x^{j_1}}{\partial y^{j_1'}} dy^{j_1'} \otimes \cdots \otimes \frac{\partial x^{j_s}}{\partial y^{j_s'}} dy^{j_s'}.
\end{aligned} \tag{3.52}$$

Comparing the first and last lines of (3.52) gives the transformation law for tensor field components:

$$\Psi^{i_1' \cdots i_r'}{}_{j_1' \cdots j_s'} = \Psi^{i_1 \cdots i_r}{}_{j_1 \cdots j_s} \frac{\partial y^{i_1'}}{\partial x^{i_1}} \cdots \frac{\partial y^{i_r'}}{\partial x^{i_r}} \frac{\partial x^{j_1}}{\partial y^{j_1'}} \cdots \frac{\partial x^{j_s}}{\partial y^{j_s'}}. \tag{3.53}$$

The components are each considered to be functions on their respective coordinate systems. If we wished to express both sides as functions of the same coordinates, say the x's, we would have to write

$$\Psi^{i_1' \cdots i_r'}{}_{j_1' \cdots j_s'}(y(x)) = \Psi^{i_1 \cdots i_r}{}_{j_1 \cdots j_s}(x) \frac{\partial y^{i_1'}}{\partial x^{i_1}} \cdots \frac{\partial y^{i_r'}}{\partial x^{i_r}} \frac{\partial x^{j_1}}{\partial y^{j_1'}} \cdots \frac{\partial x^{j_s}}{\partial y^{j_s'}}, \tag{3.54}$$

where it is understood that the derivatives also are evaluated at x or $y(x)$ as appropriate.

EXERCISE 3.22 Not every object with indices is a tensor field. Let $X = X^i \partial/\partial x^i$ be a vector field. Show that the functions $\partial X^i/\partial x^j$ cannot be the components of any tensor field, by showing that they do not obey (3.53).

Of course, the components of Ψ are different in a different basis. In order to distinguish the components of Ψ in a coordinate basis from those in a more general

---

[27] In this subsection and the next we use Ψ rather than T to denote a general tensor field, so as not to confuse Ψ_p, the tensor field at a point p, with T_p, the tangent space at p.

basis, it is traditional to use different letters for the dummy indices. Thus one writes $\{e_a\}$ for a generic frame field and $\{\theta^a\}$ for the corresponding coframe field, in which case the tensor field Ψ can be written

$$\Psi = \Psi^{a_1 \cdots a_r}{}_{b_1 \cdots b_s} e_{a_1} \otimes \cdots \otimes e_{a_r} \otimes \theta^{b_1} \otimes \cdots \otimes \theta^{b_s}. \tag{3.55}$$

Under a change of basis

$$e_a' = e_b A^b{}_a \quad \text{and} \quad \theta'^a = (A^{-1})^a{}_b \theta^b,$$

we have

$$\Psi^{a_1' \cdots a_r'}{}_{b_1' \cdots b_s'} = \Psi^{a_1 \cdots a_r}{}_{b_1 \cdots b_s} (A^{-1})^{a_1'}{}_{a_1} \cdots (A^{-1})^{a_r'}{}_{a_r} A^{b_1}{}_{b_1'} \cdots A^{b_s}{}_{b_s'}. \tag{3.56}$$

The only difference between (3.56) and (2.12) is that the tensor components and matrix elements are now smooth functions on M.

3.10.2 Tensor fields as function linear maps

If you dislike indices, you can take some comfort in the fact that a tensor field can also be viewed as a smoothly varying multilinear map. Let $\tilde{T}^r_s(p)$ be the space of all multilinear maps on

$$\underbrace{T^*_p M \times \cdots \times T^*_p M}_{r \text{ times}} \times \underbrace{T_p M \times \cdots \times T_p M}_{s \text{ times}}.$$

A tensor field Ψ of type (r, s) is then a smooth assignment of an element $\Psi_p \in \tilde{T}^r_s(p)$ to each $p \in M$. "Smooth" in this context means that, for any smooth covector fields $\alpha_1, \ldots, \alpha_r$ and smooth vector fields $X_1, \ldots, X_s$ in a neighborhood of p, the map

$$p \mapsto \Psi_p(\alpha_1(p), \ldots, \alpha_r(p), X_1(p), \ldots, X_s(p)) \tag{3.57}$$

is smooth.

The map (3.57) depends only on the values of the covector and vector fields at p. In particular, it must give the same value for any other fields that agree with the chosen ones at p. This, in turn, implies that the map Ψ must be more than just multilinear – in fact, it must be **function linear**, which means that (2.14) must hold for Ψ with the scalars a, b replaced by smooth functions.[28] More precisely, we have the following useful result.

Theorem 3.5 *Let $\Gamma(TM)$ (respectively, $\Gamma(T^*M)$) denote the space of all vector (respectively, covector) fields on M.[29] Then a tensor field Ψ of type (r, s) is a function linear map*

---

[28] Technically, we should call such maps *function multilinear* (because they are function linear in each entry).

[29] For further discussion of the objects $\Gamma(TM)$ and $\Gamma(T^*M)$, see Section 7.1.

$$\Psi : \underbrace{\Gamma(T^*M) \times \cdots \times \Gamma(T^*M)}_{r \text{ times}} \times \underbrace{\Gamma(TM) \times \cdots \times \Gamma(TM)}_{s \text{ times}} \to \Omega^0(M).$$

Proof For simplicity we will prove the theorem for the case $r = 0$. The general case is similar. Suppose that $\widetilde{\Psi}$ satisfies the conditions of the theorem. We claim that the value of $\widetilde{\Psi}$ at p depends only on the values of $X_i(p)$, $1 \le i \le s$. Let $\{e_i\}$ be a frame field defined on a neighborhood U of p. Expanding each X_i in terms of the basis $\{e_i\}$ gives $X_i = e_j X^j{}_i$ for some smooth functions $X^j{}_i$. By function linearity

$$\widetilde{\Psi}(X_1, \ldots, X_s) = X^{j_1}{}_1 \cdots X^{j_s}{}_s \widetilde{\Psi}(e_{j_1}, \ldots, e_{j_s}),$$

so we have

$$\widetilde{\Psi}(X_1, \ldots, X_s)(p) = X^{j_1}{}_1(p) \cdots X^{j_s}{}_s(p) \widetilde{\Psi}(e_{j_1}, \ldots, e_{j_s})(p).$$

Let $\{X_i'\}$ be another set of vector fields on U that satisfy $X_i'(p) = X_i(p)$, so that $X'^j{}_i(p) = X^j{}_i(p)$ for $1 \le i \le s$ and $1 \le j \le n$. Then

$$\begin{aligned}
\widetilde{\Psi}(X_1', \ldots, X_s')(p) &= X'^{j_1}{}_1(p) \cdots X'^{j_s}{}_s(p) \widetilde{\Psi}(e_{j_1}, \ldots, e_{j_s})(p), \\
&= X^{j_1}{}_1(p) \cdots X^{j_s}{}_s(p) \widetilde{\Psi}(e_{j_1}, \ldots, e_{j_s})(p),
\end{aligned}$$

which proves the claim.

Now, given vectors $Y_1, \ldots, Y_s \in T_p M$ define a map

$$\Psi : M \to \widetilde{T}^0_s$$

by

$$\Psi_p(Y_1, \ldots, Y_s) := \widetilde{\Psi}(X_1, \ldots, X_s)(p),$$

for any set of vector fields $\{X_i\}$ satisfying $X_i(p) = Y_i$. By construction the right-hand side is a smoothly varying function of p, so Ψ is a tensor field. Conversely, given a tensor field we can construct a function linear map of the above type by reversing this process. $\qquad\square$

By duality, Theorem 3.5 allows us to view other function linear maps as tensors. For example, suppose we have a function linear map $S : \Gamma(TM) \to \Gamma(TM)$. If X is a vector field then $S(X)$ is a vector field, not a function, so technically Theorem 3.5 does not apply. But S determines (and is determined by) a unique tensor field $\widetilde{S} : \Gamma(T^*M) \times \Gamma(TM) \to \Omega^0(M)$ given by $\widetilde{S}(\alpha, X) = \langle \alpha, S(X) \rangle$. In this sense, we are justified in calling S a tensor field, or more simply, **tensorial**.

3.11 Differential forms

Just as a k-vector is a special kind of tensor, namely an alternating one, a k-form is a special kind of tensor field, namely an alternating one. Viewed in this way they are not particularly interesting. But k-forms are by far the most useful kind of tensor fields, for several reasons. First, they are easy to define. Second, they are easy to use because, for the most part, one does not have to deal with all those irritating indices. Third, almost all important geometrical quantities can be expressed in terms of forms. And last, but certainly not least, differential forms are essentially the things that appear under integral signs.

Let U be a coordinate patch of M and let $p \in U$. A k-**form** (or a **form of degree** k) ω_p on U at p is an element of $\bigwedge^k(T_p^* M)$. It follows that, in local coordinates,

$$\omega_p = \frac{1}{k!} \sum a_{i_1 \cdots i_k} \, dx^{i_1} \wedge \cdots \wedge dx^{i_k} = \sum a_I \, dx^I, \tag{3.58}$$

for some constants $a_I = a_{(i_1, \dots, i_k)}$. A (**differential**) k-**form** ω on M is a smooth assignment $p \to \omega_p$. In local coordinates,

$$\omega_U = \frac{1}{k!} \sum a_{i_1 \dots i_k}(x) \, dx^{i_1} \wedge \cdots \wedge dx^{i_k} = \sum a_I(x) \, dx^I, \tag{3.59}$$

where the $a_I(x)$ are now smooth functions on U. The vector space of all k-forms on M is denoted $\Omega^k(M)$. In particular, we think of the smooth functions on M as 0-forms.

As with tensor fields, it is difficult to give explicit examples of differential forms on a general manifold M, because forms by definition are globally defined but coordinates are only local. An expression such as (3.59) only specifies the form ω restricted to a single coordinate patch U. One would have to show that ω_U extends to a global form on M. We do this in Example 3.10 below. Of course, there is one case where it is very easy to give explicit examples of differential forms, and that is the case of Euclidean space.

Example 3.9 On $\mathbb{R}^n$ we have *global* coordinates $\{x^i\}$, so it is easy to define smooth forms. For example,

$$\omega = (6xy^2 + 3z) \, dx + (3y \sin z) \, dy + (e^x y) \, dz$$

is a globally defined 1-form on $\mathbb{R}^3$, while

$$\mu = 3x^2 y \, dy \wedge dz + 7z^3 \, dz \wedge dx + 4xy \, dx \wedge dy$$

is a globally defined 2-form on $\mathbb{R}^3$.

Example 3.10 Consider again the 2-sphere with the atlas obtained from stereographic projection, as in Example 3.6. Now we do not have global coordinates, so we do not have obvious global forms. Instead, we must define everything locally and then hope that we can patch things together consistently. Typically one does this by first defining a form on a single patch and then extending it to the other patches. So, for example, suppose that we define a 2-form on the patch V by

$$\omega_V = \frac{dv^1 \wedge dv^2}{(1 + \xi)^2}.$$

First we must check that this is indeed defined everywhere on V. For example, it would do us no good to define a 2-form on V by $(dv^1 \wedge dv^2)/\xi$, because this form blows up at $(v^1, v^2) = (0, 0)$, which is a legitimate point of V. But ω_V does not suffer from this indignity, so we're ok. The next question is, can ω_V be *smoothly* extended to the rest of S^2, i.e., is there a 2-form ω_U on U that smoothly agrees with ω_V on the overlap $U \cap V$? The obvious method turns out to work: we simply *define* $\omega_U = \omega_V$ on the overlap (so that they trivially agree), reexpress ω_U in terms of the u coordinates, and define ω_U elsewhere on U using this expression. The reason why this works is that changing variables from v to u is a smooth process (although we must be careful to verify that the resulting form is well defined everywhere on U).

So, how *do* we change variables? The answer is that we just use (3.53). But that equation can be confusing because it has too many indices, so instead we just follow our instincts and use the chain rule together with the properties of the wedge product (in Section 3.14 we will formalize this process):

$$\begin{aligned}
dv^1 \wedge dv^2 &= \left(\frac{\partial v^1}{\partial u^1} du^1 + \frac{\partial v^1}{\partial u^2} du^2 \right) \wedge \left(\frac{\partial v^2}{\partial u^1} du^1 + \frac{\partial v^2}{\partial u^2} du^2 \right) \\
&= \left(\frac{\partial v^1}{\partial u^1} \frac{\partial v^2}{\partial u^2} - \frac{\partial v^1}{\partial u^2} \frac{\partial v^2}{\partial u^1} \right) du^1 \wedge du^2 \\
&= \frac{\partial(v^1, v^2)}{\partial(u^1, u^2)} du^1 \wedge du^2.
\end{aligned} \tag{3.60}$$

The appearance of the Jacobian determinant on the right-hand side of (3.60) is not coincidental; the general rule appears in Section 3.14 below. In any event, using the results of Example 3.6 and setting $\omega_U = \omega_V$ on $U \cap V$ gives

$$\begin{aligned}
\omega_U &= \frac{dv^1 \wedge dv^2}{(1 + \xi)^2} \\
&= -\frac{1}{\eta^2} \frac{du^1 \wedge du^2}{(1 + 1/\eta)^2} \\
&= -\frac{du^1 \wedge du^2}{(1 + \eta)^2}.
\end{aligned}$$

This is a 2-form defined everywhere on U and by construction it agrees with ω_V on the overlap. It follows that there is a globally defined 2-form ω on S^2 which equals ω_U on U and ω_V on V.

EXERCISE 3.23 In Exercise 3.16 we covered the 2-torus with four coordinate patches, U_i, $i = 1, \ldots, 4$. Construct two global 1-forms dx and dy on the torus by extending $dx_{U_1} = dx^1$ and $dy_{U_1} = dy^1$ to the rest of the torus. (Because the 2-torus is two dimensional, these 1-forms constitute a basis for the cotangent space.)

3.12 The exterior derivative

We now introduce a natural differential operator on k-forms, denoted simply by the letter d. It is a far reaching generalization of the ordinary gradient, curl, and divergence operators of multivariable calculus.

Theorem 3.6 *There exists a unique linear operator*

$$d : \Omega^k(M) \to \Omega^{k+1}(M)$$

*called the **exterior derivative**, satisfying the following properties. For any forms λ and μ, and for any function f, the operator d is*

(1) *linear:* $\quad d(\lambda + \mu) = d\lambda + d\mu$,
(2) *a graded derivation:* $\quad d(\lambda \wedge \mu) = d\lambda \wedge \mu + (-1)^{\deg \lambda} \lambda \wedge d\mu$,
(3) *nilpotent:* $\quad d^2\lambda = 0$, *and*
(4) *natural:* $\quad$ *in local coordinates $\{x^i\}$ about a point p,*

$$df = \sum \frac{\partial f}{\partial x^i} dx^i.$$

Two comments are in order. Property (4) makes it appear as though the definition depends on the local coordinate system but this is an illusion, for if $\{y^j\}$ is another local coordinate system at p then, by (3.37),

$$\sum \frac{\partial f}{\partial x^i} dx^i = \sum \frac{\partial f}{\partial x^i} \frac{\partial x^i}{\partial y^j} dy^j = \sum \frac{\partial f}{\partial y^j} dy^j. \tag{3.61}$$

Clearly, by property (4) the exterior derivative on $\mathbb{R}^n$ coincides with the usual notion of the differential of a function. Observe also that property (4) applied to the coordinate function x^i gives $d(x^i) = dx^i$, which is good. When this is combined with property (3) we get

$$d^2 x^i = 0, \tag{3.62}$$

another useful fact. Before proving the theorem, let's look at some examples.

Example 3.11 Let $\omega = A\,dx + B\,dy + C\,dz$ be a 1-form on $\mathbb{R}^3$. Then

$$
\begin{aligned}
d\omega &= dA \wedge dx + A \wedge d^2x \\
&\quad + dB \wedge dy + B \wedge d^2y \\
&\quad + dC \wedge dz + C \wedge d^2z &&\text{(by (1) and (2))} \\
&= dA \wedge dx + dB \wedge dy + dC \wedge dz &&\text{(by (3))} \\
&= \left(\frac{\partial A}{\partial x}\,dx + \frac{\partial A}{\partial y}\,dy + \frac{\partial A}{\partial z}\,dz \right) \wedge dx \\
&\quad + \left(\frac{\partial B}{\partial x}\,dx + \frac{\partial B}{\partial y}\,dy + \frac{\partial B}{\partial z}\,dz \right) \wedge dy \\
&\quad + \left(\frac{\partial C}{\partial x}\,dx + \frac{\partial C}{\partial y}\,dy + \frac{\partial C}{\partial z}\,dz \right) \wedge dz &&\text{(by (4))} \\
&= \left(\frac{\partial B}{\partial x} - \frac{\partial A}{\partial y} \right) dx \wedge dy + \text{cyclic}. &&\text{(3.63)}
\end{aligned}
$$

Here "cyclic" means that the subsequent terms are obtained from the first by cyclically permuting $x \to y \to z$ and $A \to B \to C$. Observe that the components of $d\omega$ are just the components of the curl of the vector field (A, B, C). We shall return to this example below.

Example 3.12 Let $\omega = A\,dy \wedge dz + B\,dz \wedge dx + C\,dx \wedge dy$ be a 2-form on $\mathbb{R}^3$. Then

$$
d\omega = \left(\frac{\partial A}{\partial x} + \frac{\partial B}{\partial y} + \frac{\partial C}{\partial z} \right) dx \wedge dy \wedge dz, \tag{3.64}
$$

which is obviously related to the divergence of the vector field (A, B, C).

Example 3.13 We can use the exterior derivative to define forms (globally) on a smooth manifold M. For example, if $f : M \to \mathbb{R}$ is a smooth function then, provided that it does not blow up anywhere, df is a (globally defined) 1-form on M.

EXERCISE 3.24 Let $f : \mathbb{R}^3 \to \mathbb{R}$ be a smooth function. By expanding terms, show explicitly that $df \wedge df = 0$.

EXERCISE 3.25 Let

$$
\alpha = 3x^2 y^3 z\,dx + xy^3 z^2\,dy + 2yz^2\,dz
$$

be a 1-form on $\mathbb{R}^3$. Compute $d\alpha$ and $d^2\alpha$.

EXERCISE 3.26 Let

$$
\omega = 6x^2\,dx \wedge dy - 3xyz^3\,dy \wedge dz + x^2 y^2\,dz \wedge dx
$$

be a 2-form on $\mathbb{R}^3$. Compute $d\omega$ and $d^2\omega$.

Proof of Theorem 3.6 Everything can be done locally in a neighborhood of p, because properties (1), (2), and (3) of the theorem are clearly coordinate independent, while we have shown that (4) holds in any system of coordinates. First we show uniqueness. Suppose there were another linear operator D satisfying properties (1) through (4). By linearity, it suffices to show that $D = d$ on a monomial of the form $a\,dx^I$. But, by (2), (3), and (4),

$$D(a\,dx^I) = \sum_k \frac{\partial a}{\partial x^k}\,dx^k \wedge dx^I = d(a\,dx^I);$$

hence $D = d$. To show existence, just define

$$d(a\,dx^I) = \sum_k \frac{\partial a}{\partial x^k}\,dx^k \wedge dx^I, \tag{3.65}$$

and extend by linearity. Then property (4) follows by setting $I = \emptyset$. Again by linearity it suffices to prove (2) and (3) on monomials. We will prove (2) and leave (3) as an exercise. So, let $\lambda = a\,dx^I$ and $\mu = b\,dx^J$. Then

$$d(\lambda \wedge \mu) = d(ab\,dx^I \wedge dx^J)$$

$$= \sum_k \left(b\frac{\partial a}{\partial x^k} + a\frac{\partial b}{\partial x^k} \right) dx^k \wedge dx^I \wedge dx^J$$

$$= \sum_k \left(\frac{\partial a}{\partial x^k}\,dx^k \wedge dx^I \right) \wedge \left(b\,dx^J \right)$$

$$+ (-1)^{|I|} \left(a\,dx^I \right) \wedge \sum_k \left(\frac{\partial b}{\partial x^k}\,dx^k \wedge dx^J \right)$$

$$= d\lambda \wedge \mu + (-1)^{\deg\lambda}\,\lambda \wedge d\mu.$$

$\square$

EXERCISE 3.27 Prove that the definition (3.65) implies that d is nilpotent (property (3)).

Example 3.14 (Maxwell's equations) Let $M = \mathbb{M}^4$ be Minkowski spacetime equipped with (global) coordinates $\{t, x, y, z\}$, so that $\{dt, dx, dy, dz\}$ is a basis for $T_p^* M$. The inner product on M induces an inner product on $T_p^* M$ given by

$$g(dt, dt) = -1,$$
$$g(dt, dx^i) = 0,$$
$$g(dx^i, dx^j) = \delta^{ij},$$

where $i, j = 1, 2, 3$. We choose

$$\sigma = dt \wedge dx \wedge dy \wedge dz,$$

so that the results of Example 2.4 apply *mutatis mutandis*. In particular, we have the following Hodge duals:

$$\star dt = dx \wedge dy \wedge dz,$$
$$\star dx = dt \wedge dy \wedge dz,$$
$$\star dy = dt \wedge dz \wedge dx,$$
$$\star dz = dt \wedge dx \wedge dy$$

and also

$$\star(dt \wedge dx) = dy \wedge dz,$$
$$\star(dt \wedge dy) = dz \wedge dx,$$
$$\star(dt \wedge dz) = dx \wedge dy,$$
$$\star(dx \wedge dy) = -dt \wedge dz,$$
$$\star(dy \wedge dz) = -dt \wedge dx,$$
$$\star(dz \wedge dx) = -dt \wedge dy.$$

We can use all this to reformulate Maxwell's equations in a particularly elegant form. If (E_x, E_y, E_z) and (B_x, B_y, B_z) are the components of the electric and magnetic field vectors $\boldsymbol{E}$ and $\boldsymbol{B}$, respectively, we first define a 2-form, sometimes called the **electromagnetic 2-form**, that encodes both fields simultaneously:[30]

$$F = -E_x \, dt \wedge dx - E_y \, dt \wedge dy - E_z \, dt \wedge dz$$
$$+ B_x \, dy \wedge dz + B_y \, dz \wedge dx + B_z \, dx \wedge dy.$$

Thus,

$$-\star F = B_x \, dt \wedge dx + B_y \, dt \wedge dy + B_z \, dt \wedge dz$$
$$+ E_x \, dy \wedge dz + E_y \, dz \wedge dx + E_z \, dx \wedge dy.$$

Applying the exterior derivative operator and collecting terms gives

$$-d \star F = \left(\frac{\partial B_x}{\partial y} - \frac{\partial B_y}{\partial x} + \frac{\partial E_z}{\partial t} \right) (dt \wedge dx \wedge dy)$$
$$+ \left(\frac{\partial B_y}{\partial z} - \frac{\partial B_z}{\partial y} + \frac{\partial E_x}{\partial t} \right) (dt \wedge dy \wedge dz)$$
$$+ \left(\frac{\partial B_z}{\partial x} - \frac{\partial B_x}{\partial z} + \frac{\partial E_y}{\partial t} \right) (dt \wedge dx \wedge dy)$$
$$+ \left(\frac{\partial E_x}{\partial x} + \frac{\partial E_y}{\partial y} + \frac{\partial E_z}{\partial z} \right) (dx \wedge dy \wedge dz). \qquad (3.67)$$

---

[30] We will work in the CGS system of units and set $c = 1$. Incidentally, we are now in a position to answer a question left open in Example 2.1. There, we wrote $F_{\mu\nu}$ for the components of the field strength tensor, but wondered what the appropriate basis was. We now see that the tensor itself is

$$F = F_{\mu\nu} \, dx^\mu \otimes dx^\nu = \tfrac{1}{2} F_{\mu\nu} \, dx^\mu \wedge dx^\nu. \qquad (3.66)$$

In other words, the basis elements are just the dx^μ.

Next, we define a 1-form that encodes the charge and current densities,

$$J = -\rho\, dt + J_x\, dx + J_y\, dy + J_z\, dz,$$

so that

$$\star J = -\rho\, dx \wedge dy \wedge dz + J_x\, dt \wedge dy \wedge dz$$
$$+ J_y\, dt \wedge dz \wedge dx + J_z\, dt \wedge dx \wedge dy. \qquad (3.68)$$

Comparing (3.67) and (3.68) we see that Maxwell's equations with sources, namely

$$\nabla \cdot \boldsymbol{E} = 4\pi\rho,$$

$$\nabla \times \boldsymbol{B} = 4\pi \boldsymbol{J} + \frac{\partial \boldsymbol{E}}{\partial t},$$

can be written compactly as

$$d \star F = 4\pi \star J. \qquad (3.69)$$

EXERCISE 3.28 Show that Maxwell's equations without sources, namely

$$\nabla \cdot \boldsymbol{B} = 0$$

$$\nabla \times \boldsymbol{E} = -\frac{\partial \boldsymbol{B}}{\partial t},$$

can be written as

$$dF = 0. \qquad (3.70)$$

Remark Note the pleasing symmetry of Maxwell's equations in the absence of sources: $dF = d \star F = 0$.

Remark The utility of writing Maxwell's equations in the forms (3.69) and (3.70) is not so much to facilitate computations of electromagnetic fields (although this can happen) but, rather, to gain an advantage in theoretical understanding. The next exercise provides one illustration of this point. Later we will see others.

EXERCISE 3.29 Show that Maxwell's equations imply the law of charge conservation,

$$\nabla \cdot \boldsymbol{J} + \frac{\partial \rho}{\partial t} = 0. \qquad (3.71)$$

where $\boldsymbol{J} = (J_x, J_y, J_z)$ is the ordinary current density vector field. *Hint:* $d^2 = 0$.

3.13 The interior product

Given a vector field X we can define a linear map $i_X : \Omega^k(M) \to \Omega^{k-1}(M)$, taking k-forms to $(k-1)$-forms, called the **interior product**, satisfying the following properties.[31] Let f be a function, ω a one-form, and λ and η arbitrary-degree forms.

---

[31] In some books the interior product $i_X \omega$ is called the "hook product" and is written $X \lrcorner \omega$.

Then we have:

(1) $i_X f = 0$;
(2) $i_X \omega = \omega(X) := \langle \omega, X \rangle$;
(3) $i_X(\lambda \wedge \eta) = i_X \lambda \wedge \eta + (-1)^{\deg \lambda} \lambda \wedge i_X \eta$.

(Property (3) says that the interior product is a graded derivation, just like the exterior derivative.) If ω is a k-form and $X_1, \ldots, X_k$ are vector fields, we *define*[32]

$$\omega(X_1, X_2, \ldots, X_k) := (i_{X_1}\omega)(X_2, \ldots, X_k) \tag{3.72}$$

or, equivalently,

$$\omega(X_1, X_2, \ldots, X_k) = i_{X_k} i_{X_{k-1}} \cdots i_{X_1} \omega. \tag{3.73}$$

Example 3.15 If α and β are 1-forms and X and Y are vector fields,

$$\begin{aligned}
(\alpha \wedge \beta)(X, Y) &= i_Y i_X (\alpha \wedge \beta) \\
&= i_Y (i_X \alpha \wedge \beta - \alpha \wedge i_X \beta) \\
&= i_Y (\alpha(X)\beta - \beta(X)\alpha) \\
&= \alpha(X)\beta(Y) - \beta(X)\alpha(Y). \tag{3.74}
\end{aligned}$$

EXERCISE 3.30

(a) Show that

$$i_X df = df(X) = Xf. \tag{3.75}$$

(b) Verify that the interior product is *function linear* on its argument. That is, show that

$$i_{fX} = f i_X, \tag{3.76}$$

for any function f and any vector field X. *Hint:* Use induction on the degree.

(c) Show that the inner product map on forms is antisymmetric:

$$i_{X_i} i_{X_j} = -i_{X_j} i_{X_i}. \tag{3.77}$$

Equivalently,

$$\omega(X_1, \ldots, X_i, \ldots, X_j, \ldots, X_k) = -\omega(X_1, \ldots, X_j, \ldots, X_i, \ldots, X_k). \tag{3.78}$$

Hint: Use another induction.

(d) Let $\omega = \lambda \wedge \mu \wedge \nu$, where λ, μ, and ν are 1-forms. Evaluate $\omega(X_1, X_2, X_3)$ as a sum of products of the form $\lambda(X_i)\mu(X_j)\nu(X_k)$.

(e) We asserted in footnote 32 that (3.72) (or (3.73)) implied a particular choice of prefactor c_p in (2.40). Generalizing from part (d), what is c_p?

---

[32] Although it does not look like it, this definition is equivalent to a particular choice of prefactor in (2.40). See Exercise 3.30(e).

EXERCISE 3.31 Suppose that $X_1, \ldots, X_n$ are n linearly dependent tangent vectors at a point p of a manifold M. Show that $\omega_p(X_1, \ldots, X_n) = 0$ for every n-form ω. *Hint:* There is a hard way and an easy way to do this problem. The hard way is to assume that $\omega = dx^1 \wedge \cdots \wedge dx^n$ and then show that $(dx^1 \wedge \cdots \wedge dx^n)(X_1, \ldots, X_n) = \det X^{ij}$, where $X^i = X^{ij} \partial_j$. The easy way is to use the function linearity and antisymmetry of the inner product together with a bit of linear algebra.

3.14 Pullback

Let $f : M \to N$ be a smooth map of manifolds. Given a smooth map $g : N \to \mathbb{R}$ we define a new map $f^*g : M \to \mathbb{R}$ by $f^*g = g \circ f$. The map f^*g is called the **pullback of g by f**, because the function g is "pulled back" from N to M. (See Figure 3.12.) We extend the pullback map to forms by requiring it to be

(1) a ring homomorphism, so that
  (i) $f^*(\lambda + \mu) = f^*\lambda + f^*\mu$,
  (ii) $f^*(\lambda \wedge \mu) = f^*\lambda \wedge f^*\mu$,
  and
(2) natural, that is, $d(f^*\lambda) = f^*(d\lambda)$.

Let ω be a 1-form on N. Let $U \subset M$ and $V \subset N$ be open sets with $f(U) \subset V$. Let $y^1, \ldots, y^n$ be local coordinates on V and $x^1, \ldots, x^m$ local coordinates on U, and suppose that $\omega = \sum a_i \, dy^i$ on V. Then applying properties (1) and (2) gives

$$
\begin{aligned}
f^*\omega &= \sum (f^*a_i) \, f^*(dy^i) \\
&= \sum (a_i \circ f) \, d(f^*y^i) \\
&= \sum (a_i \circ f) \, df^i \\
&= \sum (a_i \circ f) \frac{\partial f^i}{\partial x^j} \, dx^j.
\end{aligned}
\tag{3.79}
$$

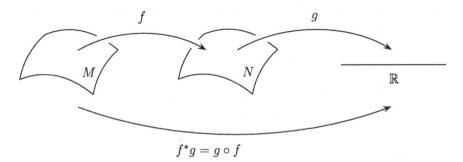

$$f^*g = g \circ f$$

Figure 3.12 The pullback map.

So far, so good. But now we employ the same convention as before (cf. (3.14)) and identify the coordinate functions f^i with y^i, so that (3.79) becomes

$$(f^*\omega)(x) = \sum a_i(y(x)) \frac{\partial y^i}{\partial x^j} \, dx^j. \tag{3.80}$$

It follows from (3.80) that property (2) is an extension of the usual chain rule from calculus, for if $g : V \to \mathbb{R}$ then (3.80) implies that

$$f^*(dg) = f^* \left(\sum \frac{\partial g}{\partial y^j} \, dy^j \right) = \sum \frac{\partial g}{\partial y^i} \frac{\partial y^i}{\partial x^j} \, dx^j$$

while

$$d(f^*g) = \sum \frac{\partial (f^*g)}{\partial x^j} \, dx^j = \sum \frac{\partial (g \circ f)}{\partial x^j} \, dx^j.$$

We see that the statement $f^*(dg) = d(f^*g)$ is just the ordinary chain rule, which is why property (2) is called "natural".

Example 3.16 Let $\alpha = 6xy \, dx + 2yz \, dy + x^2 \, dz$ be a 1-form in $\mathbb{R}^3$, and let $\gamma : I \to \mathbb{R}^3$ be a parameterized curve given by $t \mapsto (t, t^2, t^3)$. Then

$$(\gamma^*\alpha)(t) = 6(t)(t^2)(dt) + 2(t^2)(t^3)(2t \, dt) + (t^2)(3t^2 \, dt)$$
$$= (6t^3 + 3t^4 + 4t^6) \, dt.$$

Example 3.17 Let $\omega = y \, dy \wedge dz + 2y \, dz \wedge dx + xz \, dx \wedge dy$ be a 2-form in $\mathbb{R}^3$ and let $\sigma : I \times I \to \mathbb{R}^3$ be a parameterized surface given by $(u, v) \to (u \cos v, u \sin v, 2u^2)$. Then

$$(\sigma^*\omega)(u, v) = (u \sin v)(\sin v \, du + u \cos v \, dv) \wedge (4u \, du)$$
$$+ 2(u \sin v)(4u \, du) \wedge (\cos v \, du - u \sin v \, dv)$$
$$+ (u \cos v)(2u^2)(\cos v \, du - u \sin v \, dv) \wedge (\sin v \, du + u \cos v \, dv)$$
$$= (-4u^3 \sin v \cos v - 8u^3 \sin^2 v + 2u^4 \cos v) \, du \wedge dv.$$

In general, suppose that ω is an n-form on an n-dimensional manifold M and is represented in local coordinates on V by

$$\omega = a(y) \, dy^1 \wedge \cdots \wedge dy^n. \tag{3.81}$$

Pulling this back to U by means of f^* gives

$$f^*\omega = (f^*a) f^*(dy^1) \wedge \cdots \wedge f^*(dy^n)$$
$$= (a \circ f) \, d(f^*y^1) \wedge \cdots \wedge d(f^*y^n)$$
$$= (a \circ f) \, df^1 \wedge \cdots \wedge df^n$$
$$= (a \circ y) \, dy^1 \wedge \cdots \wedge dy^n$$

$$= (a \circ y) \left(\sum \frac{\partial y^1}{\partial x^{i_1}} dx^{i_1} \right) \wedge \cdots \wedge \left(\sum \frac{\partial y^n}{\partial x^{i_n}} dx^{i_n} \right)$$

$$= (a \circ y) \sum \frac{\partial y^1}{\partial x^{i_1}} \cdots \frac{\partial y^n}{\partial x^{i_n}} dx^{i_1} \wedge \cdots \wedge dx^{i_n}$$

$$= (a \circ y) \left[\frac{\partial y^i}{\partial x^j} \right] dx^1 \wedge \cdots \wedge dx^n, \tag{3.82}$$

where $[\partial y^i / \partial x^j]$ denotes the usual Jacobian determinant.[33] (For the last step, see (2.53); compare with (3.60).) It is no accident that this looks like the change of variables theorem in calculus, as the pullback operation together with the antisymmetry of the wedge product is designed to reproduce that result.

EXERCISE 3.32 Let

$$\omega = 3xz \, dx - 7y^2z \, dy + 2x^2y \, dz$$

be a 1-form on $\mathbb{R}^3$, and suppose that $\phi : \mathbb{R} \to \mathbb{R}^3$ is the map $\phi(t) = (t^2, 2t - 1, t)$. Compute the 1-form $\phi^*\omega$.

EXERCISE 3.33 Let

$$\theta = -x \, dy \wedge dz - y \, dz \wedge dx + 2z \, dx \wedge dy$$

be a 2-form on $\mathbb{R}^3$, and suppose that $\phi : \mathbb{R}^2 \to \mathbb{R}^3$ is the map $\phi(u, v) = (u^3 \cos v, u^3 \sin v, u^2)$. Compute the 2-form $\phi^*\theta$.

EXERCISE 3.34 Let $\omega = 4y \, dx + 2xy \, dy + dz$ and $\phi : \mathbb{R}^2 \to \mathbb{R}^3$ be given by $\phi(u, v) = (u, v, u^2 + v^2)$. Show that $d\phi^*\omega = \phi^*d\omega$.

3.15 Pushforward

Let $f : M \to N$ be any smooth map. We have seen how the pullback map f^* naturally pulls a form on N back to a form on M. One can ask a similar question regarding vector fields. Do they pull back as well? The answer to this question is complicated, and we shall return to it at the end of this section. Instead, we introduce a map that is very familiar already, although the language may not be so. If $g : N \to \mathbb{R}$ is a smooth function on N, we define a linear map $f_* : T_pM \to T_{f(p)}N$ of tangent spaces, called the **pushforward**, by

$$(f_*X_p)_{f(p)}(g) := X_p(f^*g). \tag{3.83}$$

---

[33] We use square brackets here and occasionally elsewhere rather than the standard vertical brackets in order to emphasize that we are referring to the determinant itself rather than its absolute value.

A hint as to the identity of the pushforward map is provided by its other name, the **derivative** or **differential** of f.[34] The next exercise provides the connection.

> **EXERCISE 3.35** If $\{x^i\}$ are local coordinates in the neighborhood of $p \in M$ and $\{y^j\}$ are local coordinates in a neighborhood of $f(p) \in N$, show that $(f_*)_{ij} = (\partial y^i / \partial x^j)_p$ relative to the local bases $\partial/\partial x^i$ and $\partial/\partial y^j$. In other words, in a local basis the derivative map is represented by the Jacobian matrix.

Remark In terms of the pushforward map, the inverse function theorem can be stated as follows: the map $f : M \to N$ is a local diffeomorphism in the neighborhood of a point p if and only if f_* is an isomorphism of $T_p M$ and $T_{f(p)} N$.

> **EXERCISE 3.36** Show that, if $f : M \to M$ is the identity map then f^* and f_* are both identity maps. Show also that if $f : U \to V$ and $g : V \to W$ then $(g \circ f)^* = f^* \circ g^*$ and $(g \circ f)_* = g_* \circ f_*$. In fancy mathematics lingo, one says that the pullback is a **contravariant functor** while the pushforward is a **covariant functor**.

Now, let $f : M \to N$ be smooth and suppose that X is a vector field on M. Furthermore, suppose that f is not injective, so that $f(p_1) = f(p_2) = q$. Then f_* maps the two tangent spaces $T_{p_1} M$ and $T_{p_2} M$ to the same image, namely $T_q N$. But there is nothing to guarantee that X_{p_1} and X_{p_2} will map to the same image and if not the image of X under f_* will not be a smooth vector field on N. This sort of pathology cannot occur if f is a diffeomorphism, because in that case f is bijective. Therefore, when f is a diffeomorphism it makes sense to define the pushforward of the vector field X in the obvious way, namely

$$(f_* X)_{f(p)} := f_* X_p. \tag{3.84}$$

Equivalently, if $g : N \to \mathbb{R}$ is any smooth function then

$$(f_* X)_{f(p)}(g) = X_p(f^* g). \tag{3.85}$$

This looks so much like (3.83) that people often get confused. Just remember that you can push *vectors* forward by any smooth map but you can only push *vector fields* forward by a diffeomorphism.

Remark There is another way to write (3.85) that is often more useful in computations. Viewing both sides of (3.85) as an equality of functions on M, we have

$$f^*[(f_* X)(g)] = X(f^* g). \tag{3.86}$$

(Evaluate both sides at $p \in M$ to recover (3.85).)

---

[34] For this reason one often sees f_* written as Df.

Theorem 3.7 *The Lie bracket of vector fields is natural with respect to pushforward by a diffeomorphism. That is, for any diffeomorphism $\varphi : M \to N$ and vector fields X and Y on M,*

$$\varphi_*[X, Y] = [\varphi_* X, \varphi_* Y]. \tag{3.87}$$

Proof It suffices to show that both sides of (3.87) yield the same result when acting on any smooth function $g : N \to \mathbb{R}$. Applying (3.86) twice gives

$$YX(\varphi^* g) = Y[\varphi^*[(\varphi_* X)(g)]] = \varphi^*[(\varphi_* Y)(\varphi_* X)(g)]. \tag{3.88}$$

Yet another application of (3.86) (with X replaced by $[X, Y]$) gives

$$\begin{aligned}
\varphi^*[\varphi_*[X, Y](g)] &= [X, Y](\varphi^* g) \\
&= (XY - YX)(\varphi^* g) \\
&= XY(\varphi^* g) - YX(\varphi^* g) \\
&= \varphi^*[(\varphi_* X)(\varphi_* Y)(g) - (\varphi_* Y)(\varphi_* X)(g)],
\end{aligned}$$

where we used (3.88) (and a similar equation with X and Y interchanged) in the last step. But φ is a bijection, so we may strip φ^* from both sides to get (3.87). $\square$

In the special case where $f : M \to N$ is a diffeomorphism, the pushforward operation can be extended to general tensor fields by recalling the definition of a tensor as a multilinear operator. Let $p \in M$ and $q := f(p) \in N$. Let $X_i \in TN$ and $\omega^i \in T^*N$, and let S be a tensor field of type (r, s) in M. As f is a diffeomorphism, the inverse function theorem guarantees that f_* is an isomorphism of tangent spaces. In particular, f_* is invertible. Thus we can write

$$\begin{aligned}
&(f_* S)(\omega^1, \ldots, \omega^r, X_1, \ldots, X_s)\Big|_q \\
&\quad := S(f^* \omega^1, \ldots, f^* \omega^r, f_*^{-1} X_1, \ldots, f_*^{-1} X_s)\Big|_p. \tag{3.89}
\end{aligned}$$

Alternatively, if R is a tensor field of type (r, s) in N then we can pull it back to M by

$$\begin{aligned}
&(f^* R)(\mu^1, \ldots, \mu^r, Y_1, \ldots, Y_s)\Big|_p \\
&\quad := R(f^{-1*} \mu^1, \ldots, f^{-1*} \mu^r, f_* Y_1, \ldots, f_* Y_s)\Big|_q, \tag{3.90}
\end{aligned}$$

where $Y_i \in TM$ and $\mu^i \in T^*M$. In this sense, pushforward and pullback are "inverse" operations.

At last we can answer the question posed at the beginning of this section: do vector fields pull back? The answer is that they naturally push forward, but by virtue of the above correspondence we can *define* the pullback of a vector field X by a diffeomorphism φ to be the pushforward via the inverse map, so that

$$\varphi^* X := (\varphi^{-1})_* X, \tag{3.91}$$

and this is what we shall do.

3.16 Integral curves and the Lie derivative

In addition to the exterior derivative of a differential form there is another intrinsic derivative on manifolds, where by "intrinsic" we mean that its definition does not require the introduction of any additional structure. If X is a vector field on M then, as we have seen, X can be thought of as a generalized directional derivative operator on functions. But we want to be able to take derivatives of tensor fields, and X does not appear to act directly on tensor fields. It turns out that the pushforward map provides the tool we need.

Given a vector field X on M, the fundamental theorem of ordinary differential equations guarantees that there is a unique maximal curve $\gamma(t) : I \rightarrow M$ through each point $p \in M$ such that $\gamma(0) = p$ and such that the tangent vector to the curve γ at $\gamma(t)$ is precisely $X_{\gamma(t)}$. The curve γ is called the **integral curve** of X through p.[35]

To see how this works, it is worth taking a moment to discuss notation. In ordinary calculus we call $\gamma'(t)$ the tangent vector to the curve. The reason is that the curve γ is a map into some Euclidean space, so $\gamma'(t)$ is naturally a vector. But, as we have previously discussed, this interpretation is no longer tenable in a general manifold that is not itself a vector space. Instead, vectors are derivations, so we really want to understand $\gamma'(t)$ as a derivation. How do we do that?

The answer is to be a bit more precise. We view I as a manifold, namely an open interval of the real line, and attach the coordinate t to the points of I. Then d/dt is a tangent vector field on I which acts on functions $f : I \rightarrow \mathbb{R}$ according to $(d/dt)f = df/dt$. Now we define $\gamma'(t) := \gamma_*(d/dt)_{\gamma(t)}$. Although this may look intimidating, it's really not. In fact, it allows us to reduce the question of the existence of an integral curve in a general manifold M to a system of ordinary differential equations.

To see why this is so, let $x^1, \ldots, x^n$ be local coordinates around p. Applying both sides of

$$\gamma'(t) = X_{\gamma(t)} \tag{3.92}$$

to the coordinate functions x^i gives

---

[35] The range of t is not specified, but "maximal" means that it is large enough that the curve exists in a neighborhood of p and no other integral curve covers more points of M. The question of how far such a curve can be extended on the manifold M is a delicate matter and the subject of much investigation. Indeed, some very beautiful results of differential geometry give conditions for the existence of global integral curves.

Figure 3.13 Flowing along the integral curves of a vector field.

$$\gamma_*(d/dt)(x^i) = \frac{d}{dt}\gamma^*(x^i) = \frac{d}{dt}(x^i \circ \gamma) = \frac{d}{dt}\gamma^i(t) = \frac{d}{dt}x^i(t), \qquad (3.93)$$

and, taking $X = X^j \partial/\partial x^j$, we have

$$X_{\gamma(t)}x^i = X^j \left.\frac{\partial x^i}{\partial x^j}\right|_{\gamma(t)} = X^i(\gamma(t)) = X^i(x^1(t), \ldots, x^n(t)). \qquad (3.94)$$

Equating (3.93) and (3.94) gives the system of differential equations

$$\frac{dx^i(t)}{dt} = X^i(x^1(t), \ldots, x^n(t)), \quad 1 \le i \le n, \qquad (3.95)$$

whose simultaneous solution yields the curve γ. We must "integrate" these equations to get the curve; hence the name "integral curve".

The existence of integral curves enables us to define the notion of a flow. At each point $p \in M$ we can imagine moving a parameter a distance t along the integral curve γ of X through p. If we do this for all points of M simultaneously we get what is called a flow along X, where you can think of this movement as the flow of a fluid. (See Figure 3.13.) More precisely, for any open set U of M and any $p \in U$, define $\gamma_p(t)$ to be the integral curve of X through p, with $p = \gamma_p(0)$. Then the map $\phi_t : U \to M$ given by $\gamma_p(0) \mapsto \gamma_p(t)$ is a **flow**, or a **one-parameter group of diffeomorphisms** of M generated by the vector field X.[36]

Using this flow, we can generalize the notion of the directional derivative to tensor fields. First observe that, in terms of the flow ϕ_t (and writing f_p for $f(p)$ and $\phi_t p$ for $\gamma_p(t)$),

$$(Xf)_p = \lim_{t \to 0} \frac{f_{\phi_t p} - f_p}{t}. \qquad (3.96)$$

If we tried to do something similar for tensors, it might look like this:

$$(\mathcal{L}_X T)_p = \lim_{t \to 0} \frac{T_{\phi_t p} - T_p}{t} \quad (???),$$

---

[36] It is a group of diffeomorphisms under composition, since $\phi_{s+t} = \phi_s \circ \phi_t = \phi_t \circ \phi_s$ with $\phi_{-t} = \phi_t^{-1}$ and $\phi_0 = id$.

where $\mathcal{L}_X$ denotes our generalized directional derivative operator. The problem with this definition is that it makes no sense, because $T_{\phi_t p}$ and T_p live in different (unrelated) vector spaces, so we cannot subtract them. But, if we were to drag $T_{\phi_t p}$ back to p by means of the flow ϕ_t then we could compare its value with T_p.

The right way to proceed is to rewrite (3.96) as

$$(Xf)_p = \lim_{t \to 0} \left. \frac{\phi_t^* f - f}{t} \right|_p , \tag{3.97}$$

for then

$$(Xf)_p = \lim_{t \to 0} \frac{f(\phi_t(p)) - f(p)}{t}$$

as before. By analogy, we define the **Lie derivative** $\mathcal{L}_X T$ of a tensor field T in the direction X at p by

$$(\mathcal{L}_X T)_p := \lim_{t \to 0} \left. \frac{\phi_t^* T - T}{t} \right|_p = \left. \frac{d}{dt} \phi_t^* T \right|_{t=0} , \tag{3.98}$$

where the pullback of a tensor field is defined by (3.90).[37] Note that a scalar field can be viewed as a type $(0, 0)$ tensor field, so that (3.90) applied to a scalar field just reduces to the ordinary pullback of functions. In particular, we have

$$(\mathcal{L}_X f)_p = (Xf)_p. \tag{3.99}$$

The following properties are immediate from the definition (3.98):

(1) $\mathcal{L}_X$ is linear;
(2) $\mathcal{L}_X$ preserves tensor type.

Also, the Lie derivative is, well, a derivative, so

(3) $\mathcal{L}_X$ is a derivation: $\mathcal{L}_X(S \otimes T) = \mathcal{L}_X S \otimes T + S \otimes \mathcal{L}_X T$.

---

[37] Other authors write this a little differently. For example, one equivalent way to write it would be

$$(\mathcal{L}_X T)_p = \lim_{t \to 0} \left. \frac{\phi_{-t*} T - T}{t} \right|_p = \lim_{t \to 0} \frac{(\phi_{-t*} T)_p - T_p}{t},$$

although in the spirit of (3.83) it is sometimes written as

$$(\mathcal{L}_X T)_p = \lim_{t \to 0} \frac{\phi_{-t*} T_{\phi_t p} - T_p}{t}$$

instead, because one imagines the first term on the right as "pushing" the tensor field T at $\phi_t p$ backwards to p. Lastly, as $\phi_{0*} = id$, the equation is sometimes written as

$$(\mathcal{L}_X T)_p = \lim_{t \to 0} \phi_{t*} \left. \left(\frac{\phi_{-t*} T - T}{t} \right) \right|_p = \lim_{t \to 0} \left. \frac{T - \phi_{t*} T}{t} \right|_p .$$

Some properties that are not so obvious include the following.

(4) $\mathcal{L}_X$ is compatible with the natural pairing of forms and vector fields:

$$\mathcal{L}_X \langle \omega, Y \rangle = \langle \mathcal{L}_X \omega, Y \rangle + \langle \omega, \mathcal{L}_X Y \rangle . \tag{3.100}$$

(5) For any vector field Y,

$$\mathcal{L}_X Y = [X, Y]. \tag{3.101}$$

(6) For vector fields X, Y, and Z,

$$[\mathcal{L}_X, \mathcal{L}_Y] Z = \mathcal{L}_{[X,Y]} Z, \tag{3.102}$$

where $[\mathcal{L}_X, \mathcal{L}_Y] := \mathcal{L}_X \mathcal{L}_Y - \mathcal{L}_Y \mathcal{L}_X$.

We offer a direct proof of property (5) here, leaving a slightly different proof to the following exercise. It suffices to show that $(\mathcal{L}_X Y)f = [X, Y]f$ for all smooth functions f. Thus, letting ϕ_t be the flow associated to X and using (3.86) in the form

$$(\phi_* Y)g = (\phi^{-1})^* Y (\phi^* g) \tag{3.103}$$

gives

$$
\begin{aligned}
(\mathcal{L}_X Y)f &= \lim_{t \to 0} \frac{\phi_t^* Y - Y}{t} f \\
&= \lim_{t \to 0} \frac{\phi_{-t*} Y - Y}{t} f \\
&= \lim_{t \to 0} \frac{\phi_t^* (Y (\phi_{-t}^* f)) - \phi_t^* (Yf) + \phi_t^* (Yf) - Yf}{t} \\
&= \lim_{t \to 0} \phi_t^* Y \left(\frac{\phi_{-t}^* f - f}{t} \right) + \lim_{t \to 0} \frac{\phi_t^* (Yf) - Yf}{t} \\
&= Y(-Xf) + X(Yf) = [X, Y]f.
\end{aligned}
$$

(We have used the facts that $\phi_0 = id$ and that ϕ_{-t} is the flow associated to the vector field $-X$.)

EXERCISE 3.37 Prove properties (4) through (6) above. *Hints:* For property (4), first show that pullback commutes with contraction, in the sense that

$$\phi_t^* \langle \omega, Y \rangle = \langle \phi_t^* \omega, \phi_t^* Y \rangle , \tag{3.104}$$

where $\phi_t^* Y$ means $(\phi_t^{-1})_* Y$. For property (5), use (4) with $\omega = df$ and recall that d commutes with pullback. For property (6), use (5) and the Jacobi identity.

Roughly speaking, the Lie derivative measures how the tensor changes as it moves along the integral curves of X. It turns out to have many uses, but it suffers from one major problem, namely that it depends on X in a neighborhood of p and not simply on X at the single point p. This necessitates the introduction of additional derivative operators, as we shall see.

Additional exercises

NOTE: *Henceforth we drop the boldface font for matrices whenever there is little or no risk of confusion between a matrix and its corresponding linear operator.*

3.38 The circle   By analogy with Example 3.6, show that $S^1 = \{(x, y) \in \mathbb{R}^2 : x^2 + y^2 = 1\}$ is an orientable manifold. (No, you cannot cheat and use Theorem 3.2.)

3.39 Cartesian products of manifolds   Show that the Cartesian product $M \times N$ of two smooth manifolds M and N (of dimensions m and n, respectively), is naturally a smooth manifold of dimension $m+n$. Furthermore, if M and N are both orientable, show that $M \times N$ is orientable. *Hint:* If M has atlas $\{(U_i, \varphi_i)\}$ and N has atlas $\{(V_j, \psi_j)\}$, show that $M \times N$ has atlas $\{U_i \times V_j, \varphi_i \times \psi_j\}$, where $(\varphi_i \times \psi_j)(p, q) := (\varphi_i(p), \psi_j(q)) \in \mathbb{R}^m \times \mathbb{R}^n$. *Remark:* From this result we immediately obtain that the *n-torus*, defined by

$$T^n = \underbrace{S^1 \times \cdots \times S^1}_{n \text{ times}},$$

is a smooth n-dimensional manifold, because the circle S^1 is a smooth manifold.

3.40 Area forms on the 2-sphere   The standard area element on the unit 2-sphere S^2 is $\sigma = \sin\theta\, d\theta \wedge d\phi$ in spherical polar coordinates (r, θ, ϕ). Let $f : \mathbb{R}^3 - \{0\} \to S^2$ be the projection map $x \mapsto x/r$, where $r = (x^2 + y^2 + z^2)^{1/2}$.

(a) Show that

$$f^*\sigma = \frac{1}{r^3}(x\, dy \wedge dz + y\, dz \wedge dx + z\, dx \wedge dy).$$

(b) Show that, in terms of the stereographic coordinates (u^1, u^2) and (v^1, v^2) described in Example 3.6,

$$\sigma_U = -4\frac{du^1 \wedge du^2}{(1 + \eta)^2} \quad \text{and} \quad \sigma_V = +4\frac{dv^1 \wedge dv^2}{(1 + \xi)^2}.$$

Compare these forms with those of Example 3.10. *Hint:* Restrict $f^*\sigma$ from part (a) to the sphere, then use the coordinate transformation between Cartesian and stereographic coordinates obtained in Example 3.6. Alternatively, you could calculate the coordinate transformation between polar and stereographic coordinates directly.

3.41 Elliptic curves The zero set Σ of the polynomial

$$f(x, y) = y^2 - x^3 - ax - b, \qquad a, b \in \mathbb{R},$$

in the plane is called an **elliptic curve** (over the reals) provided that it has no cusps or self-intersections – i.e., provided that it is an embedded smooth submanifold of $\mathbb{R}^2$. For what values of a and b does f define a real elliptic curve? *Hint:* Recall that a cubic polynomial of the form $x^3 + ax + b$ has repeated roots only if the discriminant $\Delta = -16(4a^3 + 27b^2)$ vanishes.

3.42 Regular values and tangent spaces Let $f : M \to N$ be a smooth map of manifolds with $m = \dim M \geq \dim N = n$ and let $q \in N$ be a regular value of f. Let $X = f^{-1}(q)$ and let $p \in X$. Show that $T_p X$ is just the kernel of the differential $f_* : T_p M \to T_q N$. *Hint:* X is a submanifold of M, so $T_p X$ is a subspace of $T_p M$. Consider f_* restricted to X and don't forget the rank–nullity theorem.

3.43 Lie groups A **Lie group** G is both a smooth manifold and a group such that the group operations $(g, h) \mapsto gh$ and $g \mapsto g^{-1}$ are smooth maps.

(a) The set $M_n(\mathbb{R})$ of all real $n \times n$ matrices is easily seen to be an n^2-dimensional smooth manifold isomorphic to $\mathbb{R}^{n^2}$. Each point is a matrix A, and the components of the matrix serve as coordinates for the point. In other words, the coordinate functions $\{x_{ij}\}$ are given by $x_{ij}(A) = A_{ij}$. The **general linear group** $GL(n, \mathbb{R})$ is the subset of $M_n(\mathbb{R})$ consisting of all matrices with nonzero determinant, the group operation being ordinary matrix multiplication. Prove that $GL(n, \mathbb{R})$ is an n^2-dimensional Lie group. *Hint:* Show that it is an open subset of $M_n(\mathbb{R})$ and that the group operations are smooth. *Remark:* A **matrix group** is a subgroup of $GL(n, \mathbb{R})$.

(b) Show that the set $O(n)$ of all orthogonal matrices is a Lie group. It is called the **orthogonal group**. What is its dimension? *Hint:* Let $M_n^+(\mathbb{R})$ be the set of real symmetric $n \times n$ matrices, and consider the map $\varphi : M_n(\mathbb{R}) \to M_n^+(\mathbb{R})$ given by $\varphi(A) = AA^T$. Use Theorem 3.2 with $q = I$. The only tricky thing is to prove that I is a regular value of φ. For this, show that $\varphi_{*A} B = AB^T + BA^T$ and go from there. Note that, because $M_n(\mathbb{R})$ and $M_n^+(\mathbb{R})$ are vector spaces, we have $T_p M_n(\mathbb{R}) \cong M_n(\mathbb{R})$ and $T_q M_n^+(\mathbb{R}) \cong M_n^+(\mathbb{R})$.

(c) The orthogonal group is divided into two classes of matrices: those with determinant $+1$ and those with determinant -1. The ones with determinant $+1$ constitute the **special orthogonal group**, denoted $SO(n)$. Show that $SO(n)$ is a Lie group but the other class of matrices is not. *Hint:* Use Theorem 3.2 again, then recall (2.54).

(d) Show that $SO(2)$ is diffeomorphic to the circle (1-sphere) S^1. *Hint:* Consider the map $(x, y) \rightarrow \begin{pmatrix} x & y \\ -y & x \end{pmatrix}$. You may use the following fact without proof. If $\varphi : M \rightarrow N$ is a smooth map of manifolds, and $M' \subseteq M$ and $N' \subseteq N$ are submanifolds with $\varphi(M') \subseteq N'$, then the restriction map $\varphi|_{M'} : M' \rightarrow N'$ is smooth.

3.44 Lie algebras Let G be a Lie group with identity element e. Define a map (actually a diffeomorphism) $L_g : G \rightarrow G$ given by $L_g(h) = gh$ ("left multiplication by g"). A vector field X is **left invariant** if $L_{g*}X = X$ for all $g \in G$ (which is a sloppy but convenient way to write $L_{g*}X_h = X_{gh}$ for every $g, h \in G$).

(a) If X and Y are two left-invariant vector fields, show that their Lie bracket $[X, Y]$ is also left-invariant.

(b) Show that a left-invariant vector field X is determined uniquely by its value X_e at the identity and, moreover, given a vector $Y \in T_e G$, there is a unique left-invariant vector field X with $X_e = Y$. Thus, with the obvious identification $[X_e, Y_e] := [X, Y]_e$, $T_e G$ inherits the structure of a Lie algebra, usually denoted $\mathfrak{g}$, called **the Lie algebra of the Lie group G**.[38]

(c) A **one-parameter subgroup** of G is a smooth homomorphism $\gamma : \mathbb{R} \rightarrow G$, written $t \mapsto \gamma(t)$. In other words, it's a smooth path in G satisfying $\gamma(s + t) = \gamma(s)\gamma(t)$. (Observe that this automatically implies that $\gamma(0) = e$ and $\gamma(t)^{-1} = \gamma(-t)$. Why?) Show that there is a bijection between the set of one-parameter subgroups of G and the Lie algebra $\mathfrak{g}$ of G. *Hint:* Consider the integral curves of a left invariant vector field. To show that a homomorphism is obtained, consider the derivative of $\gamma(s + t) = \gamma(s)\gamma(t)$ and use the uniqueness of integral curves. You should also show that the domain of γ can be extended to the whole of $\mathbb{R}$ (from which we conclude that left invariant vector fields are **complete**).

---

[38] The Lie algebra $\mathfrak{g}$ is therefore a (dim G)-dimensional subalgebra of the (infinite-dimensional) Lie algebra of all vector fields on G.

Remark: Let $\gamma_X(t)$ denote the one-parameter subgroup corresponding to the left invariant vector field X. By analogy with the usual exponential function, which satisfies $e^s e^t = e^{s+t}$, one writes $\gamma_X(t) = \operatorname{Exp} tX$. The map $\operatorname{Exp} : \mathfrak{g} \to G$ given by $X \mapsto \operatorname{Exp}_X(1)$ is called the **exponential map**.[39] In general the exponential map is neither injective nor surjective, but it is still quite powerful. It furnishes a link between Lie groups and Lie algebras that enables one to reduce questions about Lie groups to questions about Lie algebras. Specifically, there is an almost unique correspondence between a Lie group and a Lie algebra, and the exponential map provides a way of associating a Lie group homomorphism to a Lie algebra homomorphism and vice versa. For a nice overview, see Graeme Segal's chapter in [15] or, for more details, [88].

3.45 **The Lie algebra $\mathfrak{gl}(n, \mathbb{R})$** A **Lie algebra isomorphism** $\varphi : \mathfrak{a} \to \mathfrak{b}$ is a vector space isomorphism that is also an algebra homomorphism: $\varphi([X, Y]_\mathfrak{a}) = [\varphi(X), \varphi(Y)]_\mathfrak{b}$. The Lie algebra of the general linear group $GL(n, \mathbb{R})$ is denoted $\mathfrak{gl}(n, \mathbb{R})$. Show that there is a Lie algebra isomorphism $\mathfrak{gl}(n, \mathbb{R}) \to M_n(\mathbb{R})$, where $M_n(\mathbb{R})$ is a Lie algebra under the bracket operation $[A, B] = AB - BA$ (the commutator of A and B). *Hint:* A neat proof proceeds as follows. (i) Let $x := (x_{ij})$ be the matrix of coordinate functions on $GL(n, \mathbb{R})$, and define $\Theta := x^{-1} dx$, a matrix of 1-forms on $GL(n, \mathbb{R})$. (Incidentally, Θ is called the **Maurer–Cartan form** of $GL(n, \mathbb{R})$.) (ii) Show that Θ is left invariant, meaning that, for every $A \in GL(n, \mathbb{R})$, we have $L_A^* \Theta = \Theta$; thereby show that each entry Θ_{ij} is left invariant. (iii) Let X be a vector field on $GL(n, \mathbb{R})$ and define a map $X \mapsto B_X$, where $B_X := \langle \Theta, X \rangle$. (The latter notation means that each entry of Θ is paired with X.) Show that if X is left invariant then this map induces a vector space isomorphism $\mathfrak{gl}(n, \mathbb{R}) \to M_n(\mathbb{R})$ by demonstrating that $\Theta(X)$ is a *constant* on $GL(n, \mathbb{R})$. You will probably want to use (3.104). (iv) Given $X, Y \in \mathfrak{gl}(n, \mathbb{R})$, use (3.123) to show that $d\Theta(X, Y) = -\Theta([X, Y])$. (v) By differentiating, show that $d\Theta = -\Theta \wedge \Theta$. (The left-hand side of this expression means you are to apply d to each entry of Θ. The right-hand side of this expression is a matrix of 2-forms whose ijth entry is $(\Theta \wedge \Theta)_{ij} = \sum_k \Theta_{ik} \wedge \Theta_{kj}$. In particular, $\Theta \wedge \Theta$ does not vanish.) (vi) Use (v) to evaluate $d\Theta(X, Y)$ and compare the result with (iv) to conclude that $[X, Y] \mapsto [B_X, B_Y]$.

---

[39] For more on the exponential map in the context of Riemannian manifolds, see Exercise 8.50 and Appendix D.

3.46 The Lie algebra $\mathfrak{so}(n)$ Show that $\mathfrak{so}(n)$, the Lie algebra of $SO(n)$, can be identified with the Lie subalgebra of $M_n(\mathbb{R})$ consisting of all $n \times n$ skew symmetric matrices together with the commutator as bracket. (A **Lie subalgebra** is just a subset of a Lie algebra that is itself a Lie algebra.) *Hint:* Consider the tangent vector to a curve in $SO(n)$ at the identity e. Show that $\mathfrak{so}(n)$ is a Lie subalgebra of $M_n(\mathbb{R})$.

3.47 The exponential map   Show that, under the isomorphism $\mathfrak{gl}(n, \mathbb{R}) \cong M_n(\mathbb{R})$, the exponential map $\mathrm{Exp} : \mathfrak{gl}(n, \mathbb{R}) \to GL(n, \mathbb{R})$ is given by

$$\mathrm{Exp}\, tA = e^{tA}, \tag{3.105}$$

where the latter expression is the ordinary exponential operator applied to the matrix tA:

$$e^{tA} = 1 + tA + \frac{1}{2!}(tA)^2 + \cdots . \tag{3.106}$$

Hint: The curve $\gamma(t) = \mathrm{Exp}\, tX$ is the unique integral curve associated to the left-invariant vector field X, which means that $d\gamma/dt = X_{\gamma(t)} = L_{\gamma(t)*}X_e$. L_{A*} is just left multiplication by the matrix A. (Why?) Ignore questions of convergence, which all work out in the end.

3.48 The Lie algebra $\mathfrak{su}(n)$ All the previous results have corresponding complex analogues. The Lie group $GL(n, \mathbb{C})$ is the set of all $n \times n$ matrices with complex entries and nonzero determinant. It is a Lie group diffeomorphic to $\mathbb{R}^{2n^2}$. The Lie algebra $\mathfrak{gl}(n, \mathbb{C})$ of $GL(n, \mathbb{C})$ is isomorphic to $M_n(\mathbb{C})$, the algebra of complex $n \times n$ matrices with commutator as bracket. The analogue of the orthogonal group $O(n)$ is $U(n)$, the **unitary group**, consisting of all $n \times n$ unitary matrices. (Recall that a matrix A is *unitary* provided that $AA^\dagger = I$, where $A^\dagger$ is the conjugate transpose of A and I is the identity matrix.) The **special unitary group** $SU(n)$ comprises $n \times n$ unitary matrices with determinant equal to unity. Show that the Lie algebra $\mathfrak{su}(n)$ of $SU(n)$ can be identified with the Lie subalgebra of $M_n(\mathbb{C})$ consisting of all traceless $n \times n$ anti-Hermitian matrices. (A matrix A is **anti-Hermitian**, or **skew-Hermitian**, if $A^\dagger = -A$.) *Hint:* For the anti-Hermiticity requirement, proceed by analogy with the method of solution of Exercise 3.46. To show tracelessness, combine Exercises 3.47 and 1.48.

3.49 The adjoint representation of a Lie group   Let G be a group. A **representation** of G on a vector space V is a homomorphism $G \to \mathrm{Aut}\, V$. There is a natural representation of a Lie group G on its own Lie algebra, defined as follows. Given an element $g \in G$ we define the **conjugation** action $\varphi(g) : G \to G$ by $\varphi(g)h = ghg^{-1}$. By the properties of Lie groups this map is smooth, so we may consider its derivative at the identity $\varphi(g)_{*,e} : T_e G \to T_e G$. Identifying $T_e G$ with the Lie algebra $\mathfrak{g}$ we get a map $\text{Ad} : G \to \text{Aut}\,\mathfrak{g}$ given by $\text{Ad}\,g = \varphi(g)_{*,e}$. By the chain rule this map is a homomorphism, so it defines a representation of G called the **adjoint representation**.

(a) Show that $\text{Ad}\,g$ really is an automorphism of $\mathfrak{g}$ by showing that it is bijective and is an algebra homomorphism.

(b) Show that, when $G = GL(n, \mathbb{R})$, $A \in G$, and $B \in \mathfrak{gl}(n, \mathbb{R})$,

$$(\text{Ad}\,A)(B) = ABA^{-1}, \tag{3.107}$$

where the product operation on the right-hand side is ordinary matrix multiplication.

3.50 The adjoint representation of a Lie algebra Let $\mathfrak{g}$ be a Lie algebra. A **representation** of $\mathfrak{g}$ on a vector space V is a Lie algebra homomorphism $\mathfrak{g} \to \text{Aut}\,V$, where the Lie algebra structure on $\text{Aut}\,V$ is defined by the commutator. The **adjoint representation** of $\mathfrak{g}$ is the map $\text{ad} : \mathfrak{g} \to \text{Aut}\,\mathfrak{g}$ given by

$$\text{ad}\,X(Y) := [X, Y]. \tag{3.108}$$

As the terminology suggests, there is an intimate connection between the adjoint representation of a Lie group and its Lie algebra. Specifically, we have

$$\text{Ad}\,\text{Exp}\,X = \text{Exp}\,\text{ad}\,X. \tag{3.109}$$

The best way to understand this relation is to prove it for the special case of a matrix group such as $GL(n, \mathbb{R})$. In that case, the exponential map just becomes the ordinary matrix exponential, so (3.109) reads

$$\text{Ad}\,e^X = e^{\text{ad}\,X}. \tag{3.110}$$

Prove (3.110). *Hint:* Replace X by tX, apply both sides to a matrix Y and call the resulting equation (*). Differentiate (*) with respect to t and show that both sides of (*) satisfy the same first-order differential equation with the same initial conditions. Invoke a uniqueness theorem. *Remark:* The two kinds of adjoint map are affectionately known as **big Ad** and **little ad**, although you should be aware that some authors write "ad" for what we have called "Ad". *Caveat lector.*

3.51 Fun with matrix exponentials Show that if $A, B \in M_n(\mathbb{R})$ and $[A, [A, B]] = [B, [A, B]] = 0$ then

$$e^A e^B = e^{A+B+(1/2)[A,B]}. \tag{3.111}$$

In particular, it follows that

$$e^A e^B = e^{A+B} \tag{3.112}$$

if A and B commute. *Hint:* Write

$$e^{t(A+B)} = e^{tA} f(t), \tag{3.113}$$

then show that

$$f(t) = e^{tB} e^{-(t^2/2)[A,B]} \tag{3.114}$$

by differentiating (3.113) and demonstrating that both sides of (3.114) obey the same differential equation with the same initial conditions. *Remark:* In general, if $e^A e^B = e^C$, the **Baker–Campbell–Hausdorff formula** gives a complicated expression for C in terms of A, B, and higher-order commutators. Its utility is that it permits one to recover the Lie group locally from its Lie algebra. For more details, see *e.g.* [34].

3.52 The Killing form of a Lie algebra Every Lie algebra $\mathfrak{g}$ comes equipped with a natural symmetric bilinear form $(\cdot, \cdot)$, called the **Killing form**, given by

$$(X, Y) := \mathrm{tr}(\mathrm{ad}\, X \circ \mathrm{ad}\, Y). \tag{3.115}$$

If the Killing form is nondegenerate, $\mathfrak{g}$ is said to be **semisimple**. Show that the Killing form satisfies the following properties. (The idea is to demonstrate the properties directly from the definitions. For half credit, you may assume $\mathfrak{g}$ to be a subalgebra of $\mathfrak{gl}(n, \mathbb{R})$ – i.e., an algebra of matrices.)

(a) It is symmetric, so that

$$(X, Y) = (Y, X). \tag{3.116}$$

(b) It invariant under the natural adjoint action:

$$(\mathrm{Ad}\, g(X), \mathrm{Ad}\, g(Y)) = (X, Y). \tag{3.117}$$

Hint: Show that, for any automorphism T of $\mathfrak{g}$, $\mathrm{ad}(TX) = T \circ \mathrm{ad}\, X \circ T^{-1}$ and use Exercise 3.49(a).

(c) For every Z,

$$(\mathrm{ad}\, Z(X), Y) = -(X, \mathrm{ad}\, Z(Y)). \tag{3.118}$$

Hint: Show that $\mathrm{ad}[X, Y] = [\mathrm{ad}\, X, \mathrm{ad}\, Y]$.

3.53 The Lie algebra $\mathfrak{so}(3)$ By the result of Exercise 3.46, the Lie algebra $\mathfrak{so}(3)$ may be viewed as the vector space of all 3×3 skew symmetric matrices, so one possible basis is

$$E_1 := \begin{pmatrix} 0 & 1 & 0 \\ -1 & 0 & 0 \\ 0 & 0 & 0 \end{pmatrix}, \quad E_2 := \begin{pmatrix} 0 & 0 & -1 \\ 0 & 0 & 0 \\ 1 & 0 & 0 \end{pmatrix}, \quad \text{and } E_3 := \begin{pmatrix} 0 & 0 & 0 \\ 0 & 0 & 1 \\ 0 & -1 & 0 \end{pmatrix}.$$

Evidently, $\mathfrak{so}(3)$ is three dimensional.

(a) By explicit computation, show that

$$[E_i, E_j] = \sum_k \varepsilon_{ijk} E_k,$$

where ε_{ijk} is the sign of the permutation (i, j, k). *Remark:* The numbers ε_{ijk} are called the **structure constants** of the Lie algebra $\mathfrak{so}(3)$ relative to the given basis. Note that the structure constants also give the matrix elements of the adjoint representation of $\mathfrak{so}(3)$, relative to the same basis.

(b) Use the result of part (b) to compute the entries $g_{ij} := (E_i, E_j)$ of the Killing form and thereby show that it defines a nondegenerate inner product on $\mathfrak{so}(3)$. What is its signature? *Hint:* Although it amounts to overkill here, you could use the tensor identity (8.139).

3.54 Isomorphism between $\mathfrak{so}(3)$ and $\mathfrak{su}(2)$ Show that $\mathfrak{so}(3)$ and $\mathfrak{su}(2)$ are isomorphic Lie algebras. *Hint:* Show that one basis for $\mathfrak{su}(2)$ consists of the set $\tau_k := (-i/2)\sigma_k$, $k = 1, 2, 3$, where $i = \sqrt{-1}$ and the σ_k are the Pauli matrices:

$$\sigma_1 := \begin{pmatrix} 0 & 1 \\ 1 & 0 \end{pmatrix}, \quad \sigma_2 := \begin{pmatrix} 0 & -i \\ i & 0 \end{pmatrix}, \quad \text{and } \sigma_3 := \begin{pmatrix} 1 & 0 \\ 0 & -1 \end{pmatrix}.$$

Consider the map $\phi : \mathfrak{su}(2) \to \mathfrak{so}(3)$ given by $\sigma_i \mapsto E_i$, extended by linearity.

Remark: Lie algebras give local but not global information about Lie groups, so the fact that $\mathfrak{so}(3)$ and $\mathfrak{su}(2)$ are isomorphic tells us only that $SO(3)$ and $SU(2)$ are locally isomorphic. Actually, $SU(2)$ is the universal (double) cover of $SO(3)$. (A Lie group G is a **covering space** (or simply a **cover**) of another Lie group H if there exists a smooth surjective group homomorphism $\varphi : G \to H$. It is **universal** provided it is simply connected. A **double cover** is a cover whose fibers $\varphi^{-1}(p)$ each consist of a pair of points.) This mathematical fact about $SU(2)$ and $SO(3)$ has physical implications. In quantum theory a spin-1/2 particle is represented by a (Pauli) **spinor**, namely an element of a two-complex-dimensional vector space that transforms under rotations according to an element of $SU(2)$. Specifically, if

$\hat{n}$ denotes a unit vector in $\mathbb{R}^3$ and θ the angle of rotation about the $\hat{n}$ axis then the corresponding element of $SU(2)$ is $R := \cos(\theta/2)I - i\sin(\theta/2)(\hat{n} \cdot \sigma)$, where I is the identity matrix. After one full rotation through an angle 2π, a spinor is transformed to the negative of itself, unlike an ordinary vector, which always returns to its original value after a 2π rotation; this property has observable consequences. Mathematically, the preimage under $\varphi : SU(2) \to SO(3)$ of the identity in $SO(3)$ consists of the two maps id and $-id$ in $SU(2)$, as you can see by substituting $\theta = 2\pi$ and $\theta = 4\pi$ into R. It is not difficult to show mathematically that $SU(2)$ is simply connected, but there are also a few amusing physical models that illustrate the same thing. Look up Feynman plate trick, Dirac belt, spinor spanner, or Balinese candle dance.

3.55 Lie derivatives of forms Let ω be a k-form and let X, Y, and $\{X_i\}$, $i = 1, \ldots, k$ be vector fields. Prove the following formulae involving the Lie derivative. *Hint for all parts:* Use induction.

(a) (Cartan's formula)

$$\mathcal{L}_X \omega = (i_X d + d i_X)\omega. \tag{3.119}$$

(b) (Another formula of Cartan)

$$\mathcal{L}_X i_Y \omega - i_Y \mathcal{L}_X \omega = i_{\mathcal{L}_X Y} \omega. \tag{3.120}$$

(c) (A useful identity)

$$(\mathcal{L}_{X_0}\omega)(X_1, \ldots, X_k) = \mathcal{L}_{X_0}(\omega(X_1, \ldots, X_k))$$

$$- \sum_{i=1}^{k} \omega(X_1, \ldots, \mathcal{L}_{X_0} X_i, \ldots, X_k). \tag{3.121}$$

Extra hint: Use the previous part.

(d) (A relation between the exterior derivative and the Lie derivative)

$$d\omega(X_0, X_1, \ldots, X_k)$$

$$= \sum_{i=0}^{k}(-1)^i \mathcal{L}_{X_i}(\omega(X_0, \ldots, \widehat{X}_i, \ldots, X_k))$$

$$+ \sum_{0 \le i < j \le k} (-1)^{i+j}\omega(\mathcal{L}_{X_i} X_j, X_0, \ldots, \widehat{X}_i, \ldots, \widehat{X}_j, \ldots, X_k)$$

$$= \sum_{i=0}^{k}(-1)^i X_i(\omega(X_0, \ldots, \widehat{X}_i, \ldots, X_k))$$

$$+ \sum_{0 \le i < j \le k} (-1)^{i+j}\omega([X_i, X_j], X_0, \ldots, \widehat{X}_i, \ldots, \widehat{X}_j, \ldots, X_k). \tag{3.122}$$

In particular, when $k = 1$ we get the very useful equation

$$d\omega(X, Y) = X\omega(Y) - Y\omega(X) - \omega([X, Y]). \tag{3.123}$$

Extra hint: Use parts (a) and (c).

3.56 Lie derivatives of tensor fields Show that, in a coordinate basis, the Lie derivative along $X = X^i \partial_i$ of a tensor field T of type (r, s) has components

$$
\begin{aligned}
(\mathcal{L}_X T)^{i_1 \ldots i_r}{}_{j_1 \ldots j_s} \\
= T^{i_1 \ldots i_r}{}_{j_1 \ldots j_s, k} X^k \\
- T^{k i_2 \ldots i_r}{}_{j_1 \ldots j_s} X^{i_1}{}_{,k} - T^{i_1 k \ldots i_r}{}_{j_1 \ldots j_s} X^{i_2}{}_{,k} - \cdots - T^{i_1 \ldots i_{r-1} k}{}_{j_1 \ldots j_s} X^{i_r}{}_{,k} \\
+ T^{i_1 \ldots i_r}{}_{k j_2 \ldots j_s} X^k{}_{,j_1} + T^{i_1 \ldots i_r}{}_{j_1 k \ldots j_s} X^k{}_{,j_2} + \cdots + T^{i_1 \ldots i_r}{}_{j_1 \ldots j_{s-1} k} X^k{}_{,j_s},
\end{aligned}
\tag{3.124}
$$

where the comma followed by i in $f_{,i}$ is shorthand for a coordinate derivative $\partial f / \partial x^i$. *Hint:* Recall from (3.26) that

$$\mathcal{L}_X Y = (X^\ell Y^k{}_{,\ell} - Y^\ell X^k{}_{,\ell}) \partial_k. \tag{3.125}$$

Use the compatibility of the Lie derivative with dual pairing (contraction) to show that, for a 1-form $\alpha = \alpha_i \, dx^i$,

$$\mathcal{L}_X \alpha = (X^\ell \alpha_{k,\ell} + \alpha_\ell X^\ell{}_{,k}) dx^k, \tag{3.126}$$

then use the derivation property.

3.57 Symplectic forms and classical mechanics A smooth $2n$-dimensional manifold M equipped with a global nondegenerate 2-form ω on M satisfying $d\omega = 0$ is called a **symplectic manifold**, and ω is called the **symplectic form**. (As usual, ω is nondegenerate if $\omega(X, Y) = 0$ for all Y implies that $X = 0$.) A famous theorem of Darboux from 1884 guarantees that, on such a manifold, we can always find local coordinates $q^1, \ldots, q^n, p_1, \ldots, p_n$ such that

$$\omega = \sum_{i=1}^{n} dq^i \wedge dp_i.$$

Symplectic manifolds are the natural setting for classical mechanics.

(a) Compute ω^n (meaning the wedge product of ω with itself n times) explicitly and thereby demonstrate that it is a volume form on M, namely a global nowhere-vanishing form of degree $2n = \dim M$. In particular, this shows that M is always orientable. (See Appendix C.)

(b) If M and N are symplectic manifolds with symplectic forms μ and ν, respectively, then a smooth map $f : M \to N$ is called **symplectic** (or, in physics language, a **canonical transformation**) if $f^*\nu = \mu$. Show that f is necessarily a local diffeomorphism. For this reason, f is often called a (local) **symplectomorphism**. *Hint:* Show that f preserves the volume form ω^n and use the inverse function theorem.

(c) Let M be symplectic with symplectic form ω, and suppose that $f : M \to \mathbb{R}$ is a smooth function. As ω is nondegenerate, there exists a unique vector field X_f associated to f according to

$$i_{X_f}\omega = \omega(X_f, \cdot) = df.$$

The field X_f is called the **Hamiltonian vector field** associated to the function f. Show that, in local coordinates,

$$X_f = \sum_i \left(\frac{\partial f}{\partial p_i} \frac{\partial}{\partial q^i} - \frac{\partial f}{\partial q^i} \frac{\partial}{\partial p_i} \right).$$

(d) If M is the phase space of some dynamical system then there is a distinguished function H on M, called the **Hamiltonian** of the system. Show that $\gamma(t) = (q^1(t), \ldots, q^n(t), p_1(t), \ldots, p_n(t))$ is an integral curve of X_H if Hamilton's equations hold. They are

$$\dot{q}^i = \frac{\partial H}{\partial p_i}, \qquad \dot{p}_i = -\frac{\partial H}{\partial q^i}, \qquad i = 1, \ldots, n,$$

where $\dot{q} := dq/dt$.

(e) Show that if $\gamma(t)$ is an integral curve for X_H then $dH(\gamma(t))/dt = 0$ (i.e., the Hamiltonian is constant along the curve). When H is the energy of the system this is just the statement that energy is conserved.

(f) Let φ_t be the flow of X_H. Show that φ_t is a symplectomorphism. *Hint:* Use Cartan's formula to show that $\mathcal{L}_{X_H}\omega = 0$. *Remark:* As the volume form is preserved by φ_t, this proves **Liouville's theorem**, which says that canonical transformations preserve volume in phase space.

(g) The **Poisson bracket** of two functions f and g on M is defined by

$$\{f, g\} := \omega(X_f, X_g).$$

Show that, in local coordinates,

$$\{f, g\} = \sum_{i=1}^n \left(\frac{\partial f}{\partial q^i} \frac{\partial g}{\partial p_i} - \frac{\partial g}{\partial q^i} \frac{\partial f}{\partial p_i} \right).$$

(h) Show that the smooth functions on M equipped with the Poisson bracket form a Lie algebra. *Hint:* Clearly the Poisson bracket is bilinear and skew symmetric, so the only nontrivial part of the assertion is that the Poisson bracket obeys the Jacobi identity

$$\{f, \{g, h\}\} + \{g, \{h, f\}\} + \{h, \{f, g\}\} = 0.$$

This is a rather unpleasant computation if local coordinates are used. Instead, compute $d\omega(X_f, X_g, X_h)$ using (3.122). You will get two types of terms, which you need to simplify. For the terms of one type, the identity $\{f, g\} = -\mathcal{L}_{X_f} g$ (which you should prove) comes in handy. For terms of the second type, Cartan's formulas are useful.

4

Homotopy and de Rham cohomology

The most intuitively evident topological invariant of a space is the number of connected pieces into which it falls. Over the past one hundred years or so we have come to realize that this primitive notion admits in some sense two higher-dimensional analogues. These are the *homotopy* and *cohomology* groups of the space in question.

Raoul Bott and Loring Tu[†]

In physics one learns that a static electric field E is conservative, or curl free:

$$\nabla \times E = 0, \tag{4.1}$$

where ∇ is the usual gradient operator $(\partial/\partial x, \partial/\partial y, \partial/\partial z)$. From this, one concludes that

$$E = \nabla \Phi \tag{4.2}$$

for some scalar field Φ. In fact, there is an explicit formula for Φ, namely the line integral

$$\Phi = \int E \cdot ds. \tag{4.3}$$

In terms of differential forms, we can define a 1-form $E = E_x \, dx + E_y \, dy + E_z \, dz$, so that (4.1) is just the condition $dE = 0$ while (4.2) is the condition $E = d\Phi$ for a smooth function Φ. Obviously $E = d\Phi$ implies $dE = 0$, because mixed partials commute ($d^2 = 0$). This example shows, however, that at least in this case the converse is also true: $dE = 0$ implies $E = d\Phi$ for some Φ.

We now proceed to generalize this example. Let ω be a k-form on a smooth manifold M. If $d\omega = 0$ we say that ω is **closed**, while if $\omega = d\alpha$ for some $(k-1)$-form

---

[†] Excerpt from [12], p. 1. Reprinted by permission.

α we say that ω is **exact**. Every exact form is closed, because $d^2 = 0$. The question arises, though, whether every closed form is exact. In the above example we saw that, on $\mathbb{R}^3$ at least, every closed 1-form is indeed exact. It turns out that a similar statement holds for a form of any degree on $\mathbb{R}^n$, and this is the content of the Poincaré lemma (Theorem 4.1). But the corresponding statement is *false* for a general manifold, for subtle reasons having to do with the topology of M.

To get a feeling for what is involved, consider the following 1-form on $\mathbb{R}^2 - \{0\}$:

$$\omega = \left(\frac{-y}{x^2 + y^2}\right) dx + \left(\frac{x}{x^2 + y^2}\right) dy.$$

This form has the interesting property that it is closed but not exact. That is, $d\omega = 0$ (check!) but there is no function f such that $\omega = df$. How do we know this? Well, just integrate. If there were such a function, we would have

$$\frac{\partial f}{\partial x} = -\frac{y}{x^2 + y^2}, \qquad \frac{\partial f}{\partial y} = \frac{x}{x^2 + y^2}.$$

Solving these equations for f gives

$$f(x, y) = \tan^{-1}\left(\frac{y}{x}\right).$$

A simple calculation shows that $\omega = df$, which appears to contradict our previous assertion. But in fact, there is no contradiction, because f *is not a function*! For example, $f(1, 1) = \pi/4, 9\pi/4, 17\pi/4, \ldots$, so it is not single-valued.[1]

Notice that the example only works because we have removed the origin; ω would not be well defined otherwise. Having removed the origin, the problem is then that f is not single-valued, because going around the origin on a circle, or more generally on some simple closed curve, brings us back to a different value for f. Somehow the existence of a form that is closed but not exact in the above example seems to be due to the fact that the space $\mathbb{R}^2 - \{0\}$ has a hole in it. The content of the Poincaré lemma is that this never happens on $\mathbb{R}^n$ because $\mathbb{R}^n$ has no holes; it is topologically trivial.

4.1 Homotopy

To make this precise, we need some definitions. Let X and Y be topological spaces, and let $I = [0, 1] \in \mathbb{R}$. Two continuous maps $f, g : X \to Y$ are said to be

---

[1] Technically, f is a **multivalued function**.

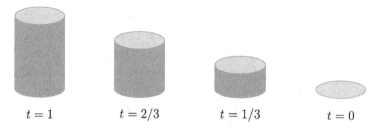

$t = 1$ $t = 2/3$ $t = 1/3$ $t = 0$

Figure 4.1 A representation of the homotopy in Example 4.1 between a cylinder and a disk.

homotopic, written $f \sim g$, if there exists a continuous map $F : I \times X \to Y$ (called a **homotopy**) such that

$$F(0, x) = f(x), \tag{4.4}$$

$$F(1, x) = g(x). \tag{4.5}$$

If there exist continuous maps $f : X \to Y$ and $g : Y \to X$ such that $g \circ f \sim id_X$ and $f \circ g \sim id_Y$, where id denotes the identity map, then X and Y have the same **homotopy type** (or, in an abuse of terminology, are homotopic), written $X \sim Y$. Intuitively, homotopic spaces can be continuously deformed into one another.

> **Example 4.1** (The cylinder construction)   Let D be a disk in $\mathbb{R}^2$. Then the cylinder $I \times D$ is homotopic to D. Let $f : I \times D \to D$ be given by $(s, y) \mapsto y$, and let $g : D \to I \times D$ be given by $y \mapsto (0, y)$. Then $f \circ g = id_D$, while $g \circ f \sim id_{I \times D}$ via the homotopy $F(t, (s, y)) = (ts, y)$. Intuitively, the cylinder can be squashed onto the disk in a continuous fashion. (See Figure 4.1.) Of course, the same thing happens if we start with any space X in place of the disk D.

A space X is **contractible** if it is homotopic to a point. In that case, there exist continuous maps $f : X \to \{p\}$ and $g : \{p\} \to X$ such that $f \circ g = id_p$ and $g \circ f \sim id_X$.

> **EXERCISE 4.1**   A map $f : X \to Y$ is **null homotopic** (or **homotopically trivial**) if is is homotopic to a **one-point map**, namely a map sending all points of the domain to a single point in the range. Show that X is contractible if and only if the identity map on X is null homotopic.

> **EXERCISE 4.2**   Let X be any space, and define the **cone over** X, denoted CX, to be the quotient space $(I \times X)/ \sim$, where $(1, x) \sim (1, y)$ for all $x, y \in X$.[2] Intuitively, construct a cylinder over X and then pinch off one end. (See Figure 4.2.) Show that a cone is always contractible.

---

[2] Be sure to distinguish the two different meanings of $\sim$: $f \sim g$ means that f and g are homotopic whereas $(1, x) \sim (1, y)$ means that the two points are identified. Sorry, but both usages are pretty standard.

Figure 4.2 A cone over a disk. (This object is solid because the disk is solid.)

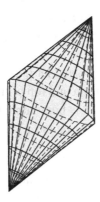

Figure 4.3 The suspension of a circle. (This object is hollow because the circle is hollow.)

Example 4.2 The **suspension** of a space X, denoted ΣX, is the quotient space $(I \times X)/ \sim$, where $(0, x) \sim (0, y)$ and $(1, x) \sim (1, y)$ for all $x, y \in X$. Intuitively, glue two cones together along the base space X. Figure 4.3 depicts the suspension of a circle. Apparently, $\Sigma S^1 \cong S^2$. Indeed, one can show that the suspension of S^n is homeomorphic to S^{n+1}.[3]

If X and Y are homeomorphic then they are homotopic, because we can take $g = f^{-1}$ in the definition, but the converse is false. One way to see this is just to note that homeomorphisms preserve open sets whereas a contractible space, such as an open disk, is homotopic to a point (a closed set). This means that homotopy is a weaker notion than homeomorphism. This is both good and bad. It is good because being a weaker notion means that homotopies are often easier to deal with than homeomorphisms (some of which can be really ugly) and because homotopy equivalence is a topological invariant. It is bad because homotopy equivalence is not a *complete* topological invariant, so knowing which spaces

---

[3] A formal proof of this fact can be found in [75], p. 334, although a more elementary proof proceeds from the notion of a join of simplicial complexes. See Exercise 5.5 at the end of Chapter 5.

are homotopy equivalent does not solve the fundamental problem of topology, which is to determine which spaces are homeomorphic.

4.2 The Poincaré lemma

With that little bit of elementary homotopy theory out of the way, we can now turn to the property of Euclidean space alluded to in the introduction.

Theorem 4.1 (Poincaré lemma) *Let $U \subset \mathbb{R}^n$ be contractible. Let $\omega \in \Omega^{k+1}(U)$ be closed. Then ω is exact, i.e., there exists an $\alpha \in \Omega^k(U)$ such that $\omega = d\alpha$.*

Proof Consider the cylinder $I \times U$, and let $s_0 : U \to I \times U$ and $s_1 : U \to I \times U$ be the inclusion maps into the bottom and top of the cylinder given by $x \mapsto (0, x)$ and $x \mapsto (1, x)$, respectively. As U is homotopic to a point, there is a homotopy[4] $F : I \times U \to U$ such that $F(0, x) = x_0$ and $F(1, x) = x$, which is the same thing as saying

$$F \circ s_0 = x_0,$$
$$F \circ s_1 = x.$$

Taking pullbacks gives

$$s_0^* \circ F^* = 0 \quad \text{(the zero map on forms)}, \tag{4.6}$$
$$s_1^* \circ F^* = 1 \quad \text{(the identity map on forms)}. \tag{4.7}$$

The theorem will follow if we can show that there exists an operator

$$h : \Omega^{k+1}(I \times U) \to \Omega^k(U),$$

called a **homotopy operator**, satisfying

$$hd + dh = s_1^* - s_0^*. \tag{4.8}$$

Given such an operator, we may apply (4.8) to the form $F^*\omega$ and use (4.6) and (4.7) to get

$$hd F^*\omega + dh F^*\omega = (s_1^* - s_0^*)F^*\omega = \omega.$$

---

[4] Technically we ought to distinguish between continuous and smooth homotopies, but fortunately the two notions are essentially equivalent for smooth manifolds. More precisely, **Whitney's approximation theorem** says that every continuous map of smooth manifolds is continuously homotopic to a smooth map (see e.g. Proposition 17.8 in [12]). Therefore, we may safely take all our homotopies to be smooth maps.

But $dF^*\omega = F^*d\omega = 0$ because ω is closed, so we conclude that

$$\omega = d(hF^*\omega),$$

i.e., ω is exact.

To show that h exists, we construct it. Any form η on $I \times U$ is a sum of terms of the following types:

$$\text{type I}, \quad a(t, x)\, dx^J;$$
$$\text{type II}, \quad a(t, x)\, dt \wedge dx^K.$$

We define the homotopy operator h by setting

$$h(a(t, x)\, dx^J) = 0,$$

$$h(a(t, x)\, dt \wedge dx^J) = \left(\int_0^1 a(t, x)\, dt \right) dx^J,$$

and extending by linearity.

On forms η of type I we have

$$d\eta = \frac{\partial a}{\partial t}\, dt \wedge dx^J + \text{terms not containing } dt,$$

so, as $h\eta = 0$ by construction,

$$hd\eta + dh\eta = \left(\int_0^1 \frac{\partial a}{\partial t}\, dt \right) dx^J$$
$$= [a(1, x) - a(0, x)]\, dx^J$$
$$= (s_1^* - s_0^*)(a(t, x)\, dx^J);$$

thus (4.8) is satisfied in this case.

On forms η of type II we have

$$hd\eta = h\left(-\sum \frac{\partial a}{\partial x^i}\, dt \wedge dx^i \wedge dx^J \right)$$
$$= -\sum \left(\int_0^1 \frac{\partial a}{\partial x^i}\, dt \right) dx^i \wedge dx^J$$

and

$$dh\eta = d\left[\left(\int_0^1 a(t, x)\, dt \right) dx^J \right]$$
$$= \sum \frac{\partial}{\partial x^i} \left[\int_0^1 a(t, x)\, dt \right] dx^i \wedge dx^J$$
$$= \sum \left(\int_0^1 \frac{\partial a}{\partial x^i}\, dt \right) dx^i \wedge dx^J.$$

But $s_0^*\eta = s_1^*\eta = 0$ (because $dt \to 0$), so again (4.8) is satisfied. $\qquad \square$

EXERCISE 4.3 Let

$$\omega = A\,dy \wedge dz + B\,dz \wedge dx + C\,dx \wedge dy \tag{4.9}$$

be a closed 2-form in $\mathbb{R}^3$, so that $d\omega = 0$. The space $\mathbb{R}^3$ is contractible via the homotopy

$$F(t, x, y, z) = (tx, ty, tz), \tag{4.10}$$

so by the Poincaré lemma we must have $\omega = dh\,F^*\omega$. We wish to verify this explicitly.

(a) Show that

$$hF^*\omega = \left(\int_0^1 A(tx, ty, tz)\,t\,dt \right)(y\,dz - z\,dy)$$

$$+ \left(\int_0^1 B(tx, ty, tz)\,t\,dt \right)(z\,dx - x\,dz)$$

$$+ \left(\int_0^1 C(tx, ty, tz)\,t\,dt \right)(x\,dy - y\,dx).$$

(b) Show that $\omega = dh\,F^*\omega$. *Hint:* It suffices to obtain only one term in ω, say the last, because the others follow similarly by cyclic permutation. At one point in the calculation you may want to use the fact (which you should justify) that

$$t\frac{d}{dt}C(tx, ty, tz) = (x \cdot \nabla)C(tx, ty, tz).$$

You may also want to integrate by parts somewhere.

Remark In Exercise 3.28 you showed that Maxwell's source-free equations can be written $dF = 0$, where F is the electromagnetic 2-form. Therefore, by the Poincaré lemma in $\mathbb{R}^3$, there exists a 1-form A such that $F = dA$; A is called the **vector potential** because its dual is the usual electromagnetic vector potential.

4.3 de Rham cohomology

Recall that $\Omega^k(M)$ denotes the vector space of all k-forms on M. It has two distinguished subspaces: $Z^k(M)$, the space of all closed k-forms on M, and $B^k(M)$, the space of all exact k-forms on M. As we have seen, all exact forms are closed, so we have the inclusions

$$B^k(M) \subseteq Z^k(M) \subseteq \Omega^k(M). \tag{4.11}$$

We seek a convenient measure of the extent to which a closed form fails to be exact. Such a measure is provided by **the kth de Rham cohomology group**[5] $H^k_{\text{dR}}(M)$, which is the quotient space of the closed k-forms modulo the exact k-forms:

$$H^k_{\text{dR}}(M) = Z^k(M)/B^k(M). \tag{4.12}$$

Put another way, two closed forms are **cohomologous** (equivalent) if they differ by an exact form, and $H^k_{\text{dR}}(M)$ is the set of all cohomology classes (equivalence classes). The remarkable thing about the cohomology groups of a manifold is that they are topological invariants, so that two manifolds with the same topology have the same cohomology groups. One simple illustration of this is provided next.

Let X be a space. If there do not exist two disjoint open sets U and V whose union is X then X is **connected**. Two points x and y are in the same **connected component** of X if there is a connected subset of X with $x, y \in X$. The number of connected components of a space is probably the simplest and most important topological invariant of a space.

Theorem 4.2 *Let M be a smooth manifold. Then*

$$\dim H^0_{\text{dR}}(M) = \textit{the number of connected components of M}. \tag{4.13}$$

Proof No 0-form f on M is exact (there are no (-1)-forms!), so every closed 0-form defines its own cohomology class. But $df = 0$ if and only if f is constant on each connected component M_i of M, and a basis for the vector space of constant functions on the connected components is provided by the unit functions $\{1_{M_i}\}$. $\square$

A slight generalization of the proof of Theorem 4.1 yields the following important result.

Theorem 4.3 *If $f, g : M \to N$ are homotopic maps then $f^* = g^* : H^\bullet_{\text{dR}}(N) \to H^\bullet_{\text{dR}}(M)$.[6] In other words, homotopic maps induce the same map in cohomology.*

---

[5] If the word "group" bothers you, you can safely replace it here with the phrase "vector space," because the de Rham cohomology group $H^k_{\text{dR}}(M)$ (over the reals) is always a vector space. In more general cohomology theories one usually works with coefficients chosen from an abelian group, in which case one obtains honest cohomology groups. For our purposes here, which is simply to illustrate the connection between differential forms and topology, the de Rham cohomology is adequate. But the reader should be aware that other cohomology theories exist that sometimes provide more detailed topological information about manifolds. (Specifically, they provide information about something called *torsion*.)

[6] We follow the standard if somewhat cavalier notation in which $H^\bullet_{\text{dR}}(M)$ stands for $\bigcup_k H^k_{\text{dR}}(M)$ and all maps respect the grading, so that cohomology in one degree is mapped to cohomology in the same degree.

Proof If $f, g : M \to N$ are homotopic then by (4.4) and (4.5) there is a homotopy $F : I \times M \to N$ such that

$$F \circ s_0 = f,$$
$$F \circ s_1 = g,$$

where $s_0, s_1 : N \to I \times M$ are the inclusion maps into the top and bottom of the cylinder $I \times M$. Hence

$$f^* = s_0^* \circ F^*,$$
$$g^* = s_1^* \circ F^*.$$

The proof of the existence of the homotopy operator in the Poincaré lemma carries over patch by patch from $\mathbb{R}^n$ to M, allowing us to conclude that there is a homotopy operator $h : I \times M \to N$ satisfying

$$hd + dh = s_1^* - s_0^*. \tag{4.14}$$

But the operator on the left of (4.14) maps closed forms to exact forms, which means that it is the zero operator on cohomology. Thus

$$f^* - g^* = (s_1^* - s_0^*)F^* \tag{4.15}$$

is also the zero operator on cohomology. $\qquad\square$

Corollary 4.4 (Homotopy invariance of de Rham cohomology) *If M and N are homotopic then $H_{\mathrm{dR}}^\bullet(M) \cong H_{\mathrm{dR}}^\bullet(N)$.*

Proof Now, if M and N are homotopic then there exist $f : M \to N$ and $g : N \to M$ such that $f \circ g \sim id_N$ and $g \circ f \sim id_M$. Hence $g^* \circ f^*$ equals the identity on $H_{\mathrm{dR}}^\bullet(N)$ and $f^* \circ g^*$ equals the identity on $H_{\mathrm{dR}}^\bullet(M)$. In other words, f^* and g^* are isomorphisms and $H_{\mathrm{dR}}^\bullet(M) \cong H_{\mathrm{dR}}^\bullet(N)$. $\qquad\square$

Homeomorphic manifolds are homotopic, so it follows that the de Rham cohomology group of a manifold is a topological invariant. But homotopic manifolds are not necessarily homeomorphic, so it is not clear at this point whether the de Rham cohomology of a manifold is a complete topological invariant. In fact, it is not; there are many distinct manifolds with the same cohomology groups. Still, the nice thing about cohomology groups is that there are many techniques with which to compute them and if two manifolds have different cohomology groups then you know that they are topologically distinct.

Example 4.3 The space $\mathbb{R}^n$ is contractible, so it has the homotopy type of a point. It follows from Theorem 4.2 and Corollary 4.4 that

$$H^k_{\mathrm{dR}}(\mathbb{R}^n) = \begin{cases} \mathbb{R}, & \text{if } k = 0, \text{ and} \\ 0 & \text{otherwise.} \end{cases} \tag{4.16}$$

Example 4.4 A **retraction** $r : X \to Y$ of a space X onto a subspace Y is a continuous map that restricts to the identity on Y. In symbols, $r \circ \iota = id_Y$, where $\iota : Y \to X$ is the inclusion map. If we also demand that $\iota \circ r \sim id_X$ then r is called a **deformation retraction** of X onto Y. A deformation retraction is a well-behaved sort of retraction, which collapses X onto Y in way that can be inverted up to homotopy. If r is a deformation retraction then r and ι are homotopy inverses, and Theorem 4.3 tells us that $r^* : H^\bullet_{\mathrm{dR}}(Y) \to H^\bullet_{\mathrm{dR}}(X)$ is an isomorphism.

To illustrate this, consider the **punctured Euclidean space** $X := \mathbb{R}^{n+1} - \{0\}$. The sphere S^n is a deformation retract of X via the map $r : X \to S^n$ given by $x \mapsto x/\|x\|$. To see this, let $\iota : S^n \to X$ be the inclusion map. Then $r \circ \iota = id_{S^n}$ and $\iota \circ r \sim id_X$, via the homotopy

$$F(t, x) = (1 - t)x + tx/\|x\|,$$

say. Therefore, by Corollary 4.4, $r^* : H^\bullet_{\mathrm{dR}}(S^n) \to H^\bullet_{\mathrm{dR}}(X)$ is an isomorphism. (In Exercise 4.7 you can compute the cohomology of the spheres.)

Remark Observe that $H^k_{\mathrm{dR}}(M) = 0$ if $k > \dim M$.

4.4 Diagram chasing*

Over the years a large number of techniques have been developed to compute various cohomology groups, of which the de Rham cohomology groups are arguably the most important. In this section and the next we describe some of the powerful machinery behind these techniques.

Let $C = \oplus_{i \in \mathbb{Z}} C^i$ be a direct sum of vector spaces. It is called a **differential complex** if there exist linear maps $d_i : C^i \to C^{i+1}$ such that $d_{i+1} d_i = 0$. The map d_i is often called the **differential** of the complex, even though it may have nothing to do with differentiation. We represent a differential complex pictorially by a sequence

$$\cdots \longrightarrow C^{i-1} \xrightarrow{d_{i-1}} C^i \xrightarrow{d_i} C^{i+1} \xrightarrow{d_{i+1}} \cdots \tag{4.17}$$

with the property that $\operatorname{im} d_i \subseteq \ker d_{i+1}$. The **cohomology** of the complex C is the direct sum $H(C) = \oplus_{i \in \mathbb{Z}} H^i(C)$, where $Z^i(C) := \ker d_i$, $B^i(C) := \operatorname{im} d_{i-1}$, and $H^i(C) = Z^i(C)/B^i(C)$. The elements of $Z^i(C)$ are called i-**cocycles** and the elements of $B^i(C)$ are called i-**coboundaries** (for reasons that will become clear in Chapter 5). In general the sequence (4.17) is *not* exact, and the degree to which it fails to be exact is measured precisely by its cohomology.

The **de Rham complex** is just $H_{dR}^\bullet(\Omega^*(M))$ equipped with the exterior derivative operator, and the de Rham cohomology is just the cohomology of this complex. The cocyles of the de Rham complex are the closed forms and the coboundaries are the exact forms. For this reason, even for an arbitrary complex, we sometimes refer to a cocycle as being closed and a coboundary as being exact.

Suppose that we have two complexes, (A, d_A) and (B, d_B).[7] A **chain map** φ : $(A, d_A) \to (B, d_B)$ is a collection of homomorphisms $\varphi_i : A^i \to B^i$ that commute with the differential operators: $\varphi_{i+1}d_{A,i} = d_{B,i}\varphi_i$. A pair of complexes related in this way by a chain map is called a **homomorphism of complexes** and is elegantly represented by the following **commutative diagram**,[8]

$$
\begin{array}{ccc}
\uparrow & & \uparrow \\
A^{i+1} & \xrightarrow{\ \varphi_{i+1}\ } & B^{i+1} \\
d_i \uparrow & & d_i \uparrow \\
A^i & \xrightarrow{\ \varphi_i\ } & B^i \\
d_{i-1}\uparrow & & d_{i-1}\uparrow \\
A^{i-1} & \xrightarrow{\ \varphi_{i-1}\ } & B^{i-1} \\
\uparrow & & \uparrow
\end{array}
\quad,
$$

$$(4.18)$$

where here and henceforth we suppress the complex labels on the differential operators to avoid cluttering the notation. This diagrammatic notation means that we get the same result no matter which way we choose to follow the arrows. The diagram is called commutative because the path independence condition is precisely equivalent to the statement that φ commutes with d.

EXERCISE 4.4 Show that a homomorphism of complexes gives rise to a homomorphism of cohomology groups. That is, show that there is a natural linear map $h_i : H^i(A) \to H^i(B)$ for all i. *Hint:* Let $a \in Z^i(A)$ be closed and denote its

---

[7] *Warning:* The quantity B is now a complex, so it consists of vector spaces B^i. Do not conflate the vector space B^i with $B^i(B)$, the vector subspace of i-coboundaries of B^i. To ensure that we do not fall into this trap, we will write $\mathrm{im}\, d_{B,i-1}$ rather than $B^i(B)$ whenever there is a danger of confusion.

[8] For the sake of future applications it is convenient to draw the differentials vertically and the chain maps horizontally.

equivalence class in $H^i(A)$ by $[a]$. Define $h_i([a]) = [\varphi_i(a)]$. Show that the map is well defined, i.e., that it maps cohomology classes to cohomology classes and is independent of class representative.

Suppose now that we have *three* complexes, A, B, and C, together with chain maps $\varphi : A \to B$ and $\psi : B \to C$. Furthermore, suppose that

$$0 \longrightarrow A^i \xrightarrow{\varphi_i} B^i \xrightarrow{\psi_i} C^i \longrightarrow 0$$

is a short exact sequence *for each i*. Then we have what is called a **short exact sequence of complexes**:

$$
\begin{array}{ccccccccc}
 & \uparrow & & \uparrow & & \uparrow & & \\
0 \longrightarrow & A^{i+1} & \xrightarrow{\varphi_{i+1}} & B^{i+1} & \xrightarrow{\psi_{i+1}} & C^{i+1} & \longrightarrow & 0 \\
 & d_i \uparrow & & d_i \uparrow & & d_i \uparrow & & \\
0 \longrightarrow & A^i & \xrightarrow{\varphi_i} & B^i & \xrightarrow{\psi_i} & C^i & \longrightarrow & 0 \\
 & d_{i-1} \uparrow & & d_{i-1} \uparrow & & d_{i-1} \uparrow & & \\
0 \longrightarrow & A^{i-1} & \xrightarrow{\varphi_{i-1}} & B^{i-1} & \xrightarrow{\psi_{i-1}} & C^{i-1} & \longrightarrow & 0 \\
 & \uparrow & & \uparrow & & \uparrow & & \\
\end{array}
$$

$$(4.19)$$

A short exact sequence of complexes gives rise to a long exact sequence in cohomology. More precisely, we have

$$(4.20)$$

The curved arrows labeled δ in (4.20) are called **coboundary operators** (or **connecting homomorphisms**) and they are somewhat tricky to define. Let $[c] \in$

$H^i(C)$. We want to define its image in $H^{i+1}(A)$ under the map δ_i. Consider (4.19) again. Choose any $c \in [c]$. By the surjectivity of ψ_i, there exists an element $b \in B^i$ such that $\psi_i b = c$. But, by commutativity, $\psi_{i+1} d_i b = d_i \psi_i b = d_i c = 0$. Hence, by the exactness of the top row, there exists an $a \in A^{i+1}$ such that $\varphi_{i+1} a = d_i b$. It turns out that a is closed, and we define $\delta_i[c]$ to be the cohomology class $[a]$ in $H^{i+1}(A)$.

We need to show two things: first that a is closed and second that the coboundary operator is well defined, in the sense that it does not depend on any of the arbitrary choices we have made at each step in the process of going from $[c]$ to $[a]$. For the first, note that $\varphi_{i+2} d_{i+1} a = d_{i+1} \varphi_{i+1} a = d_{i+1} d_i b = 0$, so, by the injectivity of φ_{i+2}, $d_{i+1} a = 0$. For the second, well, it's a useful exercise to show this.

EXERCISE 4.5 Show that the coboundary operator δ_i is well defined. *Remark:* After completing this exercise you will appreciate why it is called "diagram chasing".

EXERCISE 4.6 Show that the sequence (4.20) is exact. *Hint:* Label the maps $H_i(A) \to H_i(B)$ and $H_i(B) \to H_i(C)$ by α_i and β_i, respectively, so that $\alpha_i[a] = [\varphi_i a]$ and $\beta_i[b] = [\psi_i b]$. To show exactness at $H_i(A)$, for example, show separately that $\operatorname{im} \delta_{i-1} \subseteq \ker \alpha_i$ and $\ker \alpha_i \subseteq \operatorname{im} \delta_{i-1}$. Do something similar at $H_i(B)$ and $H_i(C)$. *Remark:* If you find yourself lost amongst the arrows, you can consult e.g. [37], pp. 113ff. (Hatcher is working in homology instead of cohomology, but the approach is the same.)

4.5 The Mayer–Vietoris sequence*

The Mayer–Vietoris sequence is an important piece of technology that allows one to compute the de Rham cohomology of a space in terms of its pieces. Like much of modern technology, however, its inner workings are somewhat complicated. Nevertheless, it is worth discussing as an illustration of how to perform simple computations in cohomology.[9]

Suppose that M is the union of two open sets U and V. Then there is a natural sequence of maps

$$M \xleftarrow{\ \varphi\ } U \coprod V \xleftarrow[\iota_V]{\iota_U} U \cap V,$$

where $\coprod$ denotes a disjoint union and ι_U and ι_V are the inclusion maps of $U \cap V$ into U and V, respectively. The map φ is just the inclusion map of each individual

---

[9] It may also be good for your health. Leopold Vietoris (1891–2002) lived to the age of 110! (By contrast, Walter Mayer (1887–1948) was only 60 at the time of his death.)

subset U and V into M. Hence, by the properties of the pullback map, we get a sequence of maps of differential forms,

$$\Omega^*(M) \xrightarrow{\ \varphi^*\ } \Omega^*(U) \oplus \Omega^*(V) \xrightarrow[\iota_V^*]{\iota_U^*} \Omega^*(U \cap V).$$

Combining the last two maps into a single map by taking their difference, we get an exact sequence.

Lemma 4.5 *The sequence*

$$0 \longrightarrow \Omega^*(M) \xrightarrow{\ \varphi^*\ } \Omega^*(U) \oplus \Omega^*(V) \xrightarrow{\ \psi^*\ } \Omega^*(U \cap V) \longrightarrow 0,$$

where $\psi^*(\mu, \nu) := \nu - \mu$, is exact. It is called the **(short) Mayer–Vietoris sequence***.

Proof To show exactness at the first term, $\Omega^*(M)$, we must show that φ^* is injective. But this is clear, because if a form vanishes on both U and V then it must vanish on $U \cup V = M$, so the only thing sent to zero by φ^* is the zero form. Next, suppose we have a form η on M. Then it maps to (η, η) under φ^* (because it is a global form on M and so is the same on U or V), whereupon it is killed by ψ^*. Hence the sequence is exact at the middle term.

To show exactness at the last term, we must show that ψ^* is surjective. That is, given a form $\omega \in \Omega^*(U \cap V)$, we must show there exist forms $\mu \in \Omega^*(U)$ and $\nu \in \Omega^*(V)$ such that $\omega = \nu - \mu$. To do this, we assume for the moment that there exist smooth functions λ_U and λ_V that have very special properties: (i) $\lambda_U, \lambda_V \geq 0$, (ii) $\lambda_U + \lambda_V = 1$, and (iii) $\mathrm{supp}(\lambda_U) \subset U$ and $\mathrm{supp}(\lambda_V) \subset V$, where $\mathrm{supp}(f)$ is the **support** of f, namely the smallest closed set containing all the points where f is nonzero. The pair of functions $\{\lambda_U, \lambda_V\}$ is called a partition of unity subordinate to the open cover $\{U, V\}$.[10] Given such a partition, define

$$\mu := \begin{cases} -\lambda_V \omega & \text{on } U \cap V, \\ 0 & \text{otherwise,} \end{cases} \quad \text{and} \quad \nu := \begin{cases} \lambda_U \omega & \text{on } U \cap V, \\ 0 & \text{otherwise.} \end{cases}$$

Then μ is a smooth form on U and ν is a smooth form on V, and $\nu - \mu = \lambda_U \omega + \lambda_V \omega = \omega$, as desired.[11]

---

[10] In general, a **partition of unity** on a smooth manifold M is an at most countable collection of nonnegative smooth functions $\{\lambda_i\}$ such that (1) $\{\mathrm{supp}\,\lambda_i\}$ is locally finite, and (2) $\sum_i \lambda_i(p) = 1$ for all $p \in M$. If $\{\mathrm{supp}\,\lambda_i\}$ is a refinement of an open cover of M we say that the partition of unity is subordinate to the open cover.

[11] At first sight it may seem a bit strange that, to define a 1-form on U, we multiply ω by λ_V instead of λ_U. The reason why we do this is rather subtle, and is best illustrated by a picture (Figure 4.4). Consider the special case in which M is one dimensional and $\omega = f$, a smooth function. For simplicity, suppose that f is the constant function 1 on $U \cap V$, as shown in Figure 4.4(a). By the properties of a partition of

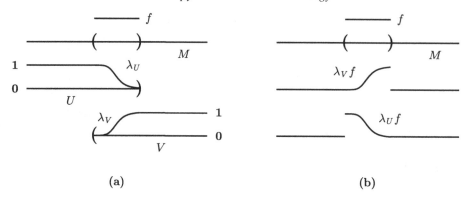

Figure 4.4 (a) A partition of unity subordinate to the cover $\{U, V\}$. (b) Weighted functions.

The only task remaining is to prove the existence of a partition of unity with the desired properties, and for this we refer the reader to [88], Theorem 1.11, or [51], Theorem 2.25. □

There is one short exact Mayer–Vietoris sequence in each degree, and the sequences are all related by the exterior derivative. Putting them all together gives a short exact sequence of complexes:

$$
\begin{array}{ccccccccc}
& & \uparrow & & \uparrow & & \uparrow & & \\
0 & \longrightarrow & \Omega^{i+1}(M) & \xrightarrow{\varphi^*} & \Omega^{i+1}(U) \oplus \Omega^{i+1}(V) & \xrightarrow{\psi^*} & \Omega^{i+1}(U \cap V) & \longrightarrow & 0 \\
& & \uparrow d & & \uparrow d & & \uparrow d & & \\
0 & \longrightarrow & \Omega^{i}(M) & \xrightarrow{\varphi^*} & \Omega^{i}(U) \oplus \Omega^{i}(V) & \xrightarrow{\psi^*} & \Omega^{i}(U \cap V) & \longrightarrow & 0 \\
& & \uparrow d & & \uparrow d & & \uparrow d & & \\
0 & \longrightarrow & \Omega^{i-1}(M) & \xrightarrow{\varphi^*} & \Omega^{i-1}(U) \oplus \Omega^{i-1}(V) & \xrightarrow{\psi^*} & \Omega^{i-1}(U \cap V) & \longrightarrow & 0 \\
& & \uparrow & & \uparrow & & \uparrow & &
\end{array}
$$

(The diagram is commutative because the exterior derivative commutes with pull-backs.) We therefore obtain a long exact sequence in cohomology, called the (**long**) **Mayer–Vietoris sequence**:

unity, the functions λ_U and λ_V must take the form shown. But then the product functions $\lambda_U f$ and $\lambda_V f$ would take the shapes displayed in Figure 4.4(b). Clearly, $\lambda_U f$ is smooth on V but not U, and vice versa for $\lambda_V f$.)

$$\xrightarrow{\quad} H^{i+1}(M) \longrightarrow H^{i+1}(U) \oplus H^{i+1}(V) \longrightarrow \quad \cdots$$

$$\delta_i$$

$$\xrightarrow{\quad} H^i(M) \longrightarrow H^i(U) \oplus H^i(V) \longrightarrow H^i(U \cap V) \xrightarrow{\quad}$$

$$\delta_{i-1}$$

$$\cdots \longrightarrow H^{i-1}(U) \oplus H^{i-1}(V) \longrightarrow H^{i-1}(U \cap V) \xrightarrow{\quad}$$

$$(4.21)$$

Remark It is sometimes useful to have an explicit expression for the coboundary map δ_i. To this end, let $\omega \in \Omega^i(U \cap V)$ be a closed form. As ψ^* is surjective, there exists an $\eta \in \Omega^i(U) \oplus \Omega^i(V)$ such that $\psi^*\eta = \omega$, namely $\eta = (-\lambda_V\omega, \lambda_U\omega)$. By commutativity $\psi^*d\eta = d\psi^*\eta = d\omega = 0$, so by exactness there exists a $\xi \in \Omega^{i+1}(M)$ such that $\varphi^*\xi = d\eta$. Again by commutativity, $\varphi^*d\xi = d\varphi^*\xi = d^2\eta = 0$ so, by the injectivity of φ^*, ξ is closed and hence it defines a cohomology class in $H^{i+1}(M)$. We therefore choose $\delta_i([\omega]) = [\xi]$. According to the result in Exercise 4.5, this map is well defined. Now, φ is the inclusion map of U and V into M, so φ^* is just the restriction map of M onto U or V. In particular, $\varphi^*\xi = d\eta = (d(-\lambda_V\omega), d(\lambda_U\omega))$ implies that ξ restricted to U is $-d(\lambda_V\omega)$ and ξ restricted to V is $d(\lambda_U\omega)$. This is consistent, because $\psi^*d\eta = 0$ implies that $-d(\lambda_V\omega) = d(\lambda_U\omega)$ on the overlap $U \cap V$. Thus ξ is globally defined and

$$\delta_i([\omega]) = \begin{cases} [-d(\lambda_V\omega)] & \text{on } U, \\ [d(\lambda_U\omega)] & \text{on } V. \end{cases} \tag{4.22}$$

Example 4.5 As a simple example of the Mayer–Vietoris machine in action, we will compute the de Rham cohomology of the circle S^1. We cover the circle with two open sets U and V, corresponding to the upper and lower "hemispheres",

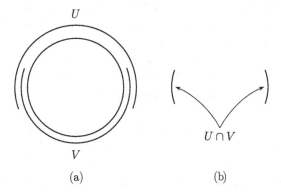

Figure 4.5 A covering of the circle.

respectively, as shown in Figure 4.5(a); then $U \cap V$ consists of two pieces, each homeomorphic to $\mathbb{R}$, as shown in Figure 4.5(b). As U and V are each homeomorphic to $\mathbb{R}$ as well, the Mayer–Vietoris sequence becomes[12]

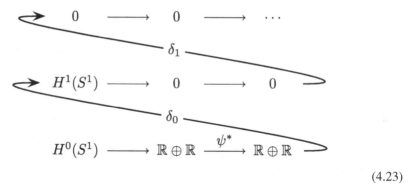

$$(4.23)$$

By Theorem 4.2 the zeroth cohomology of a manifold is just the number of connected components, and the cohomology generators are the constant functions on each piece. The circle is connected, so $H^0_{\mathrm{dR}}(S^1) = \mathbb{R}$. Next, let μ be a constant function on U and ν be a constant function on V. Then $\psi^*(\mu, \nu) = (\nu - \mu, \nu - \mu)|_{U \cap V}$, a pair of identical constant functions on the pieces of $U \cap V$, so the rank and nullity of ψ^* are both 1. (For example, if 1_U is the function that is 1 on U and 1_V is the function that is 1 on V then a basis for the kernel is $(1_U, 1_V)$, because $1_U = 1_V$ on the overlap.) By exactness $\dim \ker \delta_0 = \mathrm{rk}\, \psi^* = 1$, so as $\dim \ker \delta_0 + \mathrm{rk}\, \delta_0 = 2$ we have $\mathrm{rk}\, \delta_0 = 1$. But $H^1_{\mathrm{dR}}(S^1)$ maps to zero, so again by exactness it must be the entire image of δ_0, namely $\mathbb{R}$. In other words, $H^1_{\mathrm{dR}}(S^1) = \mathbb{R}$. Summarizing, we have

$$H^k_{\mathrm{dR}}(S^1) = \begin{cases} \mathbb{R}, & \text{if } k = 0, 1, \text{ and} \\ 0 & \text{otherwise.} \end{cases} \qquad (4.24)$$

As this differs from the cohomology of $\mathbb{R}$, it confirms the obvious fact that the circle and the line are topologically distinct.

A basis element of the vector space $H^k_{\mathrm{dR}}(M)$ is called a **generator of the cohomology of M in degree k**. There are many different types of cohomology generators, but typically one chooses a generator of the top-degree cohomology of M to be a **bump form**, which is a form with compact support and finite integral. To illustrate the construction in the case of S^1, let (λ_U, λ_V) be a partition of unity subordinate to the cover $\{U, V\}$ and let f be a function that is 1 on one component of $U \cap V$, say, the left-hand component in Figure 4.5(b), and 0 on the other. Then, by (4.22), $\delta_0 f$ (a global 1-form on S^1) generates the cohomology of S^1 in degree 1. The 1-form $\delta_0 f$ equals $-d(\lambda_V f)$ on U and $d(\lambda_U f)$ on V, and is a bump form with support in

---

[12] In the Mayer–Vietoris sequence we write $\mathbb{R} \oplus \mathbb{R}$ for $H(U \cap V)$ to emphasize that there is one cohomology generator from each piece of the overlap. But we could just as well have written $\mathbb{R}^2$ instead. When there are many generators people usually drop the unwieldy direct sum notation.

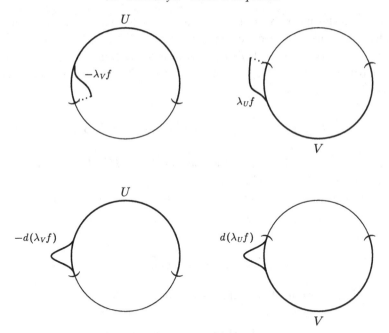

Figure 4.6 A bump form on S^1.

the left-hand component of $U \cap V$, as shown in Figure 4.6. (This 1-form clearly has compact support, because its support (a closed set) lies in $U \cap V$ (a bounded set). Also, its integral is finite because $\lambda_U f$ and $\lambda_V f$ are bounded functions.)[13]

EXERCISE 4.7 Let S^n be the n-sphere, namely $\{x \in \mathbb{R}^{n+1} : \|x\| = 1\}$, where $\|x\| = (\sum_i (x^i)^2)^{1/2}$ is the standard Euclidean norm. By analogy with the previous example, use the Mayer–Vietoris sequence to show that

$$H_{\mathrm{dR}}^k(S^n) = \begin{cases} \mathbb{R} & \text{if } k = 0 \text{ or } k = n, \text{ and} \\ 0 & \text{otherwise.} \end{cases} \qquad (4.25)$$

In particular, the n-sphere is not contractible, a result that is easy to see intuitively but difficult to prove directly. *Hint:* Cover S^n with two patches, and argue that the overlap is homotopic to a familiar space. *Remark:* As in the previous example we can choose a generator of $H_{\mathrm{dR}}^n(S^n)$ to be a bump form, although it is not necessary to know this in order to compute the cohomology.

Example 4.6 We can use the result of this last example to prove a famous theorem in topology called the **Brouwer fixed point theorem**. Let B^n denote the n-ball, namely the set $\{x \in \mathbb{R}^n : \|x\| \leq 1\}$. The Brouwer fixed point theorem asserts that any continuous map $f : B^n \to B^n$ has a fixed point, namely a point x such that

---

[13] Integration is discussed in detail in Chapter 6.

$f(x) = x$. This theorem has a large number of applications in many diverse fields, such as game theory, economics, and differential equations.[14]

We first show by contradiction that there is no retraction $r : B^{n+1} \to S^n$ of the n-ball onto its boundary. Suppose that there were. Then, by pulling back everything, we would have the composition

$$H_{dR}^n(S^n) \xrightarrow{r^*} H_{dR}^n(B^{n+1}) \xrightarrow{\iota^*} H_{dR}^n(S^n), \tag{4.26}$$

where $\iota^* \circ r^* = 1$. But the ball is contractible, so $H_{dR}^n(B^{n+1})$ vanishes whereas $H_{dR}^n(S^n)$ does not, a contradiction.

Now suppose that $f : B^n \to B^n$ has no fixed point. Then $f(x) \neq x$ for all $x \in B^n$. Let $g : B^n \to S^{n-1}$ be the map

$$g(x) = \frac{x - f(x)}{\|x - f(x)\|}. \tag{4.27}$$

But this is a retraction of the ball onto its boundary, a contradiction.

Additional exercises

4.8 Homotopic maps are equivalent   Show that homotopy is an equivalence relation on maps, i.e., that for any continuous maps $f, g, h : X \to Y$,
   (a) $f \sim f$,
   (b) $f \sim g$ implies $g \sim f$, and
   (c) $f \sim g$ and $g \sim h$ implies $f \sim h$.

4.9 Homotopy and composition of maps   Let $f, g : X \to Y$ and $h : Y \to Z$ be continuous maps. Show that if $f \sim g$ then $h \circ f \sim h \circ g$. Similarly, if $f : X \to Y, g, h : Y \to Z$, and $g \sim h$, show that $g \circ f \sim h \circ f$.

4.10 Homotopic spaces are equivalent   Show that homotopy is an equivalence relation on spaces, i.e., that for any spaces X, Y, Z,
   (a) $X \sim X$,
   (b) $X \sim Y$ implies $Y \sim X$, and
   (c) $X \sim Y$ and $Y \sim Z$ implies $X \sim Z$.

4.11 The fundamental group   Let X be a topological space, and let γ be a **path** in X from p to q, namely a (continuous) parameterized curve $\gamma : [0, 1] \to X$ with $p := \gamma(0)$ and $q := \gamma(1)$. The points p and q are the **starting point** and **ending point**, respectively. Given two paths $\gamma_1 : [0, 1] \to X$ and $\gamma_2 : [0, 1] \to X$ with the property that one ends where

---

[14] Technically, we will only prove Brouwer's theorem for smooth maps. Fortunately, a similar argument works in the continuous case by the use of singular homology (see Section 5.2) rather than de Rham cohomology.

Figure 4.7 The product of two paths.

the other starts, i.e., $\gamma_1(1) = \gamma_2(0)$, their **product** $\gamma := \gamma_1\gamma_2$ is just the concatenation of γ_1 and γ_2, rescaled so that the parameter domain remains $[0, 1]$. Explicitly,

$$\gamma(t) = \begin{cases} \gamma_1(2t), & \text{if } 0 \le t \le 1/2, \text{ and} \\ \gamma_2(2t - 1) & \text{if } 1/2 \le t \le 1. \end{cases}$$

(See Figure 4.7.) The **inverse path** γ^{-1} is just the path that goes backwards along γ from q to p: $\gamma^{-1}(t) := \gamma(1 - t)$. The **identity path** at p, id_p, is the path $id_p(t) = p$ for all $t \in [0, 1]$.

Two paths α and β from p to q are **path homotopic** (denoted $\alpha \sim \beta$) if there is a homotopy $F : [0, 1] \times [0, 1] \to X$ from one to the other preserving the endpoints, i.e., such that

$$F(0, t) = \alpha(t),$$
$$F(1, t) = \beta(t),$$

with $F(s, 0) = p$ and $F(s, 1) = q$ for all $s \in [0, 1]$. (See Figure 4.8.)
(a) Show that path-homotopy is an equivalence relation.
(b) Define the natural product on path-homotopy equivalence classes, namely $[\alpha][\beta] := [\alpha\beta]$ (whenever $\alpha\beta$ is defined). Show that this is well defined, namely, independent of class representative.
(c) Define the inverse of a path-homotopy equivalence class by $[\alpha]^{-1} := [\alpha^{-1}]$. Verify that this, too, is well defined.

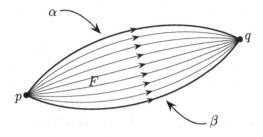

Figure 4.8 A path-homotopy from α to β.

(d) A **loop based at** p is a path that starts and ends at p. Denote by $\pi_1(X, p)$ the set of all path-homotopy equivalent loops based at p. Show that $\pi_1(X, p)$ is a group under the operations defined above, where the identity element is naturally $[id_p]$. The group $\pi_1(X, p)$ is called the **fundamental group** or **first homotopy group** of X. It is one of the most important topological invariants of a space. (For more on the fundamental group see e.g. [55].) *Hint:* Show that all four group axioms hold (cf. Appendix A). Be careful. You are probably so used to thinking that $gg^{-1} = e$ that you might just assume that $\alpha\alpha^{-1} = id$, but this is not true. If α is a nontrivial loop based at p then $\alpha\alpha^{-1}$ is a loop that travels from p around α to p then back along α^{-1} to p again. This is not the same thing as just standing in place. However, it *is* true that $[\alpha][\alpha]^{-1} = [id_p]$, because we just shrink everything back to p. Similarly, $(\alpha\beta)\gamma \neq \alpha(\beta\gamma)$ in general (because the parameterizations are different), but you should show that $([\alpha][\beta])[\gamma] = [\alpha]([\beta][\gamma])$. Associativity is actually the hardest part, because the parameterizations are a bit tricky. Here is something to get you started. Consider the map

$$F(s, t) = \begin{cases} \alpha(4t/(s + 1)) & \text{if } 0 \leq t \leq (s + 1)/4, \\ \beta(4t - (s + 1)) & \text{if } (s + 1)/4 \leq t \leq (s + 2)/4, \text{ and} \\ \gamma(?) & \text{if } (s + 2)/4 \leq t \leq 1. \end{cases}$$

Fill in the question mark, and figure out a use for F.

(e) A space X is **path connected** if any two points in X can be joined by a path in X. Show that, if X is path connected then $\pi_1(X, p) \cong \pi_1(X, q)$ for any $p, q \in X$. (In this context, $\cong$ means isomorphic as groups.) *Remark:* Topological spaces can be weird. If X is path connected then it is always connected, but the converse is not true. There are connected spaces that are not path connected. Look up "topologist's sine curve".

(f) (For extra credit) Assume that X and Y are two path connected spaces of the same homotopy type. Show that their fundamental groups are isomorphic.

4.12 A continuous map to a contractible space is null homotopic   Show that any continuous map $f : X \to Y$ is null homotopic if Y is contractible. Use this to show that any continuous but nonsurjective map $f : X \to S^n$ is null homotopic.

4.13 k-Connected spaces   A space X is **k-connected** if, for $0 \leq i \leq k$, any continuous map $f : S^i \to X$ from the i-sphere to X can be extended to a continuous map $\hat{f} : B^{i+1} \to X$ from the $(i + 1)$-ball to X, so that

$\hat{f}(p) = f(p)$ for all $p \in S^i$. (It is to be assumed that the sphere is the boundary of the ball.)

(a) Show that X is 0-connected if and only if it is path connected.
(b) Show that X is k-connected if and only if every continuous map $f : S^i \to X$ is null homotopic, for $1 \leq i \leq k$.
(c) Show that if X is k-connected and $Y \sim X$ then Y is k-connected.
(d) Show that S^n is not n-connected. *Remark:* One can also show that S^n is i-connected for $1 \leq i < n$, but the proof is more difficult. The basic idea is simple, though. One shows that any continuous map from a lower dimensional sphere to a higher dimensional one is homotopic to a map that is not surjective and so, by Exercise 4.12, the original map must be null homotopic. See e.g. [56], Theorem 4.3.2.

4.14 Cohomology of the torus Show that the cohomology of the 2-torus is given by

$$
H^k_{\mathrm{dR}}(T^2) = \begin{cases} \mathbb{R}, & k = 0, \\ \mathbb{R} \oplus \mathbb{R}, & k = 1, \\ \mathbb{R}, & k = 2, \\ 0, & k > 2. \end{cases}
$$

Can you identify natural generators in each degree? *Hint:* Try a covering by two overlapping cylinders and use Mayer–Vietoris. You will want to invoke the result discussed in Example 4.4 to show that ψ_1^* has rank 1 on cohomology. The key point is to make sure that ψ_1^* doesn't kill more cohomology classes than you might expect.

4.15 Cohomology of the punctured torus Compute the de Rham cohomology of a punctured torus, namely a torus with a point removed, and show that it is given by

$$
H^k_{\mathrm{dR}}(T'^2) = \begin{cases} \mathbb{R}, & k = 0, \\ \mathbb{R} \oplus \mathbb{R}, & k = 1, \\ 0, & k \geq 2. \end{cases}
$$

Hint: Let $U = T'^2$ be the torus minus a point p, and V an open disk on the torus containing p, and use Mayer–Vietoris. Use the fact that ψ_1^* is the zero map on cohomology, which you can prove in Exercise 6.14. Actually, there's a really easy way to see that this is the right cohomology: just show that T'^2 is homotopic to a bouquet of two circles (two circles joined at a single point) and then use the result of Exercise 4.18.

4.16 Cohomology of genus-g surfaces   Calculate the de Rham cohomology of a genus-2 surface, then guess the general result for a genus-g surface (which can be proved by induction). For the definition of "genus" see Exercise 5.9. *Hint:* Use the result of the previous exercise and Mayer–Vietoris again.

4.17 The suspension isomorphism   Prove the **suspension isomorphism** $H^k(M) \cong H^{k+1}(\Sigma M)$, where ΣM is the suspension of M and $k \geq 1$. Also, show that $H^0(M) = H^1(\Sigma M) \oplus \mathbb{R}$ and $H^0(\Sigma M) = \mathbb{R}$. Use these results and the fact that $\Sigma S^n = S^{n+1}$ to compute the cohomology of spheres by induction. *Hint:* Let p and q be the two endpoints of the suspension. Let $U = \Sigma M - \{p\}$ and $V = \Sigma M - \{q\}$. Argue that U and V are contractible and that $U \cap V$ is homotopic to M. (You needn't exhibit the homotopies explicitly.) Do not assume M is connected.

4.18 Cohomology of a wedge sum   Let X and Y be two topological spaces, and suppose that $p \in X$ and $q \in Y$ are distinguished basepoints. The **wedge sum** $X \vee Y$ of two topological spaces X and Y (along p and q) is $\{X \coprod Y\}/(p \sim q)$. Show that for two manifolds M and N with distinguished basepoints, $H^0(M \vee N) \oplus \mathbb{R} \cong H^0(M) \oplus H^0(N)$ and $H^k(M \vee N) \cong H^k(M) \oplus H^k(N)$ for $k > 0$. *Hint:* The naive thing to do is to set $U = M$ and $V = N$ and use Mayer–Vietoris. But this will not work, because neither M nor N is open in $M \vee N$. Instead, let $U_p \in M$ be an open ball around p and let $V_q \in N$ be an open ball around q. Define $U = M \cup V_q$ and $V = N \cup U_p$ and go on from there.

5

Elementary homology theory

In these days the angel of topology and the devil of abstract algebra fight
for the soul of each individual mathematical domain.

Hermann Weyl[†]

In order to gain a deeper understanding of smooth manifolds and the de Rham
theory, we travel in this chapter into the nether reaches of topology. In the process
we encounter a menagerie of topological invariants and lay the groundwork for
the following chapter, in which we acquire an understanding of that most basic of
operations, integration.

5.1 Simplicial complexes

A set $C \in \mathbb{R}^n$ is **convex** if, for every $x, y \in C$, the line segment

$$tx + (1 - t)y \qquad 0 \le t \le 1, \qquad (5.1)$$

is contained entirely in C. In other words, you can walk from any point in a convex
set to any other along a straight line and never leave the set. See Figure 5.1 for
some examples of convex and nonconvex sets. The **convex hull** of a set U, denoted
conv U, is the smallest convex set containing U. (Figure 5.2.) Equivalently, conv U
is the intersection of all convex sets containing U. A **polytope** (Figure 5.3) is the
convex hull of a finite set of points.[1]

> **EXERCISE 5.1** Let $S := \{p_1, \ldots, p_k\}$ be a finite set of points in $\mathbb{R}^n$. If $p = \sum_i t_i p_i$
> with $t_i \ge 0$ and $\sum_i t_i = 1$ then p is a **convex combination** of the points in S. Show
> that the convex hull of S is precisely the set of convex combinations of the points in
> S. *Hint:* Use induction on k.

[†] Hermann Weyl, Invariants, *Duke Math. J.* **5:3**, 489–502. Copyright 1939, Duke University Press. All rights
reserved. Republished by permission of the copyright holder, Duke University Press, www.dukeupress.edu.
[1] For a friendly and comprehensive tour of the world of polytopes, see [92].

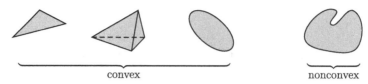

Figure 5.1 Convex and nonconvex sets.

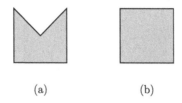

(a)   (b)

Figure 5.2 The convex hull of (a) is (b).

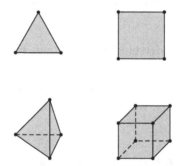

Figure 5.3 Polytopes.

Let $v \in \mathbb{R}^n$ and $a \in \mathbb{R}$, and let $g(\cdot, \cdot)$ be the usual Euclidean metric. Any set of the form $\{x \in \mathbb{R}^n : g(x, v) = a\}$ is called an **affine hyperplane**. It determines two (closed) **half-spaces**, namely $\{x \in \mathbb{R}^n : g(x, v) \geq a\}$ and $\{x \in \mathbb{R}^n : g(x, v) \leq a\}$. The affine hyperplane H is a **supporting hyperplane** of an n-dimensional polytope P if H meets P in some nonempty set and P is confined to one of the half-spaces determined by H. (See Figure 5.4.) Any set F of the form $P \cap H$, where H is a supporting hyperplane of P, is called a **face** of P. The definition allows for $F = P$; all other faces are **proper**. The **dimension** of a face is the dimension of the smallest affine subspace containing it. A zero-dimensional face is a **vertex** of P, a one-dimensional face is an **edge** of P, and an $(n - 1)$-dimensional face is a **facet**. The collection of all the facets of a polytope is the **boundary** of the polytope, and the **interior** of P is P minus its boundary. Because polytopes have facets or "seats", they are also called **polyhedra** ("many seats").[2] It seems obvious that a polytope

---

[2] Technically, a **polyhedron** is the intersection of a finite number of half-spaces, and can therefore be unbounded. To get a polytope we must insist that the polyhedron not contain a ray.

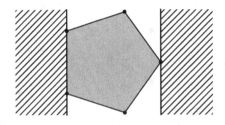

Figure 5.4 A polytope, two of its supporting hyperplanes, and two half-spaces. The intersections of the hyperplanes with the polytope determine two faces, an edge and a vertex.

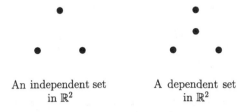

An independent set
in $\mathbb{R}^2$

A dependent set
in $\mathbb{R}^2$

Figure 5.5 Affinely independent and dependent point sets in the plane.

is the convex hull of its vertices and that every face is just the convex hull of some subset of the vertices, but the proofs are surprisingly subtle, so we will omit them.[3]

A set of points $S := \{p_0, p_1, \ldots, p_d\}$ in $\mathbb{R}^n$ is **(affinely) independent** if the vectors $p_1 - p_0, p_2 - p_0, \ldots, p_d - p_0$ are linearly independent. (see Figure 5.5.) A **simplex** is the convex hull of an independent set of points. We write $[S]$ to denote the simplex determined by the set S. If $|S| = d + 1$, the dimension of $[S]$ is d, so we called it a d-**simplex**. A triangle is a 2-simplex and a tetrahedron is a 3-simplex. (See the two polytopes on the left in Figure 5.3.) The vertices of $[S]$ are precisely the points of S. Because a subset of an independent set of points is itself independent, any proper subset $T \subset S$ determines a lower dimensional simplex $[T]$, which is necessarily a face of $[S]$. If $|T| = k + 1$, $[T]$ is a k-**face** of $[S]$. The 0-faces are the vertices, the 1-faces are the edges, and the $(n-1)$-faces are the facets.

EXERCISE 5.2 Let $S = \{p_0, \ldots, p_d\}$ be the vertices of a simplex $[S]$. Show that every point in $[S]$ can be written uniquely as $\sum_i t_i p_i$ with $t_i \geq 0$ and $\sum_i t_i = 1$. (The coefficients t_i are called the **barycentric coordinates** of p.)

Remark   As we have defined them, (geometric) simplices are closed subsets of $\mathbb{R}^n$. Given an independent set of points $S = \{p_0, \ldots, p_n\}$, one can also define the **open simplex**, (S), to be the set of all points of the form $\sum_{i=0}^{n} t_i p_i$, where $t_i > 0$

---

[3] See *e.g.* [92], Propositions 2.2 and 2.3, or [57], Proposition 5.3.2.

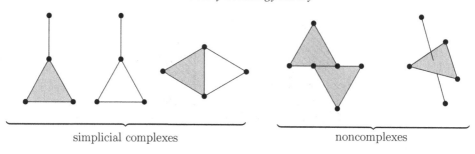

Figure 5.6 Simplicial complexes and noncomplexes.

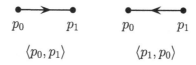

Figure 5.7 Oriented 1-simplices.

and $\sum_i t_i = 1$. If $T \subseteq S$, (T) is an **open face** of $[S]$. The open simplex (S) is the topological interior of $[S]$, and $[S]$ is the topological closure of (S). Note that vertices are both open and closed faces and that a simplex is the (disjoint) union of its open faces.

At last we come to the main definition. A **simplicial complex** K is a finite collection of simplices satisfying two properties, as follows.

(1) If $[S] \in K$ and $T \subset S$ then $[T] \in K$. (Every face of a simplex in K is a simplex in K.)
(2) For any $[S_1], [S_2] \in K$, $[S_1] \cap [S_2] = [S_1 \cap S_2]$. (If two simplices meet, they do so along a mutual face.)

The **dimension** of K is the maximum dimension of all the simplices it contains. Intuitively, a simplicial complex is a set comprising a bunch of simplices that are glued together along their faces. (See Figure 5.6.)

It turns out to be algebraically useful to assign an orientation to a simplicial complex. An **oriented simplex** $\bar{\sigma} = \langle p_0, p_1, \ldots, p_n \rangle$ is a simplex $[p_0, p_1, \ldots, p_n]$ in which we have chosen an order for the points, up to an even permutation. That is, all the simplices are divided into precisely two equivalence classes, or **orientations**, depending on whether one chooses an even or odd permutation of the vertices of $[p_0, p_1, \ldots, p_n]$.

Example 5.1   Consider the oriented 1-simplex $\langle p_0, p_1 \rangle$. The 1-simplex $\langle p_1, p_0 \rangle$ has the opposite orientation. We can represent these two simplices as in Figure 5.7.

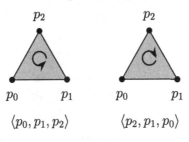

$$\langle p_0, p_1, p_2 \rangle \qquad \langle p_2, p_1, p_0 \rangle$$

Figure 5.8 Oriented 2-simplices.

Example 5.2 The ordered collections $\langle p_0, p_1, p_2 \rangle$, $\langle p_1, p_2, p_0 \rangle$, and $\langle p_2, p_0, p_1 \rangle$ represent the same oriented 2-simplex, whereas $\langle p_0, p_2, p_1 \rangle$, $\langle p_2, p_1, p_0 \rangle$, and $\langle p_1, p_0, p_2 \rangle$ represent another oriented 2-simplex. The two oriented simplices have opposite orientations. We can represent them as in Figure 5.8.

Let K be an **oriented simplicial complex**, namely a simplicial complex in which each simplex has been assigned an orientation. An *m***-chain on K with coefficients in** $\mathbb{R}$ is a formal linear combination of the oriented m-simplices in K with coefficients in $\mathbb{R}$:

$$\bar{c} = \sum_i a_i \bar{\sigma}_i \qquad (a_i \in \mathbb{R}), \tag{5.2}$$

subject to the convention that if $\bar{\sigma}_i$ and $\bar{\sigma}_j$ represent oppositely oriented simplices then $\bar{\sigma}_i = -\bar{\sigma}_j$. The vector space of all m-chains on K is denoted $C_m(K)$.

Let $\bar{\sigma} = \langle p_0, p_1, \ldots, p_n \rangle$ be an oriented n-simplex. The **boundary** $\partial \bar{\sigma}$ of $\bar{\sigma}$ is defined to be the $(n-1)$-chain

$$\partial \bar{\sigma} = \sum_{j=0}^{n} (-1)^j \langle p_0, \ldots, \widehat{p}_j, \ldots, p_n \rangle, \tag{5.3}$$

where the caret means that the vertex p_j is deleted. (See Figure 5.9.) The **boundary operator**

$$\partial : C_m(K) \to C_{m-1}(K) \tag{5.4}$$

is obtained as the unique linear extension of the boundary map on simplices. Actually, there is one boundary operator for each m. When we wish to distinguish them, we write ∂_m for the boundary operator whose domain is $C_m(K)$.

Example 5.3 We have

$$\partial^2 \langle p_0, p_1, p_2 \rangle = \partial \{ \langle p_1, p_2 \rangle + \langle p_2, p_0 \rangle + \langle p_0, p_1 \rangle \}$$
$$= p_2 - p_1 + p_0 - p_2 + p_1 - p_0$$
$$= 0.$$

We now arrive at a very simple but surprisingly powerful theorem.

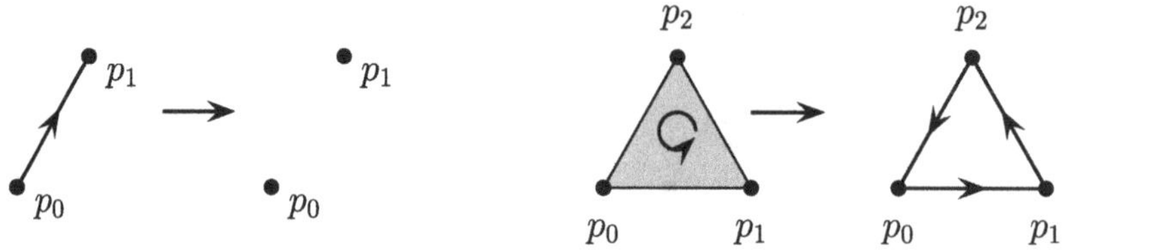

$$\partial\langle p_0, p_1\rangle = p_1 - p_0 \qquad \partial\langle p_0, p_1, p_2\rangle = \langle p_1, p_2\rangle - \langle p_0, p_2\rangle + \langle p_0, p_1\rangle$$

Figure 5.9 The boundary operator in action.

Theorem 5.1 *The operator $\partial^2 = 0$, i.e., the boundary of a boundary is zero.*

Proof The general proof is similar to that of Example 5.3, in that it uses pairwise cancellation. The only tricky thing is keeping track of the summations:

$$\partial[\partial\langle p_0, \ldots, p_{m+1}\rangle] = \partial\left[\sum_{j=0}^{m+1}(-1)^j\langle p_0, \ldots, \widehat{p}_j, \ldots, p_{m+1}\rangle\right]$$

$$= \sum_{j=0}^{m+1}(-1)^j\left[\sum_{k=0}^{j-1}(-1)^k\langle p_0, \ldots, \widehat{p}_k, \ldots, \widehat{p}_j, \ldots, p_{m+1}\rangle\right.$$

$$\left. + \sum_{k=j+1}^{m+1}(-1)^{k-1}\langle p_0, \ldots, \widehat{p}_j, \ldots, \widehat{p}_k, \ldots, p_{m+1}\rangle\right]$$

$$= \sum_{k<j}(-1)^{j+k}\langle p_0, \ldots, \widehat{p}_k, \ldots, \widehat{p}_j, \ldots, p_{m+1}\rangle$$

$$+ \sum_{j<k}(-1)^{j+k-1}\langle p_0, \ldots, \widehat{p}_j, \ldots, \widehat{p}_k, \ldots, p_{m+1}\rangle$$

$$= 0.$$

(For the last equality, flip the dummy indices j and k in the second term.) $\qquad\square$

5.2 Homology

The utility of simplicial complexes in topology is that they provide simple and tractable models of various spaces. More precisely, a **triangulation** of a topological space X is a simplicial complex K that is homeomorphic to X. For instance, $\bar{\sigma} = \langle p_0, \ldots, p_n\rangle$ is a triangulation of the $(n+1)$-ball B^{n+1}, while its boundary

$\partial\bar{\sigma}$ is a natural triangulation of the n-sphere S^n.[4] If a triangulation of X exists, X is said to be **triangulable**. It is perhaps not too surprising that there are topological spaces that are not triangulable, because general topological spaces can be quite complicated. What is really strange is that there are topological manifolds (*e.g.*, the so-called **E8-space**) that are not triangulable either. However, all smooth manifolds can be triangulated in a particularly nice way (see Chapter 6), and in that case the availability of a triangulation of a manifold makes it significantly easier to ascertain its topology. Before returning to the land of smooth manifolds, however, it is worth spending a little time in the land of merely continuous manifolds in order to better understand the idea of "homology". We begin by discussing **simplicial homology**.

An m-chain $\bar{c}$ with no boundary ($\partial\bar{c} = 0$) is called an m-**cycle**. The space of all m-cycles on K is labeled $Z_m(K)$.[5] An m-chain $\bar{c}$ that is the boundary of an $(m+1)$-chain $\bar{c}'$ ($\bar{c} = \partial\bar{c}'$) is called an m-**boundary**. Two m-cycles whose difference is an m-boundary are said to be **homologous**. We write

$$\bar{z}_1 \sim \bar{z}_2 \quad \text{when} \quad \bar{z}_1 - \bar{z}_2 = \partial\bar{c}. \tag{5.5}$$

The equivalence class of all m-cycles homologous to a fixed m-cycle $\bar{z}$ is written $[\bar{z}]$. The set of all such equivalence classes is the mth **simplicial homology group**, denoted $H_m(K)$. In other words, the mth simplicial homology group is just the quotient space of cycles modulo boundaries:

$$H_m(K) = Z_m(K)/B_m(K). \tag{5.6}$$

Although it is not obvious at this stage, the homology of K is independent of the orientation chosen for its computation; it depends only on the underlying complex K.[6]

Because we are dealing here with only real-valued quantities, $H_m(K)$ is a vector space. All vector spaces of the same dimension are the same (isomorphic), so the only relevant information contained in $H_m(K)$ is its dimension $\beta_m = \dim H_m(K)$, called the mth **Betti number** of K.[7] As we shall see, $\beta_m(K)$ basically counts the number of m-dimensional "holes" of K.

There is an evident similarity between simplicial homology and de Rham cohomology, although the precise connection between them is certainly not clear at this stage. In order to forge a link, we first discuss yet another homology theory,

---

[4] When comparing topologies we just forget about the orientation of the complex.

[5] "Z" for the German "Zykel".

[6] See e.g. ([75], Corollary 7.25).

[7] Technically, we ought to write $H_m(K, \mathbb{F})$ instead of $H_m(K)$, to emphasize that our chains are defined with coefficients chosen from a field $\mathbb{F}$. When the field $\mathbb{F}$ is replaced by a general abelian group G, $H_m(K, G)$ contains additional information called *torsion* (not to be confused with another kind of torson to be introduced in Section 8.2). For more information, consult [75].

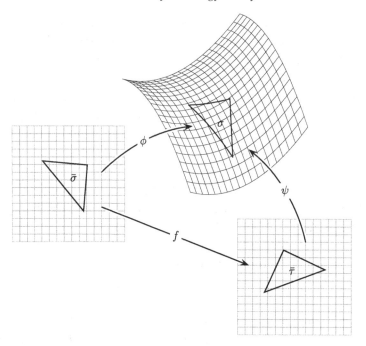

Figure 5.10 A singular 2-simplex.

called the **singular homology** of a space X and denoted $H_*(X)$. In view of our eventual destination (namely smooth manifolds), we will proceed along somewhat unconventional lines although, of course, we will end up with exactly the same theory.

Intuitively, a singular n-simplex in X is just a continuous map $\phi : \bar{\sigma}^n \to X$ from an n-simplex to X.[8] The problem is that, starting from different n-simplices we could end up with the same image, and in that case we would really like to say the two singular simplices are the same. In the conventional approach one avoids this problem by restricting $\bar{\sigma}^n$ to some kind of standard simplex, but we will find it convenient not to do this. Instead, if $\phi : \bar{\sigma}^n \to X$ is a continuous map, we define a **singular n-simplex** to be an equivalence class $(\bar{\sigma}^n, \phi)$ where two classes $(\bar{\sigma}^n, \phi)$ and $(\bar{\tau}^n, \psi)$ are the same whenever $\phi(\bar{\sigma}^n) = \psi(\bar{\tau}^n)$. Equivalently, if $f : \bar{\sigma}^n \to \bar{\tau}^n$ is the natural order-preserving affine map, then $\phi = \psi \circ f$. (See Figure 5.10.)

A **singular n-chain** with coefficients in $\mathbb{R}$ is a formal linear combination of the singular n-simplices in X, and the space of all such chains is denoted $S_n(X)$. To define a homology theory on X we need an analogue of the boundary operator. Let $\sigma = (\bar{\sigma}, \phi)$. Suppose that $\partial\bar{\sigma} = \sum_i \pm\bar{\sigma}_i$, where $\bar{\sigma}_i$ is a facet of $\bar{\sigma}$. We define the

---

[8] The map ϕ is not required to be injective, so singular simplices may not look much like their rigid counterparts. In particular, the image of a simplex could be a lower-dimensional object such as a point – hence the word "singular".

Figure 5.11 A triangulation of the circle.

boundary of σ to be $\partial\sigma := \sum_i \pm\sigma_i$, where $\sigma_i = (\bar{\sigma}_i, \phi)$, and extend it linearly to singular n-chains. We automatically obtain $\partial^2 = 0$, whereupon we simply repeat everything all over again. The **singular n-cycles** are $Z_n(X) := \ker \partial|_{S_n(X)}$, the **singular n-boundaries** are $B_n(X) := \operatorname{im} \partial|_{S_{n-1}(X)}$, and the **singular nth-homology** $H_n(X)$ is $Z_n(X)/B_n(X)$. With a little work one can show that $H_n(X)$ is a homotopy invariant and therefore a topological invariant.

As you might have guessed, there is a relation between the two kinds of homology we have discussed. If K is any triangulation of X then[9]

$$H_\bullet(K) \cong H_\bullet(X). \tag{5.7}$$

This is good, because computations in the land of singular chains are usually quite involved, whereas computations in the land of simplicial complexes are relatively straightforward (at least in principle).

Example 5.4 Here is a simple example to illustrate the computation of simplicial homology. The circle S^1 admits a triangulation K, as illustrated in Figure 5.11. (One can think of the map ϕ as the formal linear combination of three singular 1-simplices.)

We want to compute the homology of the circle, which we can do by computing the Betti numbers of K. The homologies are all of the form $H_k \cong \mathbb{R}^{\beta_k}$, where β_k is the kth Betti number. As $Z_k = \ker \partial_k$, $B_k = \operatorname{im} \partial_{k+1}$, and $H_k = Z_k/B_k$, Theorem 1.4 gives

$$\beta_k = \dim H_k = \dim \ker \partial_k - \dim \operatorname{im} \partial_{k+1}.$$

Thus, to find the homologies of the triangle K, we just need to compute the ranks and nullities of the boundary maps ∂_k.

Since K consists of a single oriented triangle $\langle p_0, p_1, p_2 \rangle$, every 0-chain on K is of the form

$$\bar{c}_0 = a_0 p_0 + a_1 p_1 + a_2 p_2,$$

for some scalars a_i. The boundary of a point is zero, so $\partial_0 \bar{c}_0 = 0$. That is, ∂_0 is the zero map. In particular, $\dim \ker \partial_0 = \dim C_0 = 3$.

The most general 1-chain is

$$\bar{c}_1 = b_0 \langle p_1, p_2 \rangle + b_1 \langle p_2, p_0 \rangle + b_2 \langle p_0, p_1 \rangle,$$

---

[9] The proof of this is not too difficult, but we shall not give it here. See e.g. [37] or [75].

for some scalars b_i. Its boundary is

$$\partial_1 \bar{c}_1 = b_0(p_2 - p_1) + b_1(p_0 - p_2) + b_2(p_1 - p_0)$$
$$= (b_1 - b_2)p_0 + (b_2 - b_0)p_1 + (b_0 - b_1)p_2.$$

Thus $\bar{c}_0 = \partial \bar{c}_1$ if

$$a_0 = b_1 - b_2,$$
$$a_1 = b_2 - b_0,$$
$$a_2 = b_0 - b_1,$$

or

$$\begin{pmatrix} 0 & 1 & -1 \\ -1 & 0 & 1 \\ 1 & -1 & 0 \end{pmatrix} \begin{pmatrix} b_0 \\ b_1 \\ b_2 \end{pmatrix} = \begin{pmatrix} a_0 \\ a_1 \\ a_2 \end{pmatrix}. \tag{5.8}$$

The matrix in (5.8) represents the boundary map ∂_1. The first two rows of that matrix are linearly independent but the sum of all three rows is zero, thus it has rank 2. Therefore

$$\beta_0 = \dim \ker \partial_0 - \dim \operatorname{im} \partial_1 = 3 - 2 = 1.$$

Observe that $\dim \ker \partial_1 = 1$, which follows from the rank–nullity theorem.

There are no 2-cycles in K, so no 1-cycle is the boundary of anything. Hence $\dim \operatorname{im} \partial_2 = 0$, which implies that

$$\beta_1 = \dim \ker \partial_1 - \dim \operatorname{im} \partial_2 = 1 - 0 = 1.$$

All the higher degree homologies vanish because there are no k-chains for $k > 1$. Summarizing, we have

$$H_k(S^1) = \begin{cases} \mathbb{R}, & k = 0 \text{ or } k = 1, \text{ and} \\ 0 & \text{otherwise.} \end{cases} \tag{5.9}$$

Example 5.5 Just for kicks, let's do one more example. This time we want to compute the homology of the two-dimensional disk (2-ball) D^2. The disk is homeomorphic to the complex K consisting of a triangle together with its interior. Proceeding as before, we find $\beta_0 = 1$ again. But β_1 is different, for now there is a nontrivial 2-chain, namely K itself: $\bar{c}_2 = \langle p_0, p_1, p_2 \rangle$. It follows that ∂_2 has rank 1, because it maps $C_2(K)$ to a one-dimensional subspace of $C_1(K)$ spanned by the chain

$$\langle p_0, p_1 \rangle + \langle p_1, p_2 \rangle + \langle p_2, p_0 \rangle.$$

Hence $\beta_1 = \dim \ker \partial_1 - \dim \operatorname{im} \partial_2 = 1 - 1 = 0$. Lastly, what of H_2? The 2-chain $a\langle p_0, p_1, p_2\rangle$ is a 2-cycle if it has no boundary, but we just saw that it does have a boundary. So there are no 2-cycles, and $H_2 = 0$. It follows that

$$H_k(D^2) = \begin{cases} \mathbb{R}, & k = 0, \text{ and} \\ 0 & \text{otherwise.} \end{cases} \tag{5.10}$$

This differs from $H_k(S^1)$, so the disk and its boundary are not homeomorphic, a result we previously obtained using de Rham cohomology.

EXERCISE 5.3 The 2-sphere S^2 can be triangulated by a simplicial complex consisting of four triangles homeomorphic to the surface of a tetrahedron. Call this complex K. By computing the simplicial homology of K, show that

$$H_k(S^2) = \begin{cases} \mathbb{R}, & k = 0 \text{ or } 2, \text{ and} \\ 0 & \text{otherwise.} \end{cases}$$

Generalize to the n-sphere S^n. (Don't do the computation for S^n – just guess the analogous answer. Of course you could always use Exercise 4.7 together with de Rham's theorem from Chapter 6.)

5.3 The Euler characteristic

As one might imagine from the previous examples, even the computation of simplicial homology can become prohibitively difficult if the approximating complex has many simplices.[10] Thus it would be nice to have an even simpler invariant at hand. The Euler characteristic is such an invariant. It is the simplest, and therefore the most fundamental, topological invariant of a space. It satisfies all kinds of intuitive properties, and it is often easy and fun to compute.

The **Euler characteristic** $\chi(X)$ of a topological space X is the integer

$$\chi(X) = \sum_{j=0}^{\dim X} (-1)^j \beta_j, \tag{5.11}$$

where $\beta_j = \dim H_j(X)$ is the jth Betti number of X.

Example 5.6 The Betti numbers of the n-sphere are

$$\beta_k(S^n) = \begin{cases} 1 & \text{if } k = 0 \text{ or } n, \text{ and} \\ 0 & \text{otherwise.} \end{cases}$$

---

[10] This is the reason why Whitehead invented the theory of CW complexes. A **CW complex** is a prescription for gluing together *cells* – essentially topological balls – according to certain rules. Typically a space needs many fewer cells than simplices for its description, so the homology theory becomes a bit simpler. See [75].

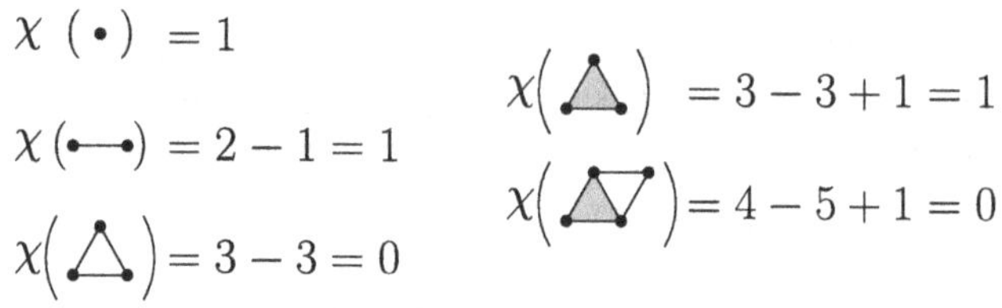

$$\chi\,(\,\bullet\,)\;=1$$

$$\chi\!\left(\triangle\right) = 3 - 3 + 1 = 1$$

$$\chi\,(\bullet\!\!-\!\!\!-\!\!\bullet)\;=2-1=1$$

$$\chi\!\left(\triangle\!\!\!\!/\right) = 4 - 5 + 1 = 0$$

$$\chi\!\left(\triangle\right)= 3 - 3 = 0$$

Figure 5.12 Euler characteristics of some simplicial complexes.

The Euler characteristic of the n-sphere is therefore

$$\chi(S^n) = 1 + (-1)^n.$$

In particular, the Euler characteristic vanishes for odd-dimensional spheres and equals 2 for even-dimensional spheres.

If X admits a triangulation by a simplicial complex K then it suffices to compute the Euler characteristic of K. For that, we use the following pretty theorem.

Theorem 5.2 *Let K be an oriented simplicial complex. For each j, with $0 \le j \le$ $\dim K$, let α_j denote the number of j-simplices of K (i.e., the number of j-faces of K). Then*

$$\chi(K) = \sum_{j=0}^{\dim K} (-1)^j \alpha_j. \tag{5.12}$$

This is the form of the Euler characteristic that you may recall from high school: vertices minus edges plus 2-faces, etc.[11] Some examples are provided in Figure 5.12.

Proof The simplicial complex K induces an exact sequence of chain spaces (called, appropriately enough, a **chain complex**),

$$\cdots \longrightarrow C_{m+1}(K) \xrightarrow{\partial_{m+1}} C_m(K) \xrightarrow{\partial_m} C_{m-1}(K) \xrightarrow{\partial_{m-1}} \cdots$$

where $C_{-1}(K)$ is the zero space, by definition. By the rank–nullity theorem,

$$\alpha_m = \dim C_m$$
$$= \dim \ker \partial_m + \dim \operatorname{im} \partial_m$$
$$= \dim Z_m + \dim B_{m-1}.$$

---

[11] In fact, you may recall that this formula applies more generally to polytopes whose faces are not simplices. That generalization follows simply by triangulating the faces of the polytope.

But, as we saw in Example 5.4,

$$\beta_m = \dim Z_m - \dim B_m.$$

Thus,

$$
\begin{aligned}
\chi(K) &= \sum_{m=0}^{\dim K} (-1)^m \beta_m \\
&= \sum_{m=0}^{\dim K} (-1)^m (\dim Z_m - \dim B_m) \\
&= \sum_{m=0}^{\dim K} (-1)^m \dim Z_m + \sum_{m=0}^{\dim K} (-1)^{m+1} \dim B_m \\
&= \sum_{m=0}^{\dim K} (-1)^m \dim Z_m + \sum_{m=1}^{\dim K} (-1)^m \dim B_{m-1} \\
&\qquad\qquad \text{(because } B_m = 0 \text{ for } m \geq \dim K) \\
&= \sum_{m=0}^{\dim K} (-1)^m (\dim Z_m + \dim B_{m-1}) \\
&= \sum_{m=0}^{\dim K} (-1)^m \alpha_m.
\end{aligned}
$$

□

Additional exercises

5.4 Perron–Frobenius theorem Let A be a matrix all of whose entries are positive real numbers. Prove that A has a positive real eigenvector (one with all positive entries) with positive real eigenvalue. *Hint:* Let $\bar{\sigma}^{n-1}$ be the $(n-1)$-simplex in $\mathbb{R}^n$, given by $\sum_{i=1}^{n} x^i = 1$ and $x^i \geq 0$. Consider the map $f : \bar{\sigma}^{n-1} \to \mathbb{R}^n$ given by

$$f(x) = \frac{Ax}{|Ax|},$$

where $x = (x^1, \ldots, x^n)$ and $|x| = \sum_i |x^i|$ is the so-called "L^1 norm". Show that $f : \bar{\sigma}^{n-1} \to \bar{\sigma}^{n-1}$ and use the Brouwer theorem. *Remark:* This theorem was originally due to Perron and Frobenius independently (with very different proofs).

Figure 5.13 The Platonic solids.

5.5 **Suspensions of spheres** Let K and L be simplicial complexes on disjoint vertex sets. Their **join** $K * L$ is the simplicial complex consisting of all simplices of the form $[S \cup T]$, where $[S] \in K$ and $[T] \in L$. The n-fold join K^{*n} is the join $K * K * \cdots * K$ of n copies of K, where we imagine the vertex sets of each copy to be disjoint. Let X be the n-fold join $(S^0)^{*n}$, where S^0 is the 0-sphere consisting of two points. In this problem we wish to show that X may be identified with a simplicial complex arising from the boundary of a certain convex polytope.

The n-dimensional **crosspolytope** is the convex hull of the points $\{e_1, -e_1, \ldots, e_n, -e_n\}$, where e_i is the ith standard basis vector of $\mathbb{R}^n$. For example, the three-dimensional crosspolytope is just an octahedron. Crosspolytopes are **simplicial**, meaning that all their faces are simplices. By the properties of polytopes, the collection of all the faces is then a simplicial complex, called the **boundary complex** of the polytope.

Show that X is isomorphic to the boundary complex of the n-dimensional crosspolytope. *Hint:* Show that $S \subset \{e_1, -e_1, \ldots, e_n, -e_n\}$ is the vertex set of a proper face of the crosspolytope if and only if, for any i, S never contains both e_i and $-e_i$. *Remark:* It is easy to see that the n-dimensional crosspolytope is homeomorphic to the n-dimensional ball B^n, and its boundary complex is homeomorphic to the $(n-1)$-dimensional sphere S^{n-1}. Furthermore, the join of a complex K with S^0 is just the suspension of K, so this problem shows why the suspension of S^k is S^{k+1}.

5.6 **The Platonic solids** A **polygon** is a two-dimensional convex polytope. If the angles between neighboring edges are all equal, it is **equiangular**. If the lengths of all edges are equal, it is **equilateral**. If it is both equiangular and equilateral, it is **regular**. The first few regular polygons are the triangle, square, and pentagon. A three-dimensional convex polyhedron is regular, or a **Platonic solid**, if its two-dimensional faces are regular polygons all of the same size and if the same number of faces meet at every vertex. In this problem we will prove that there are only five Platonic solids (so-named because they were well known to the ancient Greeks, especially Plato). They are the tetrahedron, cube (hexahedron), octahedron, dodecahedron, and icosahedron. (See Figure 5.13.)

We will use the fact that the Euler characteristic is a topological invariant. The surface of each polyhedron is topologically equivalent (homeomorphic) to a two-dimensional sphere S^2. Thus, all the polyhedral surfaces have the same Euler characteristic. Examining each figure, we see that the Euler characteristic $\chi = V - E + F$ of each surface is 2, where V is the number of vertices, E the number of edges, and F the number of two-dimensional faces.

(a) Tabulate the face numbers V, E, and F for each Platonic solid surface, and verify that $\chi = 2$ for each one. The face numbers of some of the figures are related. Can you explain why? What about the face numbers of the tetrahedron?

(b) Let p be the number of edges of each face and q the number of faces meeting at each vertex. For example, for the octahedron $p = 3$ and $q = 4$. Label each solid by its **Schläfli symbol** $\{p, q\}$. Thus, the Platonic solids are labeled $\{3, 3\}$, $\{4, 3\}$, $\{3, 4\}$, $\{5, 3\}$, $\{3, 5\}$. Using the properties of the Platonic solids together with Euler's equation, show that we must have

$$\frac{1}{p} + \frac{1}{q} > \frac{1}{2},$$

and thereby show that the only possible regular polyhedra are the Platonic solids listed above. *Hint:* Relate the face numbers to p and q.

5.7 **The Euler characteristic of the 2-torus** Using the triangulation of the two-dimensional torus T^2 shown in Figure 5.14, compute its Euler characteristic $\chi(T^2)$. *Hint:* Be careful not to overcount the vertices and edges! *Remark 1:* This seems like a lot of triangles. Actually, the number could be reduced, but only to 14, and then the triangulation would

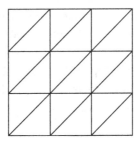

Figure 5.14 A triangulation of the torus (with opposing edges identified).

not look as nice. Why 14? That's a story for another time. *Remark 2:* Note that $\chi(T^2)$ differs from $\chi(S^2)$, thereby showing that the two surfaces are topologically distinct. We already know this from our cohomology calculations, but counting simplices is easier, don't you think?

5.8   Simplicial homology of a wedge sum   Let K be the simplicial complex consisting of two copies of the boundary of a 3-simplex glued along a single vertex. In other words, K is the wedge sum of two tetrahedra. (For the definition of a wedge sum of topological spaces, see Exercise 4.18.)

(a) Calculate $\chi(K)$ by counting faces.

(b) Compute the simplicial homology groups of K and verify (5.11). *Hint:* You may use the facts, proved for the de Rham cohomology in Exercise 4.18, that $H_0(K \vee L) \oplus \mathbb{R} \cong H_0(K) \oplus H_0(L)$ and $H_k(K \vee L) \cong H_k(L) \oplus H_k(L)$ for $k > 0$.

5.9   The Euler characteristic of a genus g surface   For most spaces X (except some pathological counterexamples), the Euler characteristic is additive. That is,

$$\chi(X) = \chi(X \backslash U) + \chi(U), \tag{5.13}$$

where U is a subspace of X and $X \backslash U$ means X with U removed.[12] This formula often provides a simple way of calculating the Euler characteristic that avoids triangulations.

(a) Let U and V be subspaces of a space X. Use (5.13) to show that the Euler characteristic is a **valuation**, namely that it satisfies the property of **inclusion–exclusion**:

$$\chi(U \cup V) = \chi(U) + \chi(V) - \chi(U \cap V). \tag{5.14}$$

(b) A two-dimensonal torus T has no boundary. (One way to see this is that the two-dimensional torus is the boundary of the solid torus and the boundary of a boundary is always zero.) But if you remove an open disk from somewhere on T you get a surface T' with boundary, namely the boundary of the disk. If you glue together two copies of T' along each of their boundaries, you get the **connected sum** of the two original tori, written $T \# T$.

   If you now cut out an open disk from the two-holed torus and glue another copy of T' along their respective boundaries you get a three-holed torus, or $T \# T \# T$. Continuing in this way you get a g-holed

---

[12] It is worth noting that U can be open or closed.

torus, where g is called the **genus** of the surface. (The two-dimensional sphere can be thought of as a 0-holed torus, or a genus-0 surface.)

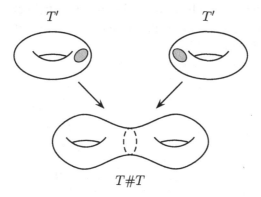

$$T' \qquad\qquad T'$$

$$T\#T$$

Show that the Euler characteristic of a genus-g surface is $\chi = 2 - 2g$. Relate this to the result of Exercise 4.16.

Remark: A compact manifold without boundary is often said to be **closed** (not to be confused with the topological meaning of "closed" introduced in Section 3.1). The **classification theorem for closed two-dimensional surfaces** asserts that any closed connected orientable-two dimensional surface is homeomorphic to a g-holed torus (including the case $g = 0$); if the surface is nonorientable then it is homeomorphic to the connected sum of real projective planes. For an elegant proof of this due to Conway, see [26].

5.10 Chain maps and homology groups   (This exercise is the homological version of Exercise 4.4.) Let K be a simplicial complex, and let $C_i := C_i(K, \mathbb{F})$ be the space of i-chains of K with coefficients in $\mathbb{F}$. Let $\partial_i : C_i \to C_{i-1}$ be the boundary operator. As $\partial_i^2 = 0$ for all i, we get a sequence

$$\cdots \xrightarrow{\partial_{i+2}} C_{i+1} \xrightarrow{\partial_{i+1}} C_i \xrightarrow{\partial_i} C_{i-1} \xrightarrow{\partial_{i-1}} \cdots .$$

For this reason, *any* sequence of vector spaces C_i with boundary maps ∂_i is called a **complex** regardless of whether the C_i arise from some simplicial complex. We adopt all the previous terminology, so that $Z_i(C) := \ker \partial_i$ is the group of i cycles, $B_i(C) := \operatorname{im} \partial_{i+1}$ is the group of i boundaries, and $H_i(C) := Z_i(C)/B_i(C)$ is the ith homology group of the complex C.

Suppose now that we have two complexes, (A_i, ∂_i) and (B_i, ∂_i). A **chain map** $\psi : (A_i, \partial_i) \to (B_i, \partial_i)$ is a sequence of linear maps that commute with the boundary maps. More precisely, we have

$$\psi_{i-1}\partial_i = \partial_i\psi_i \tag{5.15}$$

for all i. A pair of complexes related in this way by a chain map is called a homomorphism of complexes and is elegantly represented by the following commutative diagram:

$$\begin{array}{ccc}
\downarrow & & \downarrow \\
A_{i+1} & \xrightarrow{\psi_{i+1}} & B_{i+1} \\
\partial_{i+1}\downarrow & & \partial_{i+1}\downarrow \\
A_i & \xrightarrow{\psi_i} & B_i \\
\partial_i\downarrow & & \partial_i\downarrow \\
A_{i-1} & \xrightarrow{\psi_{i-1}} & B_{i-1} \\
\downarrow & & \downarrow
\end{array} \tag{5.16}$$

The statement that the diagram is commutative is precisely the statement that the maps ψ commute with the boundary maps ∂.

Show that a homomorphism of complexes gives rise to a homomorphism of homology groups. That is, show that there is a natural linear map $h_i : H_i(A) \to H_i(B)$ for all i. *Hint:* Let $[\alpha_i]$ be an element of $H_i(A)$, and consider $[\psi_i(\alpha_i)]$. Show the map is well defined, i.e., that it maps homology classes to homology classes and is independent of class representative.

5.11 Mayer–Vietoris for homology and the Euler characteristic As with cohomology, a short exact sequence of complexes gives rise to long exact sequence of homology groups. In particular, there is a homological analogue of the Mayer–Vietoris sequence for manifolds, namely[13]

$$\begin{array}{ccccc}
\longleftarrow & H_{i+1}(U \cup V) & \longleftarrow & H_{i+1}(U) \oplus H_{i+1}(V) & \longleftarrow & \cdots \\
\longleftarrow & H_i(U \cup V) & \longleftarrow & H_i(U) \oplus H_i(V) & \longleftarrow & H_i(U \cap V) & \longleftarrow \\
& \cdots & \longleftarrow & H_{i-1}(U) \oplus H_{i-1}(V) & \longleftarrow & H_{i-1}(U \cap V) & \longleftarrow
\end{array}$$

---

[13] Precisely what the maps are does not concern us here. For more details, consult [37].

Use this to obtain another proof of (5.14) (for U and V open). *Hint:* You may want to refer to Exercise 1.30.

5.12 The Hopf trace formula In Exercise 5.10 you showed that a chain map $\psi : (A_i, \partial_i) \to (B_i, \partial_i)$ between two complexes induces a natural linear map $h_i : H_i(A) \to H_i(B)$ between the homology groups of those complexes. Now suppose that $A = B = C$. Then $\psi_i : C_i \to C_i$ and $h_i : H_i(C) \to H_i(C)$ become linear endomorphisms. As such, they each have a well-defined trace. Assuming that $C_i = 0$ for $i > n$, show that

$$\sum_{i=0}^{n} (-1)^i \operatorname{tr} \psi_i = \sum_{i=0}^{n} (-1)^i \operatorname{tr} h_i. \qquad (5.17)$$

Hint: Show that there are short exact sequences

$$0 \longrightarrow Z_i \overset{\iota}{\longrightarrow} C_i \overset{\partial_i}{\longrightarrow} B_{i-1} \longrightarrow 0$$

and

$$0 \longrightarrow B_i \overset{\iota}{\longrightarrow} Z_i \overset{\pi_i}{\longrightarrow} H_i \longrightarrow 0,$$

where $Z_i = Z_i(C) = \ker \partial_i$, $B_i = B_i(C) = \operatorname{im} \partial_{i-1}$, ι is the appropriate inclusion map, ∂_i is the boundary operator, and π_i is the natural projection onto the quotient. Recall that every exact sequence of vector spaces

$$0 \longrightarrow U \overset{\iota}{\longrightarrow} V \overset{\pi}{\longrightarrow} W \longrightarrow 0$$

splits, so there exists a section $s : W \to V$ with $\pi \circ s = 1$. In particular, we have $V = U \oplus s(W)$. Thus, if ψ is a linear operator on V then $\operatorname{tr}(\psi, V) = \operatorname{tr}(\psi, U) + \operatorname{tr}(\psi, s(W))$, where $\operatorname{tr}(\psi, S)$ means the trace of ψ restricted to the subspace S. Now relate $\operatorname{tr}(\psi_i, s(B_{i-1}))$ to $\operatorname{tr}(\psi_{i-1}, B_{i-1})$ and $\operatorname{tr}(\psi_i, H_i)$ to $\operatorname{tr}(h_i, H_i)$.

Remark: A smooth map $f : M \to M$ from a manifold to itself naturally induces a map $f_* : C \to C$ (see Section 6.3) of chain complexes of the manifold. In that case, the right-hand side of (5.17) is called the **Lefschetz number** of the map f. Note that if f is the identity map and M is triangulable then the Lefschetz number is just the Euler characteristic and the Hopf trace formula reduces to Theorem 5.2. The remarkable **Lefschetz fixed point theorem** asserts that if M is compact and orientable and the Lefschetz number does not vanish then f has a fixed point, namely a point $p \in M$ with $f(p) = p$. In fact, the Lefschetz fixed point theorem is a special case of a more general theorem known as the Lefschetz–Hopf theorem. See the discussion following Theorem 9.4.

6

Integration on manifolds

> ... Stokes' Theorem has had a curious history and has
> undergone a striking metamorphosis.
>
> *Michael Spivak*[†]

After that quick jaunt into general topology we now return to the setting of smooth manifolds. Our aim in this chapter is to relate singular homology to de Rham cohomology. It will turn out that they are isomorphic, but to establish this we must first introduce a few more concepts. The key tool is integration, and we will derive some of its fundamental properties, including both the change of variables theorem and Stokes' theorem.

6.1 Smooth singular homology

Simplices (and singular simplices) tend to have sharp corners where derivatives do not exist, whereas when dealing with a smooth manifold M we would like everything to behave nicely with respect to smooth maps. Ideally we want the map $\phi : \bar{\sigma}^n \to M$ to be smooth. But smooth maps are defined only on open sets, and our simplices are closed. The way around this is to *extend* ϕ to a smooth map. More precisely, a **smooth singular m-simplex** σ is an equivalence class of triples $(\bar{\sigma}, U, \phi)$, where $\bar{\sigma}$ is an m-simplex, U is an open neighborhood of $\operatorname{Aff}\bar{\sigma}$ (the smallest affine subspace containing $\bar{\sigma}$), and $\phi : U \to M$ is a smooth map. Two triples $(\bar{\sigma}, U, \phi)$ and $(\bar{\tau}, V, \psi)$ are in the same class if $\phi(\bar{\sigma}) = \psi(\bar{\tau})$ or, equivalently, if $\phi = \psi \circ f$ where $f : \bar{\sigma} \to \bar{\tau}$ is the natural order-preserving affine map.

As before, we define $C_k^\infty(M)$, the space of **smooth singular k-chains in M**, to be the vector space generated by all the smooth singular k-simplices in M. If $\sigma = (\bar{\sigma}, U, \phi)$ and $\partial\bar{\sigma} = \sum_i \pm\bar{\sigma}_i$, the boundary of σ is defined to be $\partial\sigma :=$

[†] Excerpt from [79], p. viii. Reprinted by permission.

$\sum_i \pm \sigma_i$, where $\sigma_i = (\bar{\sigma}_i, V_i, \phi)$ and V_i is a small open neighborhood of $\bar{\sigma}_i$. The boundary operator again satisfies $\partial^2 = 0$, so the space $B_k^\infty(M)$ of smooth singular k-boundaries of M is a subspace of the space $Z_k^\infty(M)$ of smooth singular k-cycles in M and we can define the **smooth singular homology** $H_k^\infty(M)$ of M to be the quotient space

$$H_k^\infty(M) = Z_k^\infty(M)/B_k^\infty(M). \tag{6.1}$$

It is a happy fact that, for smooth manifolds, the smooth singular homology and the singular homology agree.[1]

Theorem 6.1 *For a smooth manifold M,*

$$H_k^\infty(M) = H_k(M). \tag{6.2}$$

(For this reason, one often omits the superscript ∞ in the smooth setting.) One way to prove this result is to exploit the fact that every smooth manifold admits a smooth triangulation.[2] More precisely, a **smoothly triangulated manifold** is a triple (M, K, ϕ), where M is a smooth manifold, K is a simplicial complex, and $\phi : K \to M$ is a homeomorphism satisfying the following property. For each simplex $\bar{\sigma} \in K$, the map $\phi : \bar{\sigma} \to M$ can be extended to a neighborhood U of $\bar{\sigma}$ in Aff $\bar{\sigma}$ in such a way that $\phi : U \to M$ is an *embedding*. As a smooth triangulation is also a triangulation, we obtain a smooth analogue of the continuous result that the homology of M coincides with the homology of the triangulation K.

Theorem 6.2 *If K is a smooth triangulation of M then*

$$H_k^\infty(M) = H_k(K).$$

The next step is to relate $H_k^\infty(M)$ to the de Rham cohomology. Before we can do this, however, we must first talk about integration on manifolds.

6.2 Integration on chains

First we define the integration of forms on a domain $D \subseteq \mathbb{R}^n$ in the obvious way. If

$$\omega = A(x_1, \ldots, x_n) \, dx^1 \wedge \cdots \wedge dx^n \tag{6.3}$$

---

[1] See e.g. ([51], Theorem 16.6).
[2] This result is due to Cairnes [13] and to Whitehead [90].

then we define

$$\int_D \omega := \int_D A(x_1, \ldots, x_n) \, dx^1 \cdots dx^n, \tag{6.4}$$

where the integral on the right is the ordinary Riemann integral (which has no wedge products!).[3]

> **Example 6.1** If $\omega = 3 \, dy \wedge dx \wedge dz$ and if $\bar{s}^n = \langle 0, e_1, e_2, \ldots, e_n \rangle$ is the **standard simplex** ($\{e_i\}$ being the standard basis of $\mathbb{R}^n$) then
>
> $$\int_{\bar{s}^3} \omega = -3 \int_{\bar{s}^3} dx \, dy \, dz = -3 \int_0^1 dx \int_0^{1-x} dy \int_0^{1-x-y} dz = -1/2.$$

Let M be a smooth manifold, ω a k-form on M, and c a smooth singular k-chain on M, where $k \leq n$. Then the expression

$$\int_c \omega$$

is defined by pulling it back to Euclidean space, where we know the answer. We want it to be linear on chains, so if

$$c = \sum_i a_i \sigma_i \tag{6.5}$$

then we set

$$\int_c \omega := \sum_i a_i \int_{\sigma_i} \omega. \tag{6.6}$$

Integration over simplices is carried out via pullback. If $\sigma = (\bar{\sigma}^k, U, \phi)$ then we set

$$\int_\sigma \omega := \int_{\bar{\sigma}^k} \phi^* \omega. \tag{6.7}$$

In order for this to be well defined, it must be independent of the equivalence class representative used to define it. This requirement leads us to the general change of variables theorem.

6.3 Change of variables

Let $f : M \to N$ be a smooth map of smooth manifolds. Then there is a natural induced linear map

$$f_* : C_k(M) \to C_k(N) \tag{6.8}$$

---

[3] Technically, we are defining our integrals to be **oriented integrals**, because they depend on the order of the variables in the differential forms. This is really the natural definition, as will become evident in the next section. The reader should be warned that this convention is by no means universal, although it ought to be.

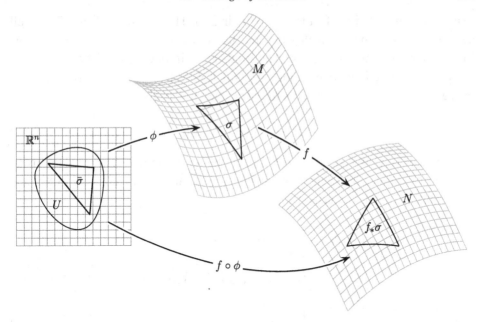

Figure 6.1 The induced map f_* on simplices.

taking k-chains on M to k-chains on N.[4] The map f_* is defined to be the linear extension of the map sending the simplex σ on M, represented by $(\bar{\sigma}, U, \phi)$, to the simplex $f_*\sigma$ on N, represented by $(\bar{\sigma}, U, f \circ \phi)$. (See Figure 6.1.) In passing we observe that, for any chain c,

$$f_*(\partial c) = \partial(f_* c), \tag{6.9}$$

while for smooth maps $f : M \to N$ and $g : N \to P$ we have

$$(g \circ f)_* = g_* \circ f_*. \tag{6.10}$$

Return now to (6.7), and suppose that σ is also represented by the triple $(\bar{\tau}^k, V, \psi)$; let $f : \bar{\sigma}^k \to \bar{\tau}^k$ be the unique order-preserving linear map of simplices, with $\phi = \psi \circ f$. Then the assumed well-definedness of (6.7) is equivalent to the chain of equalities

$$\int_{f_*\bar{\sigma}^k} \psi^*\omega = \int_{\bar{\tau}^k} \psi^*\omega = \int_\sigma \omega = \int_{\bar{\sigma}^k} \phi^*\omega = \int_{\bar{\sigma}^k} f^*\psi^*\omega \tag{6.11}$$

---

[4] There is no danger of confusing this map with the pushforward map defined in Section 3.15, even though they both use the same star subscript notation, because the former is a map of chains which live on the manifold itself while the latter is a map of tangent spaces, which do not. (In any case, the notation is now more or less standard.)

(where we have used the fact that $f = f_*$ in $\mathbb{R}^n$). Thus, to show that (6.7) is well defined we must show that the first and last terms in (6.11) are equal. But this follows from the usual change of varables theorem in Euclidean space.

Actually, something a bit more general holds. Let $f : \mathbb{R}^n \rightarrow \mathbb{R}^n$ be a diffeomorphism. Choose local coordinates, so that f is given by

$$y^i = f^i(x^1, \ldots, x^n), \tag{6.12}$$

and let $\eta = a(y) \, dy^1 \wedge \cdots \wedge dy^n$. Let $\bar{\sigma}^n$ be an n-simplex. Then, by the usual change of variables theorem from calculus (and with the assistance of (3.82) and the definition (6.4)), we have

$$\int_{f_*\bar{\sigma}^n} \eta = \int_{f_*\bar{\sigma}^n} a \, dy^1 \cdots dy^n$$
$$= \int_{\bar{\sigma}^n} (a \circ f) \det\left(\frac{\partial y^i}{\partial x^j}\right) dx^1 \cdots dx^n$$
$$= \int_{\bar{\sigma}^n} f^* \eta. \tag{6.13}$$

The same argument proves (6.11) on setting $\eta := \psi^* \omega$ and restricting f to Aff $\bar{\sigma}^k$.[5]

Equation (6.13) reveals a natural duality between f^* and f_* with respect to integration. This duality extends naturally to chains on arbitrary manifolds and thereby yields the following elegant formulation of the general change of variables theorem.

Theorem 6.3 (Change of variables) *Let $f : M \rightarrow N$ be a diffeomorphism, ω a k-form on N, and c a k-chain c on M. Then*

$$\int_c f^* \omega = \int_{f_* c} \omega. \tag{6.14}$$

Example 6.2 The change of variables theorem can also be used to recover the classical expressions for line and surface integrals. Let $\omega = A_1 \, dx^1 + A_2 \, dx^2 + A_3 \, dx^3$ be a 1-form on $\mathbb{R}^3$, and let γ be a curve. In our new terminology, a curve is just a (singular) 1-chain in $\mathbb{R}^3$. More precisely, if I is the 1-simplex in $\mathbb{R}$, the 1-chain γ is the image of I under γ given by

$$\gamma(t) = (\gamma^1(t), \gamma^2(t), \gamma^3(t)),$$

---

[5] The astute reader will notice that we have cheated here. The usual change of variables theorem employs the *absolute value* of the Jacobian, whereas we have used the Jacobian itself. But that's all right, because we defined our integrals in (6.4) precisely to account for this pesky sign.

where t is the coordinate on $\mathbb{R}$. We have

$$
\begin{aligned}
\gamma^*\omega &= \gamma^*(A_1\,dx^1 + A_2\,dx^2 + A_3\,dx^3) \\
&= (\gamma^*A_1)(\gamma^*dx^1) + (\gamma^*A_2)(\gamma^*dx^2) + (\gamma^*A_3)(\gamma^*dx^3) \\
&= (A_1 \circ \gamma)\,d\gamma^1 + \cdots + (A_3 \circ \gamma)\,d\gamma^3 \\
&= A(\gamma(t)) \cdot d\boldsymbol{\gamma} \\
&= (A(\gamma(t)) \cdot \nabla\gamma)\,dt,
\end{aligned}
$$

where A is the vector field whose components are (A_1, A_2, A_3), and ∇ is the usual gradient operator. (The dot product of two vectors is their Euclidean inner product.) Applying Theorem 6.3 gives

$$
\int_\gamma \omega = \int_{\gamma_*I} \omega = \int_I \gamma^*\omega = \int_I A(\gamma(t)) \cdot \nabla\gamma\,dt,
$$

namely the usual line integral of A along the curve γ.

EXERCISE 6.1 Let $\gamma(t) = (t, t^2, t^3)$, $0 \le t \le 2$ be a curve in $\mathbb{R}^3$. (It is called the **moment curve**.) Let $\omega = yz\,dx + zx\,dy + xy\,dz$ be a 1-form on $\mathbb{R}^3$. Evaluate $\int_\gamma \omega$.

EXERCISE 6.2 Let $\beta = B_x\,dy \wedge dz + B_y\,dz \wedge dx + B_z\,dx \wedge dy$ be a 2-form in $\mathbb{R}^3$. Let Σ be a surface in $\mathbb{R}^3$, namely the image under a map φ of a 2-chain c in $\mathbb{R}^2$. By choosing coordinates u, v on $\mathbb{R}^2$ such that Σ is given by

$$
(u, v) \mapsto (\varphi^1(u, v), \varphi^2(u, v), \varphi^3(u, v)), \tag{6.15}
$$

show that

$$
\int_\Sigma \beta = \int_c B(\varphi(u, v)) \cdot n(\varphi(u, v))\,du \wedge dv =: \int_\Sigma B \cdot dS, \tag{6.16}
$$

the ordinary flux integral of B over Σ. Here $B = (B_x, B_y, B_z)$, dS is the surface area element of Σ, and $n = \partial\varphi/\partial u \times \partial\varphi/\partial v$ is the surface normal to Σ (an ordinary cross product).

6.4 Stokes' theorem

The fundamental theorem of calculus states that

$$
\int_a^b f'(x)\,dx = f(b) - f(a). \tag{6.17}
$$

We can view the interval $[a, b]$ as an oriented chain with oriented boundary $b - a$, and write (6.17) more suggestively as

$$
\int_{[a,b]} df = \int_{\partial[a,b]} f, \tag{6.18}
$$

where the integral of a function over a point just means the evaluation of the function at the point (with the appropriate sign). In this way the fundamental theorem of calculus can be interpreted as saying that the integral of the differential of a function f over a chain equals the integral of f over the boundary of the chain (with the appropriate orientations).

In multivariable calculus the fundamental theorem is extended from $\mathbb{R}$ to $\mathbb{R}^3$ in the theorems of Stokes and Gauss. So, if A is a vector field, Σ a surface with boundary $\partial \Sigma$, and V a volume with boundary ∂V, then we have

$$\int_{\Sigma} (\nabla \times A) \cdot dS = \oint_{\partial \Sigma} A \cdot d\ell \tag{6.19}$$

and

$$\int_{V} (\nabla \cdot A) \, dV = \oint_{\partial V} A \cdot dS. \tag{6.20}$$

In both these theorems one must choose compatible orientations on the domains of integration and their respective boundaries (via the "outward pointing normal").

All these formulae are just special cases of a far-reaching theorem of integration on manifolds, called simply Stokes' theorem. Actually, there are two approaches to Stokes' theorem, which are essentially equivalent. For completeness, we discuss both of them here.

6.4.1 Stokes' theorem I

Theorem 6.4 (Stokes' theorem – chain version)   *For any k-form ω and $(k + 1)$-chain c,*

$$\int_{c} d\omega = \int_{\partial c} \omega. \tag{6.21}$$

Remark   In the next section we show how Stokes' theorem also works for "infinite chains" (i.e., noncompact manifolds with boundary), provided we restrict ourselves to differential forms with compact support. (The **support** of a form ω, denoted supp ω, is the smallest closed set that contains the set of points at which ω does not vanish.)

Proof   By linearity it suffices to prove that

$$\int_{\sigma} d\omega = \int_{\partial \sigma} \omega,$$

for some $(k+1)$-simplex σ. If $\sigma = (\bar{s}^{k+1}, U, \phi)$ then this is equivalent to

$$\int_{\bar{s}^{k+1}} d(\phi^*\omega) = \int_{\bar{s}^{k+1}} \phi^*(d\omega) = \int_\sigma d\omega = \int_{\partial\sigma} \omega = \int_{\partial\bar{s}^{k+1}} \phi^*\omega.$$

It is therefore sufficient to prove that

$$\int_{\bar{s}^{k+1}} d\lambda = \int_{\partial\bar{s}^{k+1}} \lambda \tag{6.22}$$

for λ a k-form on a neighborhood U of the standard $(k+1)$-simplex $\bar{s}^{k+1}$ in $\mathbb{R}^{k+1}$.
Again by linearity it is enough to prove (6.22) for a monomial of the form

$$\lambda = A(x^1, \ldots, x^{k+1}) \, dx^1 \wedge \cdots \wedge dx^k.$$

We have

$$d\lambda = (-1)^k \frac{\partial A}{\partial x^{k+1}} \, dx^1 \wedge \cdots \wedge dx^{k+1},$$

so that

$$\int_{\bar{s}^{k+1}} d\lambda = (-1)^k \int_{\bar{s}^{k+1}} \frac{\partial A}{\partial x^{k+1}} \, dx^1 \wedge \cdots \wedge dx^{k+1},$$

$$= (-1)^k \int_{x^i \geq 0, \sum_{i=1}^k x^i \leq 1} dx^1 \cdots dx^k \left(\int_0^{1-\sum_{i=1}^k x^i} \frac{\partial A}{\partial x^{k+1}} \, dx^{k+1} \right)$$

$$= (-1)^k \int_{x^i \geq 0, \sum_{i=1}^k x^i \leq 1} dx^1 \cdots dx^k$$
$$\times \left[A(x^1, \ldots, x^k, 1 - \sum_{i=1}^k x^i) - A(x^1, \ldots, x^k, 0) \right]. \tag{6.23}$$

Furthermore,

$$\partial \bar{s}^{k+1} = \partial \langle e_0, e_1, \ldots, e_{k+1} \rangle$$
$$= \langle e_1, e_2, \ldots, e_{k+1} \rangle + (-1)^{k+1} \langle e_0, e_1, \ldots, e_k \rangle \pm \text{other faces},$$

giving

$$\int_{\partial \bar{s}^{k+1}} \lambda = \int_{\langle e_1, \ldots, e_{k+1} \rangle} \lambda + (-1)^{k+1} \int_{\langle e_0, e_1, \ldots, e_k \rangle} \lambda. \tag{6.24}$$

(The form λ vanishes on the "other faces", because on those faces at least one of $x^1, \ldots, x^k$ is constant.) But $\langle e_0, e_1, \ldots, e_k \rangle$ is just the standard simplex $\bar{s}^k$ with $x^{k+1} = 0$, so the second term in (6.24) becomes

$$(-1)^{k+1} \int_{\bar{s}^k} A(x^1, \ldots, x^k, 0) \, dx^1 \cdots dx^k. \tag{6.25}$$

By projecting down along the x^{k+1} axis, the first term in (6.24) becomes

$$\int_{\langle e_1, \ldots, e_{k+1} \rangle} \lambda = \int_{\langle e_1, \ldots, e_k, 0 \rangle} A(x^1, \ldots, x^k, 1 - \textstyle\sum_{i=1}^{k} x^i) \, dx^1 \cdots dx^k$$

$$= (-1)^k \int_{\langle e_0, \ldots, e_k \rangle} A(x^1, \ldots, x^k, 1 - \textstyle\sum_{i=1}^{k} x^i) \, dx^1 \cdots dx^k. \tag{6.26}$$

Now compare (6.23) with (6.24), (6.25), and (6.26) to obtain (6.22). □

Example 6.3 Let $\omega = A_1 \, dx^1 + A_2 \, dx^2 + A_3 \, dx^3$ be a 1-form and let Σ be a closed surface in $\mathbb{R}^3$. Then, using the results of Example 6.2 and Exercise 6.2, Stokes' theorem

$$\int_c d\omega = \int_{\partial c} \omega$$

becomes

$$\int_\Sigma (\nabla \times A) \cdot dS = \int_{\partial \Sigma} A \cdot d\phi.$$

EXERCISE 6.3 Let $\omega = -yz \, dx + xz \, dy$ be a 1-form in $\mathbb{R}^3$, and let U be an open subset of the (u, v) plane given by $1 \le u \le 2$ and $0 \le v \le 2\pi$. Let $\Sigma := \varphi(U)$ be a parameterized surface (2-chain) in $\mathbb{R}^3$, where

$$\varphi(u, v) = (u^3 \cos v, u^3 \sin v, u^2).$$

Using the change of variables theorem, verify that Stokes' theorem holds, i.e.,

$$\int_\Sigma d\omega = \int_{\partial \Sigma} \omega.$$

EXERCISE 6.4 Show that the integral of a closed form over the boundary of any chain is zero. Show that the integral of an exact form over a boundaryless chain is zero.

EXERCISE 6.5 Using Stokes' theorem, show that the property $d^2 = 0$ (mixed partials commute) is equivalent to the property $\partial^2 = 0$ (the boundary of a boundary vanishes).

6.4.2 Stokes' theorem II

The second version of Stokes' theorem is perhaps more familiar to modern readers. It is a statement about the integral of a differential form over an oriented manifold with boundary, so it looks a little different from the chain version. However, the distinction is illusory, as one can obtain the second version from the first (see e.g. [88], Section 4.8). But the second version can also be derived *ab initio* without recourse to the machinery of chains, which is what we do here. First we need some preparation.

A (**smooth**) n-**dimensional manifold** M **with boundary** is defined much like an ordinary smooth manifold, except that the coordinate patches $\{U_i\}$ belonging to an atlas $\{U_i, \varphi_i\}$ now come in two varieties: either U_i is homeomorphic to $\mathbb{R}^n$, as before, or else U_i is homeomorphic to the upper half space $\mathbb{H}^n = \{(x^1, \ldots, x^n) : x^n \geq 0\}$.[6] The boundary $\partial \mathbb{H}^n$ of $\mathbb{H}^n$ is the set $\{(x^1, \ldots, x^n) : x^n = 0\}$, which is homeomorphic to $\mathbb{R}^{n-1}$. The **boundary** ∂M of M consists of all the points $p \in M$ such that $\varphi_i(p) \in \partial \mathbb{H}^n$. It is an embedded $(n-1)$-dimensional submanifold of M, and it inherits an orientation from that of M as follows. Let $\mathbb{H}^n$ be given the standard orientation $dx^1 \wedge \cdots \wedge dx^n$.[7] Then by definition the induced orientation of $\partial \mathbb{H}^n$ is the orientation class of $\lambda = (-1)^n \, dx^1 \wedge \cdots \wedge dx^{n-1}$. If η is a nowhere vanishing $(n-1)$-form on ∂M then ∂M has the **induced orientation** if $\varphi_i^* \lambda$ and η (restricted to $U_i \cap \partial M$) are in the same orientation class.

Let ω be a top-dimensional form on an oriented manifold M and let $\{\rho_i\}$ be a partition of unity subordinate to the open cover $\{U_i\}$. Then we define

$$\int_M \omega = \sum_i \int_{U_i} \rho_i \omega. \tag{6.27}$$

That is, we break up the integral into little integrals over each coordinate neighborhood U_i, integrate the fraction of ω belonging to each U_i according to the partition of unity, then sum all the contributions.

> **EXERCISE 6.6** Show that the definition (6.27) makes sense, i.e., that it is independent of the coordinate covering and the partition of unity.

Theorem 6.5 (Stokes' theorem – manifold version)   *Let M be an oriented smooth n-dimensional manifold with boundary, and let ω be an $(n-1)$-form with compact support. Then*

---

[6] Note that $\mathbb{H}^n$ is an open set in the subspace topology even though it is closed as a subset of $\mathbb{R}^n$. Therefore there is no problem with demanding that the transition maps be smooth.

[7] See Appendix C.

$$\int_M d\omega = \int_{\partial M} \omega, \tag{6.28}$$

where ∂M is given the induced orientation.

Proof Let $\{U_i\}$ be a locally finite covering of M by coordinate neighborhoods with subordinate partition of unity $\{\rho_i\}$. Clearly $\omega = \sum_i \rho_i \omega$. Thus, by the linearity of integration, it suffices to prove Stokes' theorem for each $\rho_i \omega$. (Note that $\rho_i \omega$ has compact support because $\operatorname{supp} \rho_i \omega \subset \operatorname{supp} \rho_i \cap \operatorname{supp} \omega$ is a closed subset of a compact set.) As the support of $\rho_i \omega$ is contained in U_i, and as each U_i is homeomorphic to a subset of either $\mathbb{R}^n$ or $\mathbb{H}^n$, it suffices to prove the result assuming either $M = \mathbb{R}^n$ or $M = \mathbb{H}^n$.

We have

$$\omega = \sum_{i=1}^n a_i(x) \, dx^1 \wedge \cdots \wedge dx^{i-1} \wedge dx^{i+1} \wedge \cdots \wedge dx^n,$$

so that

$$d\omega = \left(\sum_{i=1}^n (-1)^{i+1} \frac{\partial a_i}{\partial x^i} \right) dx^1 \wedge \cdots \wedge dx^n.$$

If $M = \mathbb{R}^n$ then

$$\int_{\mathbb{R}^n} d\omega = \sum_{i=1}^n (-1)^{i+1} \int_{\mathbb{R}^n} \frac{\partial a_i}{\partial x^i} \, dx^1 \cdots dx^n$$

$$= \sum_{i=1}^n (-1)^{i+1} \int \left(\int_{-\infty}^\infty \frac{\partial a_i}{\partial x^i} \, dx^i \right) dx^1 \cdots dx^{i-1} dx^{i+1} \cdots dx^n.$$

But ω has compact support so $a(x)$ has compact support, and

$$\int_{-\infty}^\infty \frac{\partial a_i}{\partial x^i} \, dx^i = a_i(\ldots, \infty, \ldots) - a_i(\ldots, -\infty, \ldots) = 0.$$

As $\mathbb{R}^n$ has no boundary, Stokes' theorem holds for this case.

If $M = \mathbb{H}^n$ then everything goes as before, except that the limits of integration on the x^n integral are now 0 to ∞, so we get

$$\int_{\mathbb{H}^n} d\omega = (-1)^n \int_{\mathbb{H}^n} a_n(x^1, \ldots, x^{n-1}, 0) \, dx^1 \cdots dx^{n-1}.$$

But ω restricted to $\partial \mathbb{H}^n$ is

$$\omega = a_i(x^1, \ldots, x^{n-1}, 0) \, dx^1 \wedge \cdots \wedge dx^{n-1},$$

so using the induced orientation on $\partial \mathbb{H}^n$ gives

$$\int_{\partial \mathbb{H}^n} \omega = (-1)^n \int_{\mathbb{H}^n} a_n(x^1, \ldots, x^{n-1}, 0) \, dx^1 \cdots dx^{n-1}.$$

Hence Stokes' theorem holds in all cases. $\square$

EXERCISE 6.7 Let f be a function, η an $(n-1)$-form, and M an n-dimensional manifold. Derive the general integration by parts formula,

$$\int_M f \, d\eta = \int_{\partial M} f\eta - \int_M df \wedge \eta.$$

Remark   Let $\iota : \partial M \to M$ be the inclusion map of the boundary of M into M. Technically, Stokes' theorem should be written

$$\int_M d\omega = \int_{\partial M} \iota^*\omega, \tag{6.29}$$

because ω really lives on M, whereas to evaluate the integral on the right-hand side we must evaluate ω on the boundary. Most of the time this formulation is needlessly pedantic, but occasionally it is useful. (A similar remark applies to the first version of Stokes' theorem.)

6.5 de Rham's theorem

Having developed the theory of integration on manifolds, we are now in a position to fulfill our promise of relating smooth singular homology and de Rham cohomology. As you might have guessed, the key tool that ties them together is integration. The basic idea is that integration provides a nondegenerate pairing between de Rham cohomology and smooth singular homology, so that the two are dual to one another in a precise sense. There are two ways to state de Rham's theorem, and each is worth knowing. We first discuss de Rham's original formulation, as it fits nicely with our previous discussion of integration on chains, and then turn to the slightly more modern statement.

Given a form ω and a chain c we get a real number, namely $\int_c \omega$. In this sense, the space of forms can be thought of as a kind of dual to the space of chains. But this duality is not too useful because each space is infinite dimensional and, as we have said, infinite-dimensional dualities are treacherous. But we know that both the de Rham cohomology and the smooth singular homology of a space are finite dimensional, so we might expect that restricting ourselves to elements of those spaces might be more profitable, and this is indeed the case.

If ω is a closed form and z is a cycle then $\int_z \omega$ is called the **period** of ω on z. Two important facts about periods follow immediately from Stokes' theorem:[8] (i) the period of an exact form on any cycle is zero, and (ii) the period of a closed form on any boundary is zero. This begs the question of whether the converses are true. Put another way, if all the periods of a closed form vanish, is the form exact? Given a map from cycles to real numbers that vanishes on all boundaries, is it defined by the periods of some closed form? The answer to both questions is yes, and this is the content of de Rham's two theorems.[9]

Theorem 6.6 (de Rham's first theorem) *A closed form is exact if and only if all its periods vanish.*

Theorem 6.7 (de Rham's second theorem) *Let* per *be a linear map that assigns a real number to every cycle. If* per *vanishes on all boundaries then there exists a closed form ω whose periods are given by* per:

$$\int_z \omega = \mathrm{per}(z).$$

But what does this have to do with homology and cohomology? Well, everything, but to see this we first introduce another kind of cohomology. Given a chain complex $(C_\bullet, \partial)$,

$$\cdots \longrightarrow C_{\ell+1} \xrightarrow{\;\partial_{\ell+1}\;} C_\ell \xrightarrow{\;\partial_\ell\;} C_{\ell-1} \xrightarrow{\;\partial_{\ell-1}\;} \cdots \;,$$

we obtain a dual chain complex $(C^\bullet, \partial^*)$,

$$\cdots \longleftarrow C^{\ell+1} \xleftarrow{\;\partial_\ell^*\;} C^\ell \xleftarrow{\;\partial_{\ell-1}^*\;} C^{\ell-1} \xleftarrow{\;\partial_{\ell-2}^*\;} \cdots \;,$$

where $C^\ell = C_\ell^*$ and ∂^* is the dual map, defined by

$$(\partial_\ell^* f)(c) = f(\partial_{\ell+1} c), \tag{6.30}$$

where $f \in C^\ell$ and $c \in C_{\ell+1}$. (The only thing one needs to check is that ∂^* is nilpotent, but that follows immediately from the nilpotence of ∂.) By definition, the **smooth singular cohomology** of M is the homology of $(C^\bullet(M), \partial^*)$ where $C_\bullet(M)$ is the complex of smooth singular chains. Explicitly, we have

$$Z^\ell(M) = \{ f \in C^\ell(M) : \partial^* f = 0 \}$$

$$= \text{the space of (smooth singular) } \ell\text{-cocycles of } M, \tag{6.31}$$

---

[8] Compare Exercise 6.4.
[9] For proofs, see [21] or [42].

$$B^{\ell}(M) = \{\partial^* f : f \in C^{\ell-1}(M)\}$$
$$= \text{the space of (smooth singular) } \ell\text{-\textbf{coboundaries} of } M, \qquad (6.32)$$
$$H^{\ell}(M) = Z^{\ell}(M)/B^{\ell}(M)$$
$$= \text{the } \ell\text{th-\textbf{smooth singular cohomology group} of } M. \qquad (6.33)$$

(Note that, in general, $Z^{\ell} \not\cong Z^*_{\ell}$ and $B^{\ell} \not\cong B^*_{\ell}$. As for H^{ℓ} and H^*_{ℓ}, see Theorem 6.9 below.)

The theorems of de Rham say that the de Rham cohomology and the smooth singular cohomology of a manifold are one and the same. To see this, observe that, given any ℓ-form ω, the integral of ω can be thought of as a linear map

$$\int \omega : C_{\ell}(M) \to \mathbb{R},$$

given by

$$\left(\int \omega\right)(c) = \int_c \omega. \qquad (6.34)$$

In other words, $\int \omega$ is an ℓ-cochain.

If ω is closed then $\int \omega$ is actually an ℓ-cocycle because, by (6.30), (6.34), and Stokes' theorem, for any $(\ell+1)$-chain c,

$$\left(\partial^* \int \omega\right)(c) = \left(\int \omega\right)(\partial c) = \int_{\partial c} \omega = \int_c d\omega = 0. \qquad (6.35)$$

Hence, integration provides a homomorphism

$$\int : Z^{\ell}_{\mathrm{dR}}(M) \to Z^{\ell}(M), \qquad \omega \mapsto \int \omega, \qquad (6.36)$$

from the space of closed ℓ-forms to the space of smooth singular ℓ-cocycles on M.

EXERCISE 6.8 Show that if $\omega = d\tau$ is an exact ℓ-form then $\int \omega$ is an ℓ-coboundary. That is, the integration map provides a natural homomorphism

$$\int : B^{\ell}_{\mathrm{dR}}(M) \to B^{\ell}(M). \qquad (6.37)$$

It follows[10] that the integration map provides a homomorphism

$$\int : H^{\ell}_{\mathrm{dR}}(M) \to H^{\ell}(M), \qquad [\omega] \mapsto \left[\int \omega\right] \qquad (6.38)$$

of cohomologies. The two theorems of de Rham tell us that this map is actually an isomorphism.

---

[10] Compare Exercise 1.15.

EXERCISE 6.9 Prove that the map (6.38) is an isomorphism. *Hint:* As a map between cohomologies, de Rham's first theorem says that $\int$ is injective and de Rham's second theorem says that $\int$ is surjective.

We therefore obtain the following repackaging of de Rham's two theorems.

Theorem 6.8 (de Rham) *Let M be a smooth manifold. Then the de Rham cohomology coincides with the smooth singular cohomology of M:*

$$H_{\mathrm{dR}}^{\ell}(M) \cong H^{\ell}(M).$$

The obvious question is: what is the relation between smooth singular cohomology and smooth singular homology? To our great relief, they are isomorphic. More precisely, we have the following theorem, which relates the cohomology and the homology (with coefficients taken from a field) of any complex.[11]

Theorem 6.9 *For any complex C and for any coefficient field $\mathbb{F}$,*

$$H^{\ell}(C, \mathbb{F}) \cong H_{\ell}^{*}(C, \mathbb{F}). \tag{6.39}$$

Proof First, we show there is a natural linear surjection $h : H^{\ell} \to H_{\ell}^{*}$. Let $[f] \in H^{\ell}$ and $[z] \in H_{\ell}$. Define h by

$$h([f])([z]) := f(z).$$

This is well defined, for if $[f] = [g]$ and $[z] = [y]$ then $g = f + \partial^{*}b$ for some b and $y = z + \partial x$ for some x, and

$$h([g])([y]) = g(y) = (f + \partial^{*}b)(z + \partial x) = f(z),$$

because $f(\partial x) = \partial^{*}f(x) = 0$, $\partial^{*}b(z) = b(\partial z) = 0$, and $\partial^{*}b(\partial x) = b(\partial^{2}x) = 0$.

Let $g \in H_{\ell}^{*}$. This extends naturally to an element $\widetilde{g} \in Z_{\ell}^{*}$ by setting

$$\widetilde{g}(z) := \begin{cases} 0 & \text{if } z \in B_{\ell}, \text{ and} \\ g([z]) & \text{otherwise.} \end{cases}$$

Let $\pi : C_{\ell} \to Z_{\ell}$ denote the natural projection map. Composing $\widetilde{g}$ with the projection π yields an element $\widetilde{g}\pi \in C_{\ell}^{*}$. By the definition of $\widetilde{g}$, $\partial^{*}(\widetilde{g}\pi) = (\widetilde{g}\pi)\partial = 0$ so $\widetilde{g}\pi \in Z^{\ell}$. Therefore $[\widetilde{g}\pi] \in H^{\ell}$.

---

[11] If there were any justice in the world, cohomology would always be isomorphic to the dual of homology. Unfortunately, the world is not always just. In general, the issue depends on the coefficient groups one uses to define these spaces. For homology and cohomology over general abelian groups (with a corresponding more general definition of "dual"), the situation is much more subtle and leads to something called the **universal coefficient theorem**. (See e.g. [37].)

We must verify that h takes $[\widetilde{g}\pi]$ to g. If z is a boundary then

$$h([\widetilde{g}\pi])([z]) = h([\widetilde{g}\pi])([0]) = 0 = g([0]) = g([z]),$$

while if z is a nontrivial cycle, we have

$$h([\widetilde{g}\pi])([z]) = \widetilde{g}\pi(z) = \widetilde{g}(z) = g([z])$$

yet again, so all is well.

We conclude the proof by showing that H^ℓ and H_ℓ have the same dimensions. Consider the chain complex

$$\cdots \xrightarrow{\partial_{\ell+2}} C_{\ell+1} \xrightarrow{\partial_{\ell+1}} C_\ell \xrightarrow{\partial_\ell} C_{\ell-1} \xrightarrow{\partial_{\ell-1}} \cdots ,$$

and the corresponding cochain complex

$$\cdots \xleftarrow{\partial_{\ell+1}^*} C^{\ell+1} \xleftarrow{\partial_\ell^*} C^\ell \xleftarrow{\partial_{\ell-1}^*} C^{\ell-1} \xleftarrow{\partial_{\ell-2}^*} \cdots .$$

By Exercise 1.53,

$$Z^\ell = \ker \partial_\ell^* = \operatorname{Ann} \operatorname{im} \partial_{\ell+1} = \operatorname{Ann} B_\ell,$$
$$B^\ell = \operatorname{im} \partial_{\ell-1}^* = \operatorname{Ann} \ker \partial_\ell = \operatorname{Ann} Z_\ell.$$

However, for any subspace W of a vector space V, $\dim V = \dim W + \dim \operatorname{Ann} W$ (Exercise 1.21), so

$$\dim Z_\ell + \dim B^\ell = \dim C_\ell = \dim B_\ell + \dim Z^\ell,$$

from which we obtain

$$\dim Z_\ell - \dim B_\ell = \dim Z^\ell - \dim B^\ell$$

or $\dim H_\ell = \dim H^\ell$. $\qquad\square$

Combining Theorems 6.8 and 6.9 enables us to conclude the following.

Corollary 6.10

$$H_{\mathrm{dR}}^\ell(M) \cong H_\ell(M).$$

From Theorem 6.2 and Corollary 6.10 we get the following.

Corollary 6.11 (de Rham triangulated version)   *Let K be a smooth triangulation of M. Then*

$$H_{\mathrm{dR}}^\ell(M) \cong H_\ell(K).$$

We already knew that de Rham cohomology was a topological invariant, but this theorem shows that we get the same topological information from de Rham cohomology as we get from simplicial homology, which is reassuring.[12]

For a fun application of homology and cohomology to electrical circuit theory, see Appendix G.

Additional exercises

6.10 Integration over the sphere Evaluate $\int_{S^2} \omega$, where ω is the (global) 2-form defined in Example 3.10. *Hint:* This exercise is actually a bit tricky, because of questions about integration domains and orientations. One way to do the integral would be to integrate ω_U over the patch U that covers everything except the north pole, because a single point is a set of measure zero and therefore irrelevant to the integral. Alternatively, you could integrate ω_U over U_0, the part of U corresponding to the southern hemisphere, and ω_V over V_0, the part of V corresponding to the northern hemisphere, and add the results together (ignoring the fact that you are integrating twice over the equator, because it is also a set of measure zero). Either way, you must choose an orientation of S^2. If you decide to integrate over both U_0 and V_0, you must make sure that you choose the same orientation on each patch. Answer: $\pm\pi$, depending on the orientation chosen. Compare this result with that of Exercise 3.40.

6.11 Integration over the torus Let $dx \wedge dy$ be the global 2-form on the torus defined in terms of the global 1-forms dx and dy. (See Exercise 3.23.) Evaluate $\int_{T^2} dx \wedge dy$. *Remark:* This is not the area of the familiar torus in $\mathbb{R}^3$ because, as we defined it, our torus does not actually live in $\mathbb{R}^3$.

6.12 Integration using stokes' theorem Let $V = \{(x, y, z) \in \mathbb{R}^3 : x^2 + y^2 = (az/h)^2, -1 \leq x, y \leq 1, 0 \leq z \leq h\}$ be a three-dimensional right circular (upside-down) cone of height h and base radius a. Let $\sigma = x\,dy \wedge dz + y\,dz \wedge dx$.
(a) Evaluate $\int_{\partial V} \sigma$. *Hint:* Parameterize, and watch out for orientations.
(b) Use Stokes' theorem to find the volume of the cone.

6.13 Stokes' theorem illustrated Let $\omega = y\,dx + z\,dy + x\,dz$ and let Σ be the disk obtained as the intersection of the plane $z = \sqrt{3}a/2$ with the sphere $x^2 + y^2 + z^2 = a^2$. Verify that Stokes' theorem holds in this case. That is, show that $\int_{\partial\Sigma} \omega = \int_{\Sigma} d\omega$.

---

[12] For a direct proof of Corollary 6.11 see e.g. [64], Section 3.4, or [77], Section 6.2. For other proofs of de Rham's theorem see e.g. [51], Chapter 16, or [88], Chapter 5.

6.14 A cohomological exercise Let M be a compact oriented surface without boundary, and let $p \in M$. Let $C \in M$ be a small circle surrounding p, and let $\iota : C \to M - \{p\}$ be the inclusion map. Show that the induced map $\iota^* : H^1_{dR}(M - \{p\}) \to H^1_{dR}(C)$ is zero. *Hint:* Stretch the hole caused by the puncture at p until it fills the inside of C and use Stokes' theorem. At one point you will need to use the fact that $\int_C : H^1_{dR}(C) \to \mathbb{R}$ is an isomorphism, a fact that you can prove either directly or else by wielding the hammer of de Rham's theorem.

6.15 Boundary manifolds Prove that the boundary of a manifold with boundary is a manifold without boundary.

7

Vector bundles

The Editor is convinced that the notion of a connection in a vector bundle will soon find its way into a class on advanced calculus, as it is a fundamental notion and its applications are wide-spread. His chapter "Vector Bundles with a Connection" hopefully will show that it is basically an elementary concept.

S. S. Chern[†]

7.1 The definitions

Let M be a differentiable manifold. Crudely put, a vector bundle is just a collection, or bundle, of vector spaces, one for each point p of M, that vary smoothly as p varies. For example, if M is a differentiable manifold and if T_pM is the tangent space to M at a point p then the union TM of all the T_pM as p varies over M is a vector bundle called the tangent bundle of M. Similarly, the cotangent bundle T^*M of M is just the union of all the cotangent spaces T_p^*M as p varies over M.

Essentially, a vector bundle over M is a space E that looks locally like the Cartesian product of M with a vector space. As we are primarily concerned with the local properties of vector bundles, we do not lose much by limiting ourselves to product bundles. But, for those who insist on knowing all the gory details, the official definition is provided here. The beginner should skim over the next two paragraphs and revisit them only as needed.

A **vector bundle** (E, M, Y, π) over a manifold M is a manifold E and a projection map $\pi : E \to M$ satisfying the following three axioms. (The axioms may make a bit more sense if you refer to Figure 7.1.)

[†] S. S. Chern, Vector bundles with a connection, in *Global Differential Geometry*, Studies in Mathematics Vol. 27, ed. S. S. Chern (Mathematical Association of America, Washington, DC, 1989), p. ii. Copyright The Mathematical Association of America 2013. All rights reserved. Reprinted by permission.

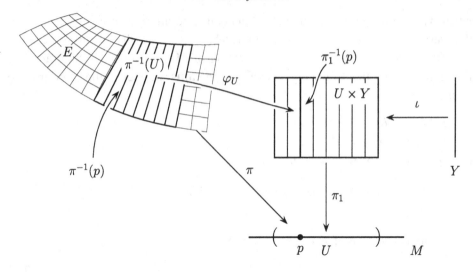

Figure 7.1 Vector bundle maps.

(B1) *All the fibers are isomorphic.* Each **fiber** $\pi^{-1}(p)$, $p \in M$ is isomorphic to a fixed vector space Y of dimension m (m is called the **rank** of the vector bundle).

(B2) *The manifold E is locally a product.* For all $p \in M$ there is a neighborhood U containing p and a diffeomorphism

$$\varphi_U : \pi^{-1}(U) \to U \times Y \tag{7.1}$$

(called a **local trivialization** of E over U).

(B3) *The local trivialization φ_U carries fibers to fibers and the restriction is linear,* i.e.,

$$\varphi_U : \pi^{-1}(p) \to \pi_1^{-1}(p) \tag{7.2}$$

is linear for all p. The map $\pi_1 : U \times Y \to U$ is a projection onto the first factor: $(p, g) \mapsto p$.

The manifold E is called the **total space** or the **bundle space**, the manifold M is called the **base space**, and π is called the **bundle projection** of the vector bundle (E, M, Y, π). For brevity, one usually writes $E \to M$ to denote the bundle (E, M, Y, π) if the projection map π is obvious or understood.

Suppose that U and V are two overlapping neighborhoods of p on M. Then we get a smooth map

$$\varphi_V \circ \varphi_U^{-1} : (U \cap V) \times Y \to (U \cap V) \times Y. \tag{7.3}$$

By virtue of (B2) we have

$$(\varphi_V \circ \varphi_U^{-1})(p, y_U) = (p, y_V), \tag{7.4}$$

where $y_V = g_{VU}(p)y_U$ for some nondegenerate endomorphism $g_{VU}(p) : Y \to Y$ that varies smoothly with p. The endomorphism g_{VU} is called a **bundle transition function**. If $\{U, V, W, \ldots\}$ is a covering of M, it follows that the transition functions satisfy the following **cocycle conditions** whenever the overlaps are nontrivial:[1]

(1) $g_{UU}(p) = id$,
(2) $g_{UV}(p)g_{VU}(p) = id$,
(3) $g_{UV}(p)g_{VW}(p)g_{WU}(p) = id$.

> **EXERCISE 7.1** Prove that the transition functions satisfy properties (1), (2), and (3) above.

A (smooth) **section** of E is a (smooth) map

$$s : M \to E \tag{7.5}$$

such that

$$\pi \circ s = id. \tag{7.6}$$

The definition implies that s can be viewed as a vector-valued map on M. In particular, the space $\Gamma(E)$ of all sections of E inherits a natural vector space structure. (Vector addition and scalar multiplication are defined pointwise.)

> **Example 7.1** The simplest vector bundle is the **trivial bundle**, which is *globally* a product: $E = M \times Y$.

It is useful to have a criterion for triviality. This is provided by the next result. A (smooth) **basis of sections** over a subset $U \subseteq M$ is a collection $\{e_1, \ldots, e_n\}$ of n sections that is linearly independent at each point of U. Such a smooth basis of sections is just what we have called a frame field. As E is locally a product, we can always find a basis of sections for any sufficiently small neighborhood of M. But what about larger neighborhoods?

Theorem 7.1 *A vector bundle E is trivial if and only if it has a global smooth basis of sections.*

Proof Assume E to be trivial. Then there is a *global* trivialization map $\varphi : E \to M \times Y$ which is smooth and preserves the linear structure on fibers. Pick a basis $\{f_i\}$ for Y, and define $e_i(p) := \varphi^{-1}(p, f_i)$. This gives n smooth sections on E, and they

---

[1] There is a reverse theorem that says roughly that any product structure glued together with transition functions satisfying the cocycle conditions defines a vector bundle.

form a basis for $\pi^{-1}(p)$ because φ is a linear isomorphism of fibers. Conversely, suppose that $\{e_i\}$ is a basis of sections over M. Then any element of $\pi^{-1}(p)$ can be written $v_p = \sum e_i(p)v^i$. Pick a basis $\{f_i\} \in Y$ and define $v = \sum f_i v^i$. Then we get a map $\psi : E \to M \times Y$ given by $\psi(v_p) = (p, v)$. This is smooth and linear on fibers by construction. $\qquad\square$

EXERCISE 7.2 Show that one triviality implies another: if E is trivial, we can choose $g_{UV} = id$ for all U and V.

Remark It can be shown (e.g. [38], Corollary 1.8) that a vector bundle over a contractible manifold is always trivial.

Remark If TM is trivial then M is called **parallelizable**. Most manifolds are not parallelizable. For example it is known that, of all the n-spheres, only S^0, S^1, S^3 and S^7 are parallelizable, a fact that is intimately related to the fact that there are only four normed division algebras: the real numbers, the complex numbers, the quaternions, and the octonions. (See e.g. [20].)

EXERCISE 7.3 (For readers who have completed the problems on Lie groups in Chapter 4.) Show that every Lie group is parallelizable, i.e., $TG = G \times T_eG$.

Of course, not all bundles are trivial.

Example 7.2 The Möbius band B that we encountered in Example 3.7 can be viewed as a nontrivial smooth vector bundle over the circle with fiber $\mathbb{R}$. The coordinate charts used to define the manifold structure can be modified slightly to provide a local trivialization. Recall that B was defined to be the quotient $\mathbb{R}^2/\sim$, where $(x, y) \sim (x + 1, -y)$. The circle S^1 is just $\mathbb{R}/\sim$ where $x \sim x + 1$. Let $[x, y]$ denote the equivalence class of (x, y) in B as before, and let $[x]$ denote the equivalence class of x on the circle. Then the projection map $\pi : B \to S^1$ is given by $[x, y] \mapsto [x]$.

Consider the open cover of S^1 given by

$$U_1 = \{[x] : 0 < x < 2/3\},$$
$$U_2 = \{[x] : 1/3 < x < 1\},$$
$$U_3 = \{[x] : 2/3 < x < 4/3\}.$$

Define local trivializations $\varphi_i : \pi^{-1}(U_i) \to U_i \times \mathbb{R}$ by $\varphi_i([x, y]) = ([x], y)$, where (x, y) is the unique representative of $[x, y]$ in the parameter range corresponding to U_i. For example, $\varphi_1([1/3, y]) = ([1/3], y)$ but $\varphi_1([4/3, y]) = \varphi_1([1/3, -y]) = ([1/3], -y)$.

Now consider the transition maps. Following in our old footsteps, we find that $(\varphi_1 \circ \varphi_2^{-1})([x], y) = ([x], y) = (\varphi_2 \circ \varphi_3^{-1})([x], y)$ but that $(\varphi_3 \circ \varphi_1^{-1})([x], y) = ([x], -y)$. Thus the bundle transition functions are $g_{12} = g_{23} = 1$ and $g_{31} = -1$.

(Observe that they satisfy the cocycle conditions.) This confirms the obvious fact that B is nontrivial.

Example 7.3   The archetypical example of a vector bundle is the **tangent bundle** $E = TM$ of a manifold M, which consists of the union of all the tangent spaces T_pM together with the natural projection $\pi : TM \to M$ that sends $T_pM \mapsto p$. A point of TM may be thought of as a pair (p, Y_p) where $p \in M$ is a point and $Y_p \in T_pM$ is a tangent vector. Since $\dim T_pM = \dim M = n$, the rank of TM is n. Sections of TM are precisely vector fields on M. In general, TM is nontrivial.

If $\dim M = n$, TM is a $2n$-dimensional manifold whose bundle transition functions are precisely the Jacobian matrices relating two local charts of M. To see this, first observe that we may cover TM by the countable collection of open sets $\pi^{-1}(U_i)$, where $\{U_i\}$ is a countable cover of M. Next, suppose that q and r are distinct points in TM. If on the one hand $\pi(q) \neq \pi(r)$ then choose a neighborhood U of $\pi(q)$ and a neighborhood V of $\pi(r)$ that are disjoint. (This is possible because M is Hausdorff.) Then $\pi^{-1}(U)$ and $\pi^{-1}(V)$ are also disjoint. On the other hand, suppose that $\pi(q) = \pi(r) = p$. Then q and r must be two distinct elements of T_pM, so they have disjoint neighborhoods U_q and U_r in T_pM (because vector spaces are Hausdorff). If U is a neighborhood of p in M then $\pi^{-1}(U) \cap U_q$ and $\pi^{-1}(U) \cap U_r$ are disjoint open sets in TM. It follows that TM is Hausdorff.

We coordinatize TM as follows. Pick a point $q \in TM$ and set $p := \pi(q)$. Let (U, ψ_U) and (V, ψ_V) be overlapping charts of M containing p, with corresponding local coordinates $u^1, \ldots, u^n$ and $v^1, \ldots, v^n$, respectively. Let Y be a vector field on U and set $Y_U^i = Y(u^i)$. Define $\chi_U : \pi^{-1}(U) \to \mathbb{R}^n \times \mathbb{R}^n$ by $q \mapsto (u^1(p), \ldots, u^n(p), Y_U^1(p), \ldots, Y_U^n(p))$. Define χ_V similarly. Writing

$$Y_V^i = \sum_j \frac{\partial v^i}{\partial u^j} Y_U^j = \sum_j (g_{VU})_{ij} Y_U^j, \tag{7.7}$$

we see that the manifold transition functions are

$$\chi_V \circ \chi_U^{-1} = (\psi_V \circ \psi_U^{-1}, g_{VU}).$$

As these are diffeomorphisms, TM is indeed a $2n$-dimensional manifold, as claimed. Moreover, the bundle transition functions g_{UV} are the Jacobian matrices of the coordinate charts of M, as promised.

Example 7.4   Another important example is $T = T^*M$, the **cotangent bundle** on M, whose fibers are just the cotangent spaces to M. The transition functions are just the inverses of the Jacobian matrices introduced in Example 7.3. The rank of T^*M is the same as that of TM. Sections of T^*M are 1-form fields on M.

Example 7.5   Let $E \to M$ and $F \to M$ be two vector bundles over M. For simplicity, write E_p to denote the fiber over p. Then the **tensor product bundle**

$(E \otimes F) \to M$ is the vector bundle over M whose fibers are just the tensor products of the individual fibers $(E \otimes F)_p = E_p \otimes F_p$. If the transition functions of $E \to M$ are g_{UV} and the transition functions of $F \to M$ are h_{UV}, the transition functions of $(E \otimes F) \to M$ are just $g_{UV} \otimes h_{UV}$. By iterating this construction we can obtain the tensor product bundle

$$\underbrace{TM \otimes \cdots \otimes TM}_{r \text{ times}} \otimes \underbrace{T^*M \otimes \cdots \otimes T^*M}_{s \text{ times}} \to M,$$

sections of which are just tensor fields of type (r, s) on M.

7.2 Connections

Now we want to introduce an important device that will allow us to differentiate sections on vector bundles in a useful way. Consider for a moment the tensor product bundle $TM^{\otimes r} \otimes TM^{*\otimes s}$ introduced in Example 7.5. A section of this bundle is just a tensor field, and ideally we want some kind of derivative operator that will act on a tensor field and give back another tensor field. As we have seen, one way to do this is to use the Lie derivative $\mathcal{L}_X$. However, although the Lie derivative is useful for some purposes, it suffers from a serious flaw: it is in some sense *nonlocal*. It depends, not just on X at a point, but on X in the neighborhood of the point.

It turns out to be much more useful to have an analogue of the ordinary directional derivative that depends only on the direction of X at a point. But, as we saw in Exercise 3.22, the ordinary directional derivative itself will not do because typically the partial derivative of a tensor is not a tensor. As we have already exhausted all the reasonable definitions of differentiation that use only the manifold structure itself, we are compelled to introduce some additional structure to yield the right sort of derivative operator. On the tensor product bundle we have been discussing this is accomplished by means of something called a "covariant derivative", which is a connection on a vector bundle.

Although one does not need the notion of a tensor product bundle to discuss covariant derivatives, these are not the only bundles and connections that are important in mathematics and physics. For example, the standard model of particle physics is based upon the notion of gauge fields, which can be viewed as a connection on a special kind of vector bundle called an associated bundle.[2] It therefore behoves us to broaden our perspective slightly and discuss the general notion of a vector bundle connection from the start.

---

[2] Unfortunately, a discussion of gauge fields from the point of view of modern differential geometry would require a chapter in itself. Instead, we refer the reader to [7] or [67]. See also Exercise 7.10.

A (**linear**) **connection** on a vector bundle E is essentially the simplest kind of operator that allows us to differentiate sections.[3] More precisely, a connection is a map[4]

$$D : \Gamma(E) \to \Gamma(E \otimes T^*M) \qquad (7.8)$$

satisfying the following two properties. For $s, s_1, s_2 \in \Gamma(E)$ and f a smooth function on M, D obeys

(D1) (linearity) $D(s_1 + s_2) = Ds_1 + Ds_2,$ and
(D2) (Leibniz rule) $D(sf) = Ds \cdot f + s \otimes df.$

Surprisingly, this simple definition has powerful consequences although it takes a while to see them.

When E is the tensor product bundle $TM^{\otimes r} \otimes TM^{*\otimes s}$ we want the connection to satisfy certain compatibility properties beyond the basic ones. This gives rise to what is usually called a **covariant derivative**, for which one uses slightly different notation. As a section of E is a tensor field, we write T instead of s. Then DT is a tensor valued 1-form and, for any vector field X, we write

$$\nabla_X T := i_X DT, \qquad (7.9)$$

where i_X is the usual interior product. The result is another tensor field, the covariant derivative of T in the direction X. The axioms for a connection imply that for a function f, vector fields X and Y, and tensor fields S and T the covariant derivative operator satisfies

(C1) (linearity) $\nabla_X(S + T) = \nabla_X S + \nabla_X T,$ and
(C2) (Leibniz rule) $\nabla_X(fT) = f\nabla_X T + (Xf)T.$

In addition, by virtue of the definition of the covariant derivative and the function linearity of the interior product, we also have, for functions f and g,

(C3) (function linearity on subscript) $\nabla_{fX+gY} T = f\nabla_X T + g\nabla_Y T.$

It is the property (C3) that distinguishes the covariant derivative from the Lie derivative and which allows us to build tensorial objects from the covariant derivative.

We will make further demands of a covariant derivative on a tensor product bundle, however. In particular, by analogy with the Lie derivative, we insist that

---

[3] The word "connection" is meant to suggest a device that "connects" the fibers above different points of a manifold. One can define more general bundles and more general connections on them, but we will not do so here; see e.g. [10] or [47].

[4] Elements of $\Gamma(E \otimes T^*M)$ can be thought of as "section-valued 1-forms" (or "1-form valued sections"). Viewing a 1-form as a linear functional on vector fields via the natural dual pairing, one can think of an element of $\Gamma(E \otimes T^*M)$ as an object that eats vector fields and spits out sections.

it be "natural" with respect to dual pairings and tensor products. Thus, for any function f and 1-form ω, we add the axioms

(C4) (reduction to ordinary derivative on functions) $\nabla_X f = Xf$,
(C5) (compatibility with dual pairing) $\nabla_X \langle Y, \omega \rangle = \langle \nabla_X Y, \omega \rangle + \langle Y, \nabla_X \omega \rangle$, and
(C6) (Leibniz rule II) $\nabla_X (S \otimes T) = \nabla_X S \otimes T + S \otimes \nabla_X T$.

In Chapter 8 we will play around further with covariant derivatives. For now we content ourselves with the simple but important observation that properties (C4) to (C6) ensure that the covariant derivative on $TM^{\otimes r} \otimes TM^{*\otimes s}$ is completely determined by the choice of covariant derivative on TM. (See Exercise 8.31.)

7.3 Cartan's moving frames and connection forms

Let us temporarily restrict our attention to a single neighborhood U of M where E is locally trivial. Because the bundle over U is trivial, Theorem 7.1 assures us that we can find a smooth basis of sections for the fiber $\pi^{-1}(p)$ at each point.

The connection maps each e_i to a section-valued 1-form. Because the e_i form a basis for each fiber, De_i must be a linear combination of the e_i with 1-form coefficients, written[5]

$$De_i = \sum_j e_j \otimes \omega^j{}_i. \tag{7.10}$$

The elements $\omega^j{}_i$ are 1-forms, and the entire collection of m^2 1-forms constitutes a matrix $\omega = (\omega^j{}_i)$, called the **connection matrix** relative to the frame field $\{e_i\}$. For ease of writing, we introduce the following matrix notation (cf. Section 1.8):

$$e = (e_1\ e_2\ \cdots\ e_m) \qquad \text{and} \qquad \omega = (\omega^j{}_i). \tag{7.11}$$

(In other words, e is a $1 \times m$ matrix of sections while ω is a matrix of 1-forms.) Following Cartan, we then write equations (7.10) as[6]

$$De = e\omega. \tag{7.12}$$

The connection matrix determines the connection completely. To see this, let

$$s = \sum_i e_i \sigma^i \tag{7.13}$$

---

[5] Note that some authors write $De_i = \sum_j \omega_i{}^j \otimes e_j$ instead. This has the unfortunate consequence of flipping signs in some of the formulae that follow.

[6] In this equation the tensor product symbol is suppressed. It should really be written $De = e \otimes \omega$ but people are lazy, so you may as well get used to it now. You just have to remember that the tensor product symbol is still there although invisible, otherwise you may be tempted to think that, when $E = TM$, e acts on ω as a derivation, which would be nonsense. (Recall the discussion at the end of Section 3.7.)

be an arbitrary section. (Each σ^i is a smooth function on M.) Then, by properties (D1) and (D2),

$$Ds = \sum_j D(e_j \sigma^j)$$

$$= \sum_j De_j \cdot \sigma^j + e_j \otimes d\sigma^j$$

$$= \sum_i e_i \otimes \left(\sum_j \omega^i{}_j \sigma^j + d\sigma^i \right). \qquad (7.14)$$

In other words, if we know the connection matrix then we know how D acts on arbitrary sections and vice versa. In shorthand notation (7.13) can be written

$$s = e\sigma, \qquad (7.15)$$

where

$$\sigma = \begin{pmatrix} \sigma^1 \\ \vdots \\ \sigma^m \end{pmatrix} \qquad (7.16)$$

is an $m \times 1$ matrix of smooth functions on M. Similarly, (7.14) becomes

$$Ds = e\omega\sigma + ed\sigma. \qquad (7.17)$$

7.4 Curvature forms and the Bianchi identity

Define a matrix of 2-forms by the shorthand formula

$$\Omega = d\omega + \omega \wedge \omega, \qquad (7.18)$$

which means

$$\Omega^i{}_j = d\omega^i{}_j + \sum_k \omega^i{}_k \wedge \omega^k{}_j. \qquad (7.19)$$

(Observe that the wedge product of the matrix of 1-forms with itself does not vanish!) The matrix elements of Ω are called **curvature 2-forms**. The significance of the term "curvature" will not become apparent until Chapter 8.

Differentiating (7.18) using the ordinary derivative operator d and using (7.18) again gives

$$d\Omega = d^2\omega + d\omega \wedge \omega - \omega \wedge d\omega = (\Omega - \omega \wedge \omega) \wedge \omega - \omega \wedge (\Omega - \omega \wedge \omega).$$

Simplifying, we get the **Bianchi identity**

$$d\Omega = \Omega \wedge \omega - \omega \wedge \Omega. \qquad (7.20)$$

Equation (7.12) and (7.18) are known as **Cartan's equations of structure**, and (7.20) is sometimes called an integrability condition.[7]

7.5 Change of basis

Suppose that we had chosen a different frame field on U, related to the old one by a nondegenerate matrix $A = (a^i{}_j)$ of smooth functions. Then

$$e'_i = \sum_j e_j a^j{}_i \tag{7.21}$$

or, in our matrix notation,

$$e' = eA. \tag{7.22}$$

In the new basis e' the connection D is represented by a new matrix ω', where

$$De' = e'\omega', \tag{7.23}$$

meaning, of course,

$$De'_i = \sum_j e'_j \otimes \omega'^j{}_i. \tag{7.24}$$

By virtue of (7.12), (7.22), (7.23), and property (D2), we have

$$eA\omega' = e'\omega' = De' = D(eA) = (De)A + e\,dA = e\omega A + e\,dA. \tag{7.25}$$

As the basis e is arbitrary, we can strip it from both sides to conclude that

$$A\omega' = dA + \omega A, \tag{7.26}$$

---

[7] The phrase "integrability condition" is applied rather loosely here. Strictly speaking an integrability condition is the statement that a specific Pfaffian system is a differential ideal (see the discussion of Frobenius' theorem in Appendix F.2). But the phrase is also used informally to refer to a situation in which differentiating a partial differential equation gives rise to a nontrivial necessary condition for the solution of the original equation. (This use of the word "integrable" arises by analogy with the Poincaré lemma, which guarantees that, in any local neighborhood, $d\lambda = 0$ implies that $\lambda = d\eta$ for some η; in other words, $d\lambda = 0$ is the integrability condition ensuring that λ can be "integrated".) In our case, we begin with the structure equation $d\omega + \omega \wedge \omega = \Omega$, which is a set of first-order differential equations for the connection 1-forms $\omega^i{}_j$ in terms of the curvature 2-forms $\Omega^i{}_j$. Differentiating this equation yields the Bianchi identity, which provides necessary (but, as can be shown, not sufficient) zeroth-order conditions that the connection 1-forms $\omega^i{}_j$ must satisfy in order to constitute a solution of the structure equation. (My thanks go to Robert Bryant for a discussion on this point.) It is interesting to observe that there are no additional integrability equations here, because applying the differential operator d to the Bianchi identity gives $0 = 0$ identically. (The Bianchi identities play an important role in gauge theory and in Einstein's theory of general relativity, where they can be interpreted as generalized conservation laws.)

or, left multiplying by A^{-1},

$$\omega' = A^{-1}dA + A^{-1}\omega A. \tag{7.27}$$

This important formula illustrates how the connection matrix changes under a change of frame field. By patching together connection matrices via this formula and a partition of unity it can be shown that connections exist.[8]

Under the change of basis, the curvature 2-form matrix changes in a particularly nice way. Differentiate the left-hand side of (7.26) and use (7.26) again to get

$$dA \wedge \omega' + Ad\omega' = (A\omega' - \omega A) \wedge \omega' + Ad\omega'$$
$$= A(d\omega' + \omega' \wedge \omega') - \omega \wedge A\omega'$$
$$= A\Omega' - \omega \wedge (dA + \omega A), \tag{7.28}$$

where

$$\Omega' = d\omega' + \omega' \wedge \omega'. \tag{7.29}$$

Differentiate the right-hand side of (7.26) to get

$$d\omega A - \omega \wedge dA. \tag{7.30}$$

Now equate (7.28) and (7.30) to obtain the transformation law

$$A\Omega' = \Omega A, \tag{7.31}$$

or

$$\Omega' = A^{-1}\Omega A. \tag{7.32}$$

7.6 The curvature matrix and the curvature operator

Let X and Y be two vector fields on M. From the curvature 2-forms we obtain the **curvature matrix** by taking inner products:

$$\Omega(X, Y) = i_Y i_X \Omega. \tag{7.33}$$

Let $s = e\sigma$ and $X, Y \in \Gamma(TM)$. Define the **curvature operator** $R(X, Y)$: $\Gamma(TM) \to \Gamma(TM)$ by

$$R(X, Y)s = e\Omega(X, Y)\sigma. \tag{7.34}$$

---

[8] For a proof see [19], p. 106. For the existence of a specific connection, see Theorem 8.4.

Note that it is linear by construction. The question, though, is whether it is well defined. The problem is that it appears to depend on the basis e. Suppose instead we had $s = e'\sigma'$. Would we still get the same image? First we note by (7.22) that $s = e\sigma = e'\sigma' = eA\sigma'$ implies that $\sigma = A\sigma'$ or $\sigma' = A^{-1}\sigma$. Thus, using (7.32),

$$e'\Omega'\sigma' = (eA)(A^{-1}\Omega A)(A^{-1}\sigma) = e\Omega\sigma, \tag{7.35}$$

so we do get the same image after all.

Although (7.34) is elegant, in order to do certain computations it is useful to have an expression for it in terms of covariant derivatives.

Theorem 7.2 *As an operator on vector fields,*

$$R(X, Y) = \nabla_X\nabla_Y - \nabla_Y\nabla_X - \nabla_{[X,Y]}. \tag{7.36}$$

Proof Let $s = e\sigma$ be a section. Then from (7.9) and (7.17) we have

$$\nabla_X s = i_X Ds = i_X D(e\sigma) = i_X(e\omega\sigma + ed\sigma) = e\omega(X)\sigma + eX\sigma, \tag{7.37}$$

where $\omega(X)$ and $X\sigma$ are defined entrywise. (That is, $\omega(X)^i{}_j = \omega^i{}_j(X)$ and $(X\sigma)^j = X\sigma^j$, the latter meaning that X acts on the function σ^j as a derivation.) Similarly, we have

$$\nabla_Y s = e\omega(Y)\sigma + eY\sigma = e[\omega(Y)\sigma + Y\sigma]. \tag{7.38}$$

It follows that

$$\begin{aligned}
\nabla_X\nabla_Y s &= e\omega(X)[\omega(Y)\sigma + Y\sigma] + eX[\omega(Y)\sigma + Y\sigma] \\
&= e\omega(X)\omega(Y)\sigma + e\omega(X)Y\sigma \\
&\quad + e[X\omega(Y)]\sigma + e\omega(Y)X\sigma + eXY\sigma \\
&= e\{\omega(X)\omega(Y) + \omega(X)Y + [X\omega(Y)] + \omega(Y)X + XY\}\sigma. \tag{7.39}
\end{aligned}$$

Hence

$$\begin{aligned}
\nabla_X\nabla_Y &- \nabla_Y\nabla_X - \nabla_{[X,Y]} \\
&= e\{\omega(X)\omega(Y) + \omega(X)Y + [X\omega(Y)] + \omega(Y)X + XY \\
&\quad - \omega(Y)\omega(X) + \omega(Y)X + [Y\omega(X)] + \omega(X)Y + YX \\
&\quad - \omega([X, Y]) - [X, Y]\}\sigma \\
&= e\{\omega(X)\omega(Y) - \omega(Y)\omega(X) \\
&\quad + [X\omega(Y)] - [Y\omega(X)] - \omega([X, Y])\}\sigma. \tag{7.40}
\end{aligned}$$

But, appealing to (3.123),

$$\begin{aligned}
\Omega(X, Y) &= (d\omega + \omega \wedge \omega)(X, Y) \\
&= X\omega(Y) - Y\omega(X) - \omega([X, Y]) \\
&\quad + \omega(X)\omega(Y) - \omega(Y)\omega(X).
\end{aligned} \tag{7.41}$$

Comparison of (7.40) and (7.41), together with (7.34), yields the theorem.   □

Informally, one hears the statement that curvature is a measure of the degree to which covariant derivatives (as opposed to simple partial derivatives) fail to commute, but the last term in (7.36) shows that this statement is not precisely true. Of course, it *is* true if one considers covariant derivatives along two distinct coordinate directions, because in that case $X = \partial_i$ and $Y = \partial_j$ so the last term in (7.36) disappears.

Recall that the interior product operator is function linear. It follows immediately from (7.33) and (7.34) that the curvature operator is function linear as well. Specifically, for any functions f, g, and h and vector fields X, Y, and Z,

$$R(fX, gY)hZ = fghR(X, Y)Z. \tag{7.42}$$

We are therefore justified in referring to $R(X, Y)$ as the **curvature tensor**.

EXERCISE 7.4 Prove (7.42) directly using the expression (7.36) together with the properties of the covariant derivative operator. (The point of this exercise is just to convince some doubters that an abstract definition can sometimes be more useful than a concrete one.)

Additional exercises

7.5 The unit tangent bundle of the 2-sphere Show that the bundle space of the unit tangent bundle of the 2-sphere S^2 is homeomorphic to $SO(3)$. *Remark:* It is actually diffeomorphic, but you need not show this. *Hint:* As usual, view S^2 as a submanifold of $\mathbb{R}^3$. Let $x \in S^2$ and let $y \in T_x S^2$ be a unit tangent vector. Consider $z = x \times y$, where $\times$ is the ordinary cross product.

7.6 Parallelizability of the 3-sphere Show that the tangent bundle of the 3-sphere S^3 is trivial by finding three linearly independent nowhere vanishing tangent vector fields. *Hint:* Define S^3, as usual, by

$$\{(x^1, x^2, x^3, x^4) \in \mathbb{R}^4 : (x^1)^2 + (x^2)^2 + (x^3)^2 + (x^4)^2 = 1\}.$$

Identify a tangent vector field with its components in the standard basis $\partial/\partial x^i$. Show that $(-x^2, x^1, -x^4, x^3)$, $(-x^3, x^4, x^1, -x^2)$, and $(-x^4, -x^3, x^2, x^1)$ do the trick.

7.7 Direct sum of vector bundles Let $E \to M$ and $F \to M$ be two vector bundles over M. The **Whitney sum** $E \oplus F \to M$ is the vector bundle whose fibers are the direct sums of the individual fibers $(E \oplus F)_p = E_p \oplus F_p$. Show this indeed a vector bundle, and find the transition functions.

7.8 Affine connections Let D and $\tilde{D}$ be two connections on a vector bundle E with connection matrices ω and $\tilde{\omega}$, respectively. Let D_t be an **affine combination** of D and $\tilde{D}$, so that

$$D_t = (1-t)D + t\tilde{D}, \qquad t \in \mathbb{R}.$$

Show that D_t is again a connection, with connection matrix

$$\omega_t = \omega + t\eta,$$

where $\eta := \tilde{\omega} - \omega$. This exercise explains why D is often referred to as an **"affine connection"** (although this terminology is sometimes reserved for a connection on the tangent bundle).

7.9 Characteristic classes Let E be a vector bundle over a manifold M with curvature form Ω.

(a) Show that tr Ω^k is invariant under a frame change, where[9]

$$\Omega^k := \underbrace{\Omega \wedge \cdots \wedge \Omega}_{k \text{ times}}$$

and "tr" denotes the trace. It follows that $b_k := \text{tr } \Omega^k$ is a globally defined $2k$-form.

(b) Show that b_k is closed. *Hint:* You will need to use the Bianchi identity, the properties of the trace, and the commutation properties of forms.

(c) Because b_k is closed, it defines a cohomology class of the vector bundle E over M.[10] Show that b_k is an invariant of E by showing that it does not depend on the connection used to define it. *Hint:* In the notation of Section 7.5 let Ω, $\tilde{\Omega}$, and Ω_t be the curvature forms associated to the connections ω, $\tilde{\omega}$, and ω_t, respectively. We want to show that tr Ω^k

---

[9] Written out explicitly, we have

$$(\Omega^k)^i{}_j = \sum_{i_1,\ldots,i_{k-1}} \Omega^i{}_{i_1} \wedge \Omega^{i_1}{}_{i_2} \wedge \cdots \wedge \Omega^{i_{k-1}}{}_j.$$

[10] These cohomology classes are called **characteristic classes**, and they play an important role in geometry and topology. (See Section 9.3 below.) On a complex vector bundle these classes are related, by Newton's identities, to the celebrated Chern classes of M (see [18] or [64], Chapter 5).

and tr $\widetilde{\Omega}^k$ differ by an exact form. So, show that $\alpha := \mathrm{tr}(\eta \wedge \Omega_t^{k-1})$ is a globally defined form on M and that $d(\mathrm{tr}\,\Omega_t^k)/dt = k\,d\alpha$. Integrate this to obtain the desired conclusion.

7.10 Gauge theory The nucleus of an atom consists of protons and neutrons. Experimentally one observes that, aside from the obvious difference of their electric charges, protons and neutrons behave pretty much alike inside a nucleus. For this reason, Werner Heisenberg proposed that the two types of particles could be thought of as two faces of one type of particle called a **nucleon**. By analogy with spin-1/2 particles, which can be either up or down relative to some axis, Eugene Wigner suggested that the nucleon could be thought of as possessing a new kind of property called **isospin** and that the two isospin states are related by a unitary transformation in a two-dimensional complex vector space called isospin space. To account for the observations, the law governing the interactions of two nucleons must therefore be *form invariant*, or *covariant*, under the action of the Lie group $SU(2)$, meaning that the law looks the same regardless of how the isospin state is transformed.

But one person's proton is another person's neutron. To allow for the possibility that different observers might disagree on the identity of a given nucleon, we permit the symmetry to be *local*, meaning that we are free to transform the nucleon state by different group elements at different points. To retain the idea of continuity (actually, differentiability) we demand that the group elements vary smoothly over the manifold. In other words, we *gauge* the $SU(2)$ symmetry.[11]

As you may have surmised by now, all this is intimately related to the theory of vector bundles. The nucleon field is modeled by a section of a vector bundle over a four-dimensional Lorentzian manifold (see Chapter 8), where the fibers are two-dimensional (complex) vector spaces, each of which carries the same representation of $SU(2)$. As with all laws of physics, a differential equation governs the time evolution of the nucleon field, and the requirement that it be form invariant under local group transformations means that we must use covariant derivatives instead of ordinary ones. As we have seen, the introduction of a covariant derivative leads to the appearance of a connection form (traditionally denoted by the letter "A"). This form is called a gauge potential, and it turns out here

---

[11] In the period 1918–19 Hermann Weyl wrote three papers on the possibility of local scale invariance in physics. In the last of these he called his idea *Eichinvarianz*. This was initially translated as "calibration invariance" and later as "gauge invariance", probably because the English word "gauge" is employed to describe the different scales of such things as wires and railroad tracks.

to have a physical interpretation as the field that mediates the interaction between two proximate nucleons. It is now understood that nucleons are not fundamental particles but are instead composed of quarks. Generalizing the construction above leads to a gauge theory in which the matter fields are quarks and the gauge fields are gluons. The gauge group in that case is $SU(3)$. Other gauge theories describe other fields.

Formally, we may define a **gauge field theory** to be a vector bundle (E, M, V, π) over a smooth manifold M, where V is a vector space carrying a representation of a Lie group G (called the **gauge group**). Sections of (E, M, V, π) are called **matter fields**. A connection on the bundle gives rise to local connection forms A, called (local) **gauge potentials**,[12] obeying (7.10), (7.22), and (7.27), namely

$$De = eA, \qquad e' = eg, \qquad \text{and} \qquad A' = g^{-1} dg + g^{-1} Ag. \qquad (7.43)$$

The last two equations are referred to as **gauge transformations** in this context.[13]

(a) Let $\Psi = e\psi = \sum_a e_a \psi^a$ be a matter field. A gauge transformation is just a local change of basis, so it leaves Ψ invariant. From $\Psi' = e'\psi' = e\psi = \Psi$ and $e' = eg$ we get $\psi' = g^{-1}\psi$. This is what a physicist calls a gauge transformation of the matter field. Continuing in (awful) physics notation, (7.14) reads

$$D\psi^a = d\psi^a + A^a{}_b \psi^b, \qquad (7.44)$$

where $e_a(D\psi^a) := e_a(D\Psi)^a = D\Psi$. Given local coordinates $\{x_i\}$ on M we have $A^a{}_b = A_i{}^a{}_b dx^i$. Defining $(A_i)^a{}_b := A_i{}^a{}_b$ and taking the inner product of (7.44) with $\partial/\partial x_i$ gives

$$D_i \psi = \partial_i \psi + A_i \psi, \qquad (7.45)$$

where $D_i := i_{\partial/\partial x^i} D$.

---

[12] There are at least three different ways to view gauge potentials. The first two involve something called a **principal bundle**, which is beyond the scope of this book. (For details see [7] or [69].) The last approach, in terms of local connection forms, given here, is probably the easiest to understand. Mathematicians generally prefer the principal bundle approach, because it gives a very geometric way of understanding the meaning of a connection, while physicists generally prefer the local, more analytic, approach. Fortunately, all the approaches are ultimately equivalent.

[13] We are following traditional physics notation, so there is some unavoidable sloppiness. For example, we really ought to write $e' = e\rho(g)$, where ρ is the representation of G on V. But fixing the representation means that G can be thought of as a matrix Lie group, so often one conflates the group element g with the matrix $\rho(g)$ representing it. Note that $g^{-1}dg$ is a Maurer–Cartan form that naturally lives in the Lie algebra $\mathfrak{g}$ of G. Also, we may interpret the last term in (7.43) as $\mathrm{Ad}\,g(A)$, the adjoint action of the group on its Lie algebra. In this way we see that A', and therefore A, can be thought of as a Lie-algebra-valued 1-form.

Show that

$$D_i(g\psi) = g D_i'\psi, \tag{7.46}$$

where D_i' is defined using A_i'. This equation is the motivation for calling D_i a **gauge covariant derivative**.

(b) The **field strength** F is the curvature of the connection A. Show that the (matrix-valued) components of F in local coordinates are

$$F_{ij} = \partial_i A_j - \partial_j A_i - [A_i, A_j], \tag{7.47}$$

where $F = \frac{1}{2} F_{ij}\, dx^i \wedge dx^j$ and the bracket $[\cdot, \cdot]$ is defined by

$$[A, B] = AB - BA.$$

Show that (7.47) can also be written as

$$F_{ij} = [D_i, D_j]. \tag{7.48}$$

(c) Show that the Bianchi identity (7.20) is equivalent to the Jacobi identity

$$[D_i, [D_j, D_k]] + [D_j, [D_k, D_i]] + [D_k, [D_i, D_j]] = 0. \tag{7.49}$$

or, equivalently,

$$[D_i, F_{jk}] + [D_j, F_{ki}] + [D_k, F_{ij}] = 0. \tag{7.50}$$

Remark: Although we do not do so here (see instead [6] or [7]), one can define a covariant analogue of the ordinary differential d, often denoted d_A, in terms of which the Bianchi identity (7.50) can be written

$$d_A F = 0. \tag{7.51}$$

Moreover, when sources are present we also have

$$d_A \star F = \star J. \tag{7.52}$$

The resulting equations (7.51) and (7.52) are called the **Yang–Mills equations**. They are a nonabelian generalization of Maxwell's equations.

8

Geometric manifolds

Une géométrie ne peut pas être plus vraie qu'une autre; elle peut seulement être plus commode. (One geometry cannot be more true than another; it can only be more convenient.)

Henri Poincaré, La science et l'hypothèse

As we have seen, a differentiable or smooth manifold M is just a topological space on which we have the ability to differentiate stuff. This is adequate if one is interested only in differential topology, but to do more we need to introduce additional structure. This additional structure is a **geometry**. Basically, we want to be able to measure distances on M as well as determine the angles between vectors. Of course, the vectors don't live on M – they live on the tangent spaces to M. What we need is an inner product on the tangent spaces. But we want to be able to differentiate things, so we want the inner product to vary smoothly.

This leads us to the following definition. A **smooth inner product** or **metric** g on M is just a smooth map $p \mapsto g_p$, where g_p is an inner product on T_pM. A **geometric manifold** (M, g) is a manifold M equipped with a smooth inner product g. Two geometric manifolds (M, g) and (N, h) are considered equivalent if they are **isometric**, meaning there exists a diffeomorphism $\varphi : M \to N$ with $\varphi_* g = h$. This amounts to saying that lengths of vectors and angles between pairs of vectors are preserved under φ, because

$$g(X, Y)_p = (\varphi_* g)(\varphi_* X, \varphi_* Y)_{\varphi(p)} = h(\varphi_* X, \varphi_* Y)_{\varphi(p)}. \tag{8.1}$$

As a symmetric bilinear form on T_pM, the inner product g_p has a definite signature.[1] If the signature is $(+, +, \ldots, +)$ then the metric is positive definite, and M equipped with such a metric is called a **Riemannian manifold**. If the signature is anything else (except if it is $(-, -, \ldots, -)$) then the metric is indefinite, and M

---

[1] Although the inner product itself will vary smoothly from point to point, its signature cannot change unless the inner product becomes degenerate at some point. This is a consequence of Sylvester's law of inertia.

equipped with such a metric is called a **pseudo-Riemannian manifold**. Of particular interest for physics are **Lorentzian manifolds**, which are manifolds equipped with a metric with signature $(-, +, +, \ldots, +)$ (or equivalently, but less aesthetically, $(+, -, -, \ldots, -)$). With a metric at our disposal we can do a lot of pretty geometry.

For the lawyers, we should probably clear up one more point.

Theorem 8.1 *Every smooth manifold admits a Riemannian metric.*

Proof Let $\{\rho_i\}$ be a partition of unity subordinate to a locally finite open cover $\{U_i\}$ on M. Let h be the usual Euclidean metric on $\mathbb{R}^n$ and define $g_i = \varphi_i^* h$, where φ_i is the coordinate map on U_i. Then $g := \sum \rho_i g_i$ is a Riemannian metric on M.   □

Remark The analogue of Theorem 8.1 does not hold for arbitrary signatures – there may be topological obstructions. (Can you see where the proof of Theorem 8.1 breaks down?) In fact, it is known that a compact manifold admits a Lorentzian metric if and only if its Euler characteristic vanishes. However, any noncompact manifold admits a Lorentzian metric ([39], p. 40).

8.1 Index gymnastics

In order to do some honest work, we have to (re-)introduce indices. The reason is that we need to manipulate components, and components require a definite frame field. We begin by discussing the two types of frame field on a manifold.

8.1.1 Holonomic and nonholonomic frames

A **holonomic** or **coordinate** frame field is one that is derived from a system of local coordinates. Specifically, if $\{x^1, \ldots, x^n\}$ are coordinates on a patch U then

$$e_i = \partial_i := \frac{\partial}{\partial x^i} \tag{8.2}$$

is called a (local) coordinate frame on U. We are already familiar with how these work. They are local derivations in the coordinate directions. For example, ∂_1 is a derivative in the direction in which the x^1 coordinate is increasing. Any vector field X can be written $X = X^i \partial_i$, where the functions X^i are the components of X with respect to the coordinate frame.

A **nonholonomic** or **noncoordinate** frame field is one that *cannot* be derived from a local system of coordinates. Such fields are usually just written $\{e_a\}$, where one chooses early Latin letters rather than middle Latin letters to indicate that one is dealing with a general frame. The vector field X is now written $X = X^a e_a$, where the functions X^a are the components of X with respect to the given frame.

It is important to have a good working criterion for when a frame is holonomic. This is provided by the next theorem.[2]

Theorem 8.2 *A frame field $\{e_a\}$ is locally holonomic (i.e., a coordinate frame field) if and only if*

$$[e_a, e_b] = 0 \qquad \text{for all a and b.} \tag{8.3}$$

Proof If $e_a = \partial_a$ for some local coordinates $\{x^a\}$ then

$$[e_a, e_b] = [\partial_a, \partial_b] = 0, \tag{8.4}$$

because mixed partials commute. Conversely, suppose that we have a frame field $\{e_a\}$ satisfying the hypothesis of the theorem. Let θ^a be the dual frame field, so that $\theta^a(e_b) = \delta_b^a$. By (3.123),

$$d\theta^a(e_b, e_c) = e_b \theta^a(e_c) - e_c \theta^a(e_b) - \theta^a([e_b, e_c]). \tag{8.5}$$

The first two terms vanish because the derivations e_a give zero when acting on constants, and by hypothesis (and linearity) the last term vanishes. Hence θ^a is closed. By the Poincaré lemma θ^a is locally exact, so there exist functions x^a such that $\theta^a = dx^a$. Hence $dx^1 \wedge \cdots \wedge dx^n = \theta^1 \wedge \cdots \wedge \theta^n \neq 0$ (because the θ^a are linearly independent), so by the inverse function theorem the x^a form a good set of local coordinates. By duality, $e_a = \partial_a$. □

8.1.2 The line element and the metric tensor components

Now that we know what a tensor field is, we can see that the metric g is one because it is a smoothly varying bilinear function on the tangent spaces to M. In a local coordinate basis the metric tensor can be written

$$g = g_{ij} \, dx^i \otimes dx^j, \tag{8.6}$$

where the components are given by

$$g_{ij} = g(\partial_i, \partial_j). \tag{8.7}$$

(Note that g is a *symmetric* tensor field because $g_{ij} = g_{ji}$, which follows from the symmetry of the inner product.)

---

[2] Theorem 8.2 is closely related to an important theorem known as Frobenius' theorem; we discuss this in detail in Appendix F.

Example 8.1 In Cartesian coordinates in Euclidean (respectively, Minkowski) space, the metric tensor components are δ_{ij} (respectively, η_{ij}); cf. (1.48) and (1.50). For this reason δ_{ij} (respectively, η_{ij}) is often sloppily called the **Euclidean** (respectively, **Minkowski**) **metric tensor**.

In a general basis $\{e_a\}$ and cobasis $\{\theta^a\}$ we have

$$g_{ab} = g(e_a, e_b) \tag{8.8}$$

and

$$g = g_{ab}\,\theta^a \otimes \theta^b. \tag{8.9}$$

Note that the nondegeneracy of the inner product is equivalent to the nonsingularity of the matrices whose entries are the components of the metric tensor in the various frames. The frame $\{e_{\hat{a}}\}$ is **orthonormal** if

$$g_{\hat{a}\hat{b}} = \pm\delta_{\hat{a}\hat{b}}, \tag{8.10}$$

where we adopt the convention that indices with carets denote orthonormal bases or components. Orthonormal frame fields always exist locally.

Theorem 8.3 *In the neighborhood of any point of a geometric manifold (M, g) there exists an orthonormal frame field.*

Proof Pick a point $p \in M$ and let $\{x^i\}$ be local coordinates on a neighborhood U of p. Then $\{\partial_i\}$ is a frame field on U. If (M, g) is Riemannian then Gram–Schmidt applied to this frame field yields an orthonormal frame field at every point. The only question is whether this can be done smoothly, and the answer is "yes". Define the smooth vector field $e_{\hat{1}} = \partial_1/\sqrt{|g(\partial_1, \partial_1)|}$. Next, define

$$e_2' = \partial_2 - \frac{g(e_{\hat{1}}, \partial_2)}{g(e_{\hat{1}}, e_{\hat{1}})}e_{\hat{1}} \quad \text{and} \quad e_{\hat{2}} = \frac{e_2}{\sqrt{|g(e_2', e_2')|}},$$

and observe that e_2 is also a smooth vector field because it is constructed out of smooth quantities by smooth processes. Continuing in this way gives a smooth orthonormal frame field on U.

If (M, g) is pseudo-Riemannian, the only problem that one might encounter is that at the ith stage one of the e_i' might have zero length. In that case, just replace ∂_i by $\partial_i' = \sum_j a^j{}_i \partial_j$, where the $a^j{}_i$ are smooth functions and such that ∂_i' does not lie in the subspace spanned by $\{e_1, \ldots, e_{i-1}\}$. By the nondegeneracy of the metric (see the first proof of Theorem 1.5) some choice of $a^j{}_i$ will ensure that e_i' has nonzero length. $\qquad\square$

A metric tensor allows us to calculate the distance s along any path $\gamma : I \to M$ by adding up all the infinitesimal distances along the path. This is easiest to understand if you think of a particle traveling along a curve $\gamma(t)$, where t denotes time. Its velocity vector is $\dot{\gamma}(t)$ so its speed is $ds/dt = |\dot{\gamma}(t)|$ the length of the velocity vector. The distance it travels in a time t is therefore

$$s(t) = \int \frac{ds}{dt} dt = \int |\dot{\gamma}(t)| \, dt. \tag{8.11}$$

On a curved surface the length of a tangent vector is determined by the metric, so

$$s(t) = \int \sqrt{g_{ij}(t)\dot{\gamma}^i(t)\dot{\gamma}^j(t)} \, dt. \tag{8.12}$$

But the length of the path cannot depend on how fast it is traversed, and indeed if $\gamma(t')$ denotes a curve with the same image as $\gamma(t)$ but a different parameterization then we have

$$s(t') = \int \frac{ds}{dt'} dt' = \int \frac{ds}{dt}\frac{dt}{dt'} dt' = \int \frac{ds}{dt} dt = s(t).$$

For this reason we define the **line element** ds^2 by

$$ds^2 = g = g_{ij} \, dx^i \, dx^j \tag{8.13}$$

(where the tensor product symbol is usually suppressed). Then the distance between any two points along a curve is defined by pulling everything back to the domain of the curve:

$$\int_\gamma ds = \int_I \gamma^* ds = \int_I \sqrt{(\gamma^* g_{ij})(\gamma^* dx^i)(\gamma^* dx^j)} = \int_I \sqrt{g_{ij}(t)\frac{dx^i}{dt}\frac{dx^j}{dt}} \, dt. \tag{8.14}$$

You can think of $ds := \sqrt{ds^2}$ as the distance between two infinitesimally separate points. For example, in three-dimensional Euclidean space $ds = \sqrt{dx^2 + dy^2 + dz^2}$.

8.1.3 Raising and lowering indices

The metric is not just a passive object for measuring distances and angles. It also provides a means of tying together the vector space V and its dual V^* via the isomorphism $\psi : T_p M \to T_p^* M$ taking X to $g(X, \cdot)$.[3] In other words, to every vector field X there exists a 1-form $g(X, \cdot)$. What does this 1-form do? As we have seen, we may view a 1-form as a machine that eats vector fields and spits

---

[3] This is just the Riesz lemma.

out numbers (in a smooth way). The 1-form $g(X, \cdot)$ acts on, say, Y, and spits out $g(X, Y)$.[4]

The map ψ is both a blessing and a curse. It is a blessing because it allows us to simplify certain formulae, but it is also a curse because it blurs the distinction between covariant and contravariant indices, which can sometimes lead to confusion. To see how it works, consider a coordinate frame. Then, under ψ we have, from (8.13),

$$X^j \partial_j = X \to g(X, \cdot) = g_{ij} \, dx^i(X) \otimes dx^j(\cdot) = g_{ij} X^i dx^j. \tag{8.15}$$

Next, we *define* the covariant components X_j of the vector field X by

$$X_j := g_{ij} X^i, \tag{8.16}$$

in which case (8.15) can be written

$$X^j \partial_j \to X_j dx^j. \tag{8.17}$$

We say (with reference to (8.16)) that the index i has been **lowered** by the metric.

The isomorphism ψ induces a unique metric $\hat{g}$ on $T_p^* M$, namely the one that makes the following diagram commute:

$$
\begin{array}{ccc}
T_p M \times T_p M & \xrightarrow{\;\psi\;} & T_p^* M \times T_p^* M \\
{\scriptstyle g}\big\downarrow & & \big\downarrow{\scriptstyle \hat{g}} \\
g(\cdot, \cdot) & =\!=\!= & \hat{g}(\cdot, \cdot).
\end{array}
$$

In other words, one gets the same answer following the arrows along either path. If we define

$$g^{ij} := \hat{g}(dx^i, dx^j), \tag{8.18}$$

then the fact that the diagram commutes means that

$$g_{ij} = g(\partial_i, \partial_j) = \hat{g}(g_{ik} dx^k, g_{j\ell} dx^\ell) = g_{ik} g_{j\ell} g^{k\ell}. \tag{8.19}$$

Using the symmetry of the metric we can rewrite (8.19) as

$$g_{ij} = g_{ik} g^{k\ell} g_{\ell j}. \tag{8.20}$$

Viewing the components of g as the entries of a matrix, (8.20) shows that g^{ij} is the matrix inverse to g_{ij} because this relation can only hold if

$$g_{ik} g^{k\ell} = \delta_i^\ell \quad \text{and} \quad g^{k\ell} g_{\ell j} = \delta_j^k. \tag{8.21}$$

---

[4] Because the metric is symmetric, we could have written the map as $X \to g(\cdot, X)$; it makes no difference here although it does make a difference for complex manifolds equipped with a Hermitian inner product.

This in turn, makes it possible to use the inverse metric to raise indices, undoing what was done by (8.16). In particular, we have

$$X^i = g^{ij} X_j \tag{8.22}$$

because

$$g^{ij} X_j = g^{ij} g_{jk} X^k = \delta^i_k X^k = X^i. \tag{8.23}$$

Similarly, we can raise and lower indices on any tensor using the metric in this way. For example,

$$T^{ij}{}_k = g^{i\ell} g^{jm} T_{\ell mk} \quad \text{and} \quad T_{ijk} = g_{im} g_{jn} g_{kp} T^{mnp}. \tag{8.24}$$

It is worth emphasizing that, whereas we can always change the vertical positions of the indices in this way by using the metric, the horizontal positions are much less flexible. In fact, a statement like $T_{ij} = T_{ji}$ is only possible if the tensor has a special property (namely symmetry).

8.2 The Levi-Civita connection

8.2.1 The Koszul formula

Let M be a manifold. As we saw in Section 7.2, any covariant derivative ∇ on TM satisfies three basic properties (C1)–(C3). By demanding a few more natural compatibility properties, (C4)–(C6), we can extend the connection on TM to tensor product bundles of TM and TM^*. In general there are infinitely many covariant derivatives on TM. To narrow down the possibilities, one usually imposes some additional restrictions on ∇. It is not obvious at this stage why one should impose the following conditions but they turn out to be useful in many applications, especially Einstein's theory of general relativity.

The first additional condition has to do with something called **torsion**,[5] which is a generalization to manifolds of the twisting of curves in $\mathbb{R}^3$. Define a map

$$\tau : \Gamma(TM) \times \Gamma(TM) \to \Gamma(TM)$$

given by

$$\tau(X, Y) = \nabla_X Y - \nabla_Y X - [X, Y]; \tag{8.25}$$

τ is called the **torsion tensor**.

> **EXERCISE 8.1** Verify that τ is indeed a tensor by verifying that it is function linear in both arguments.

---

[5] The idea of torsion is subtle and has engendered much confusion. For a discussion, see e.g. [40].

We now insist that this tensor vanish everywhere, so we add the condition of

(C7) (torsion-freeness) $\tau(X, Y) = 0$.

The second additional condition, whose significance becomes apparent later, is

(C8) (metric compatibility) $Xg(Y, Z) = g(\nabla_X Y, Z) + g(Y, \nabla_X Z)$.

Theorem 8.4 *There is a unique connection on TM, satisfying (C1)–(C3), (C7), and (C8), called the* **Levi-Civita connection** *(or, if the metric is positive definite, the* **Riemannian connection***).*

First we prove a little lemma.

Lemma 8.5 (Koszul) *A connection satisfying (C1)–(C3), (C7), and (C8) obeys the* **Koszul formula***,*

$$2g(\nabla_X Y, Z) = Xg(Y, Z) + Yg(X, Z) - Zg(X, Y)$$
$$+ g([X, Y], Z) - g([X, Z], Y) - g([Y, Z], X). \qquad (8.26)$$

Proof By metric compatibility

$$Xg(Y, Z) = g(\nabla_X Y, Z) + g(Y, \nabla_X Z),$$
$$Yg(X, Z) = g(\nabla_Y X, Z) + g(X, \nabla_Y Z),$$
$$Zg(X, Y) = g(\nabla_Z X, Y) + g(X, \nabla_Z Y),$$

so the right-hand side of (8.26) becomes

$$g(\nabla_X Y + \nabla_Y X + [X, Y], Z)$$
$$+ g(\nabla_X Z - \nabla_Z X - [X, Z], Y)$$
$$+ g(\nabla_Y Z - \nabla_Z Y - [Y, Z], X).$$

But this equals $2g(\nabla_X Y, Z)$ by torsion-freeness.   □

Proof of Theorem 8.4 To prove the theorem, we *define* ∇ by the Koszul formula. We must then show that it exists (i.e., that the operator ∇ so defined is indeed a connection) and is unique. Uniqueness is obvious, because Z is arbitrary so any other covariant derivative ∇' satisfying (8.26) would have to equal ∇. To prove existence, we must show that ∇ satisfies axioms (C1)–(C3), (C7), and (C8). Actually, we need only show that it satsifies axioms (C1)–(C3) because if it does then, by construction, since it satisfies (8.26) it must also be torsion free and metric compatible.

The linearity of ∇ is obvious from (8.26), because both the Lie bracket and the metric are bilinear. So (C1) holds. To prove (C2) it suffices to prove function linearity in the form $\nabla_{fX} Y = f \nabla_X Y$, because the same argument as before shows that $\nabla_{X+Y} Z = \nabla_X Z + \nabla_Y Z$. By (8.26) we have

$$
\begin{aligned}
2g(\nabla_{fX} Y, Z) =& f X g(Y, Z) + Y g(fX, Z) - Z g(fX, Y) \\
&+ g([fX, Y], Z) - g([fX, Z], Y) - g([Y, Z], fX). \\
=& f X g(Y, Z) + Y[fg(X, Z)] - Z[fg(X, Y)] \\
&+ g(fXY - (Yf)X - fYX, Z) \\
&- g(fXZ - (Zf)X - fZX, Y) - fg([Y, Z], X) \\
=& f X g(Y, Z) + (Yf)g(X, Z) + fYg(X, Z) \\
&- (Zf)g(X, Y) - fZg(X, Y) \\
&+ fg([X, Y], Z) - (Yf)g(X, Z) \\
&- fg([X, Z], Y) + (Zf)g(X, Y) - fg([Y, Z], X) \\
=& 2fg(\nabla_X Y, Z).
\end{aligned}
$$

Property (C3) can be proved similarly; this is left to the reader. □

EXERCISE 8.2 Let $\{e_a\}$ be a frame field and let $\{\theta^a\}$ be its dual frame field on M. Show that, in terms of these bases, the torsion tensor can be written

$$
\tau(e_b, e_c) = \sum \tau^a(e_b, e_c) e_a, \tag{8.27}
$$

where the τ^a are 1-forms given by

$$
\tau^a = d\theta^a + \omega^a{}_b \wedge \theta^b. \tag{8.28}
$$

In a more compact notation one writes

$$
\tau = d\theta + \omega \wedge \theta, \tag{8.29}
$$

where τ is called the **torsion form** of the connection, relative to the basis θ.[6] Clearly the torsion tensor and the torsion form embody the same information. *Remark:* One sometimes sees the even shorter shorthand notation

$$
\tau = D\theta, \tag{8.30}
$$

where the action of the connection D on 1-forms is defined as follows:

$$
D\lambda := d\lambda + \omega \wedge \lambda. \tag{8.31}
$$

In this notation the curvature form becomes

$$
\Omega = D\omega. \tag{8.32}
$$

However, this notation can be a bit tricky so we shall eschew its use. If you're curious about how it works, look at [7], pp. 250–252.

---

[6] Strictly speaking, τ is not a 1-form but, rather, a collection of 1-forms τ^a.

8.2.2 *The Christoffel symbols and structure functions*

Let $\{e_a\}$ be an arbitrary frame field and $\{\theta^a\}$ the dual frame field, and let ∇ be the Levi-Civita connection. We define the **Christoffel symbols** $\Gamma^c{}_{ab}$ by the formula

$$\nabla_{e_a} e_b = \Gamma^c{}_{ab} e_c. \tag{8.33}$$

The Christoffel symbols are intimately related to the connection forms. We have

$$\nabla_{e_a} e_b = i_{e_a} D e_b = i_{e_a}(e_c \otimes \omega^c{}_b) = \omega^c{}_b(e_a)e_c, \tag{8.34}$$

so, comparing (8.33) and (8.34), we get

$$\Gamma^c{}_{ab} = \omega^c{}_b(e_a) \qquad \text{or} \qquad \omega^c{}_b = \Gamma^c{}_{ab}\theta^a. \tag{8.35}$$

The Christoffel symbols play an important role in geometry, because they tell you how vectors twist and turn as you move about the manifold.

EXERCISE 8.3 Show that

$$\nabla_{e_a} \theta^c = -\Gamma^c{}_{ab}\theta^b. \tag{8.36}$$

We now derive an explicit expression for the Christoffel symbols in terms of simpler quantities. Associated to any frame field $\{e_a\}$ are its **structure functions** $c^c{}_{ab}$, defined by

$$[e_a, e_b] = c^c{}_{ab} e_c \tag{8.37}$$

where the expression on the left is the usual Lie bracket of vector fields.

Theorem 8.6 *The Christoffel symbols are determined uniquely by the metric tensor and the structure functions.*

Proof Let $X = e_a$, $Y = e_b$, and $Z = e_c$ and apply the Koszul formula (8.26) to get

$$2g(\nabla_{e_a} e_b, e_c) = e_a g(e_b, e_c) + e_b g(e_a, e_c) - e_c g(e_a, e_b)$$
$$+ g([e_a, e_b], e_c) - g([e_a, e_c], e_b) - g([e_b, e_c], e_a).$$

In terms of components this reads (using the bilinearity of the metric)

$$2\Gamma^d{}_{ab} g_{dc} = e_a g_{bc} + e_b g_{ac} - e_c g_{ab} + c^d{}_{ab} g_{dc} - c^d{}_{ac} g_{db} - c^d{}_{bc} g_{da}.$$

This equation is the result we seek. It can be written more compactly by using the metric tensor to lower all the indices:

$$\Gamma_{cab} = \frac{1}{2}(e_a g_{bc} + e_b g_{ac} - e_c g_{ab} + c_{cab} - c_{bac} - c_{abc}). \tag{8.38}$$

$\square$

As an important special case of the theorem, we observe by Theorem 8.2 that the structure functions vanish in a coordinate (holonomic) basis. In that case (8.38) simplifies to give

$$\Gamma_{kij} = \frac{1}{2}(\partial_i g_{jk} + \partial_j g_{ik} - \partial_k g_{ij}) = \frac{1}{2}(g_{jk,i} + g_{ik,j} - g_{ij,k}). \tag{8.39}$$

Notice a handy feature of the Christoffel symbols in a coordinate basis, namely that they are symmetric in the last two indices:

$$\Gamma^k{}_{ij} = \Gamma^k{}_{ji}. \tag{8.40}$$

One word of warning. *The Christoffel symbols look like the components of a tensor, but they are not.* That is, they would appear to be tensor components because they have all those indices, but they fail to satisfy the correct transformation property (3.53). This fact follows from (7.27) and (8.35), but it is also instructive to see it directly.

EXERCISE 8.4 Starting from the definition

$$\nabla_{\partial_i} \partial_j = \Gamma^k{}_{ij} \partial_k \tag{8.41}$$

of the Christoffel symbols in a coordinate basis (with $\partial_i := \partial/\partial x^i$), verify that the $\Gamma^k{}_{ij}$ are not the components of any tensor field by showing that, under a coordinate transformation $x^i \to y^{i'}$,

$$\Gamma^{k'}{}_{i'j'} = \frac{\partial y^{k'}}{\partial x^k} \frac{\partial x^i}{\partial y^{i'}} \frac{\partial x^j}{\partial y^{j'}} \Gamma^k{}_{ij} + \frac{\partial y^{k'}}{\partial x^j} \frac{\partial^2 x^j}{\partial y^{i'} \partial y^{j'}}. \tag{8.42}$$

EXERCISE 8.5 Consider the unit 2-sphere S^2 equipped with the standard line element

$$ds^2 = d\theta^2 + \sin^2\theta \, d\phi^2, \tag{8.43}$$

where θ and ϕ are the usual spherical polar coordinates in $\mathbb{R}^3$. (As previously noted, these coordinates are not well defined at the poles, so their use implies that we are standing somewhere else on the sphere.) Show that the only nonzero Christoffel symbols for the 2-sphere metric are

$$\Gamma^\theta{}_{\phi\phi} = -\sin\theta\cos\theta \qquad \text{and} \qquad \Gamma^\phi{}_{\theta\phi} = \Gamma^\phi{}_{\phi\theta} = \cot\theta.$$

Remark: For a derivation of (8.43), see Exercise 8.47.

EXERCISE 8.6 Let $\{e_{\hat{a}}\}$ be an orthonormal frame field on some neighborhood U of $p \in M$, and let ω and Ω be the connection and curvature matrices, respectively, relative to this frame. Show that both matrices are skew symmetric *when their indices are lowered with the metric*. That is, show that $\omega_{\hat{a}\hat{b}} := g_{\hat{a}\hat{c}}\omega^{\hat{c}}{}_{\hat{b}}$ satisfies $\omega_{\hat{a}\hat{b}} = -\omega_{\hat{b}\hat{a}}$, with a similar result holding for Ω. *Hint:* Use metric compatibility. *Remark:* This result holds irrespective of the signature of the metric, but a similar statement is false if M is pseudo-Riemannian and the indices are not both lowered. For example, $\omega^{\hat{a}}{}_{\hat{b}} \neq -\omega^{\hat{b}}{}_{\hat{a}}$ in general; see Exercise 8.40. However, if M is Riemannian (so that the metric is positive definite) then we have both $\omega^{\hat{a}}{}_{\hat{b}} = -\omega^{\hat{b}}{}_{\hat{a}}$ and $\Omega^{\hat{a}}{}_{\hat{b}} = -\Omega^{\hat{b}}{}_{\hat{a}}$. These last equations refer to individual components and, as should be clear, are not meant to imply a tensor equality, so in the Riemannian case it is better simply to write $\omega^T = -\omega$ and $\Omega^T = -\Omega$. Note that all these results apply only to *orthonormal* frames.

EXERCISE 8.7 As before, let $\{e_{\hat{a}}\}$ be a local orthonormal frame field around $p \in M$. Assume the metric to be positive definite. By virtue of Exercise 8.6, ω is a skew symmetric matrix (of 1-forms). Show that if $\omega = A^{-1} dA$ for some matrix A of smooth functions with $A(p) = I$ (the identity matrix) then A must be an orthogonal matrix.

8.3 The Riemann curvature tensor

Exercise 8.4 shows that the Christoffel symbols are nontensorial and consequently appear to be of limited utility. However, they can be combined in a special way to form an important tensor, namely the Riemann curvature tensor, an object of great utility.

The **Riemann curvature tensor** is just the curvature tensor of the Levi-Civita connection. The significance of the Riemann curvature tensor is certainly not obvious at this stage but, intuitively, it measures how much the manifold curves in various directions. Here we discuss a few simple consequences of the definition and later we investigate its properties more thoroughly.

In a general basis $\{e_a\}$ the components of the Riemann curvature tensor are defined by

$$R(e_c, e_d)e_b = R^a{}_{bcd} e_a. \tag{8.44}$$

EXERCISE 8.8 Let $\{e_a, \theta^b\}$ be a pair of dual bases. Starting from (7.34), show that the curvature 2-forms can be written

$$\Omega^a{}_b = \tfrac{1}{2} R^a{}_{bcd}\, \theta^c \wedge \theta^d. \tag{8.45}$$

EXERCISE 8.9 Using (7.36) and (8.44) show that, in a coordinate (holonomic) basis, the components of the Riemann curvature tensor are

$$R^i{}_{jk\ell} = \Gamma^i{}_{\ell j,k} - \Gamma^i{}_{kj,\ell} + \Gamma^i{}_{km}\Gamma^m{}_{\ell j} - \Gamma^i{}_{\ell m}\Gamma^m{}_{kj}, \tag{8.46}$$

where $\Gamma^i{}_{jk}$ are the Christoffel symbols.

The Riemann curvature tensor satisfies various symmetry properties, some of which are obvious and some of which are not. They are

$$R(X, Y)Z = -R(Y, X)Z, \tag{8.47}$$

$$g(W, R(X, Y)Z) = -g(Z, R(X, Y)W), \tag{8.48}$$

$$R(X, Y)Z + R(Y, Z)X + R(Z, X)Y = 0, \tag{8.49}$$

and

$$g(W, R(X, Y)Z) = g(X, R(W, Z)Y). \tag{8.50}$$

Equation (8.47) is clear from (7.36). To prove (8.48) we simplify things by observing that, by tensoriality, it suffices to prove the result in a local coordinate basis. In particular, we may assume that all brackets such as $[X, Y]$ vanish. By metric compatibility

$$g(W, \nabla_X \nabla_Y Z) = Xg(W, \nabla_Y Z) - g(\nabla_X W, \nabla_Y Z),$$

so that

$$g(W, R(X, Y)Z) = Xg(W, \nabla_Y Z) - Yg(W, \nabla_X Z)$$
$$- g(\nabla_X W, \nabla_Y Z) + g(\nabla_Y W, \nabla_X Z).$$

Now add this to the same expression with W and Z interchanged. After obvious cancellations and another appeal to metric compatibility the right-hand side becomes

$$XYg(W, Z) - YXg(W, Z) = [X, Y]g(W, Z) = 0.$$

For (8.49), we again assume that the brackets vanish. Expanding the left-hand side gives

$$\nabla_X \nabla_Y Z - \nabla_Y \nabla_X Z + \nabla_Y \nabla_Z X - \nabla_Z \nabla_Y X + \nabla_Z \nabla_X Y - \nabla_X \nabla_Z Y.$$

By torsion-freeness (e.g. $\nabla_Y Z = -\nabla_Z Y$) the terms in this expression all cancel pairwise. The last property is left as the following exercise.

EXERCISE 8.10 Prove (8.50) using (8.47)–(8.49). *Hint:* Take the inner product of (8.49) with a fixed vector, then add and subtract various permutations of it and use the other properties.

It is useful to translate the symmetry conditions (8.47)–(8.50) into statements about the (downstairs) components of the Riemann curvature tensor, say in a coordinate basis. Defining $R_{ijk\ell} := g_{im} R^m{}_{jk\ell}$ we have:

(1) $R_{ijk\ell} = -R_{ij\ell k}$ (antisymmetry in the last two indices);
(2) $R_{ijk\ell} = -R_{jik\ell}$ (antisymmetry in the first two indices);
(3) $R_{i[jk\ell]} = 0$ (total antisymmetry in the last three indices);

(4) $R_{ijk\ell} = R_{k\ell ij}$ (symmetry under the interchange of the first and last pair of indices).

These follow directly from (8.47)–(8.50) on choosing $W = \partial_i$, $X = \partial_k$, $Y = \partial_\ell$, and $Z = \partial_j$. The only exception is property (3), which requires the use of property (1).

> **EXERCISE 8.11** Prove property (3) above.

> **EXERCISE 8.12** By counting the number of constraints, show that the Riemann tensor has only
>
> $$\frac{n^2(n^2 - 1)}{12} \tag{8.51}$$
>
> (rather than n^4) independent components in n dimensions. *Hint:* By property (1) there are only $N := \binom{n}{2} = n(n-1)/2$ choices for the last two indices, because there are that many ways of choosing two distinct objects from a list of n objects. By property (4) the Riemann tensor can be viewed as a symmetric $N \times N$ matrix (which means that the conditions in property (2) are redundant and can be discarded). Argue that property (3), after accounting for properties (1) and (4), imposes $\binom{n}{4}$ conditions.

> **EXERCISE 8.13** Using the standard coordinates on the 2-sphere, show that $R^\theta{}_{\phi\theta\phi} = \sin^2\theta$ and $R^\phi{}_{\theta\phi\theta} = 1$ and that all the other independent components vanish. *Hint:* Use Exercise 8.5 and (8.46).

8.4 More curvature tensors

From the Riemann curvature tensor we can construct various other curvature tensors by contraction. In a coordinate basis the components of the **Ricci tensor** R_{ij} are defined by

$$R_{ij} := R^k{}_{ikj}. \tag{8.52}$$

> **EXERCISE 8.14** Show that the Ricci tensor is symmetric.

> **EXERCISE 8.15** Compute the components of the Ricci tensor for the 2-sphere in standard coordinates.

> **EXERCISE 8.16** Define
>
> $$\mathrm{Ric}(X, Y) = \mathrm{tr}(Z \to R(Z, Y)X). \tag{8.53}$$
>
> Show that Ric is precisely the Ricci tensor. That is, show that $\mathrm{Ric}(\partial/\partial x^i, \partial/\partial x^j) = R_{ij}$.

The **Ricci curvature scalar** is

$$R := g^{ij} R_{ij}. \tag{8.54}$$

Note that this number is the same in all coordinate systems. It is therefore called a **curvature invariant**.[7] Note that $R^{ij} R_{ij}$ is also a curvature scalar built from the Ricci tensor, but it has no standard name. Many other curvature invariants exist.

EXERCISE 8.17 Show that $R = 2$ for the unit 2-sphere.

Although this is not a book about general relativity, it would be gross negligence to have reached this point and yet not to write down Einstein's equations; for completeness we will do so here. The components of the **Einstein tensor** G_{ij} are defined by

$$G_{ij} := R_{ij} - \frac{1}{2} R g_{ij}. \tag{8.55}$$

Einstein's famous equations of general relativity can then be written (in natural units) as

$$G_{ij} = 8\pi T_{ij}, \tag{8.56}$$

where T_{ij} are the components of something called the **stress–energy tensor**, which contains all the information about the distribution of mass–energy and momentum in spacetime.

Einstein's equations (8.56), together with the constraints imposed on the Einstein tensor by the Bianchi identity, comprise $n(n-1)/2$ coupled, nonlinear partial differential equations for the metric tensor components in terms of the stress–energy tensor components. Despite their complexity, in all cases in which they can be solved, either exactly or approximately, they successfully predict the behavior of every observable large scale phenomenon in the universe; they also predict the shape of the universe itself. Einstein's equations are sometimes summarized by the (entirely too cavalier) slogan "geometry = mass", because they predict that spacetime is curved by the presence of matter.

---

[7] **Remarks** (i) A scalar is an element of the underlying field $\mathbb{F}$, but a scalar field is a coordinate invariant function on a manifold. (How could a function not be coordinate invariant? The answer is, if it is a pseudoscalar. See the discussion in the footnote indicated after (8.130).) Sometimes people are sloppy and call a scalar field a scalar. *Mea culpa.* (ii) Some authors would say that R is a diffeomorphism invariant, but this is misleading. Recalling the discussion in the footnote indicated before (3.14), we could indeed say that R is a diffeomorphism invariant in the sense that "diffeomorphism" is just the active way of stating "coordinate transformation". But if we think of a diffeomorphism as a smooth deformation of the manifold, it would seem to engender a contradiction. After all, a squashed sphere has a different geometry from a standard sphere. But this is the point: when those authors say "diffeomorphism invariant" they mean that both the manifold and the metric are changed simultaneously by the diffeomorphism. In that case, R is indeed invariant. If we were to just squash the sphere but not change the metric accordingly (by, for example, embedding both the standard sphere and the squashed sphere in Euclidean space and demanding that they each inherit their respective metrics from the ambient Euclidean metric) then R would certainly *not* be invariant. For this reason, it is best not to call R a diffeomorphism invariant. However, certain *integrals* of curvature scalars are indeed diffeomorphism invariants in the stronger sense because they are invariant under arbitrary smooth deformations of the manifold, no matter what metric is used. In other words, they are *smooth topological invariants*. See Theorem 9.6 below.

Remark   Another striking application of the Ricci tensor is to a fundamental question of topology. A topological space is **simply connected** if it is path connected and if every simple closed curve can be contracted to a point while remaining in the space. Simple connectivity can be used to distinguish two topologically distinct manifolds. For example, both the 2-sphere and the 2-torus are compact and without boundary, but the 2-sphere is simply connected while the 2-torus is not. The question then becomes, if a manifold (of any dimension) is compact, boundaryless, and simply connected is it a (topological) sphere?

The famous mathematician Henri Poincaré showed that this is true for two-dimensional manifolds, and in 1904 he conjectured that the answer was also true in three dimensions.[8] For almost one hundred years Poincaré's notorious conjecture resisted all attempts to solve it, until Grigory Perelman used Richard Hamilton's technique of **Ricci flow** to construct a proof around 2003. The basic idea of Ricci flow is to treat the curvature of a manifold as something that flows from one point to another (akin to the flow of heat) while preserving the original topology.[9] The hope was that one could then show that a 3-manifold with the desired properties would flow into a manifold with constant positive curvature (which, given the other properties, has to be a 3-sphere). The details are extremely delicate, but ultimately Perelman's argument was judged successful. For a good popular discussion of the Poincaré conjecture and its history, see [72] or [83]. For the technical details, see [63].

8.5 Flat manifolds

In the next few sections we briefly investigate the significance of curvature. We begin by trying to characterize spaces that are not curved, so we start with Euclidean space. Intuitively, Euclidean space is "flat" in all directions. Thus we would expect the Riemann curvature tensor of Euclidean space to vanish. Fortunately, it does. The Euclidean metric tensor in Cartesian coordinates has constant components $g_{ij} = \delta_{ij}$ so, by (8.39), all the Christoffel symbols vanish whereupon by (8.46) $R_{ijk\ell} = 0$. But a tensor that vanishes in one coordinate system vanishes in all coordinate systems, because by construction tensors are coordinate independent.[10]

---

[8] In dimensions four and above, the analogous conjecture (that any compact, boundaryless, simply connected and homologically trivial manifold is a sphere) was affirmed in the 1960s through the efforts of mathematicians such as Smale, Stallings, Wallace, Zeeman, and Freedman. For more about this story, see [59].

[9] The basic equation of Ricci flow is $dg_{ij}/dt = -2R_{ij}$.

[10] The same thing holds for Minkowski space, as well as for other spaces with globally constant metric tensor components. To avoid having to discuss spaces of every possible signature we mostly restrict our attention in this section to Riemannian spaces.

By definition, a space whose Riemann curvature tensor vanishes everywhere is called a **flat space**.[11] We have just seen that Euclidean space is flat. Is this the only flat space there is? Interestingly, the answer is no. For example, you will hear people say that the cylinder and the torus are also flat, which sounds absurd on the face of it but is true in a certain sense. We really have to be a bit more precise.

The question of whether a geometric manifold is flat depends on its *metric* and only indirectly on its topology. A topological space equipped with one metric may be flat, but equipped with another it may not be. Consider, for example, the torus. If we think of torus as a doughnut in $\mathbb{R}^3$ with its metric inherited from the ambient Euclidean metric, then it is *not* flat. (See Exercise 8.49). But, if the torus is defined as $\mathbb{R}^2/\mathbb{Z}^2$ and equipped with the metric $ds^2 = dx^2 + dy^2$, where x and y are the standard coordinates defined in Exercise 3.16 (see also Exercise 3.23), then it is definitely flat.[12]

For this reason it would be better to avoid saying that a manifold is or is not flat and instead say that a manifold is or is not equipped with a flat metric. In fact, it would be best to speak of a geometric manifold (M, g) as being flat or not. But everyone speaks of flat manifolds, so we will do so as well, remembering that the statement always refers to a manifold with a metric.

This begs two questions. First, how could a manifold that *looks* curved actually be flat? For example, the standard torus looks curved, and in fact is curved. But the standard right circular cylinder in $\mathbb{R}^3$ looks curved, and yet is flat (see Exercise H.8). The resolution of this conundrum lies in distinguishing two kinds of curvatures, intrinsic and extrinsic. The intrinsic curvature of a manifold is the curvature that an observer living in the manifold would measure (see Section 8.6 below for exactly how one might do this), whereas the extrinsic curvature depends on how the manifold is embedded in a higher-dimensional space. The cylinder has zero intrinsic curvature but nonzero extrinsic curvature. In this chapter we focus exclusively on intrinsic curvature, so when we speak of "curvature" we are always referring to the intrinsic curvature. But both types are important, and there are deep connections between them. For more about this, see Appendix H.

The second question is, when does a manifold admit a flat metric? For example, no matter how hard you try, you cannot put a flat metric on a topological sphere. To answer this question one needs the following result, which is of sufficient importance that we present it here.

---

[11] Note that the Riemann tensor is built from the metric via the (torsion free, metric compatible) Levi-Civita connection, so our definition of flat space *assumes* the use of the Levi-Civita connection. More generally one may speak of a vector bundle equipped with a **flat connection** (a connection with zero curvature), but we do not do so because we are focused here on geometric manifolds.

[12] This flat torus can only be embedded *isometrically* in $\mathbb{R}^4$.

Theorem 8.7 *A manifold M is flat if and only if every point of M has a neighborhood isometric to an open subset of Euclidean space. Put another way, M is flat if and only if every point has a coordinate neighborhood U such that $g_{ij} = \delta_{ij}$ everywhere on U.*[13]

Basically this theorem says that the only way to get a flat manifold is to patch together undistorted bits of Euclidean space without kinks or bends. How could you do this and get anything but Euclidean space? We basically gave it away above. Observe that the flat cylinder may be viewed as the quotient $\mathbb{R}^2/\mathbb{Z}$, while the flat 2-torus was defined as $\mathbb{R}^2/\mathbb{Z}^2$ (both with the induced metric). It turns out that these examples are typical, in the sense that every flat manifold can be obtained from Euclidean space by identifying various points appropriately.[14] Thus, flat manifolds are locally identical to, but globally distinct from, Euclidean space.[15]

Remark Because a flat manifold is only locally isometric to Euclidean space, one sometimes defines a manifold whose Riemann tensor vanishes to be "locally flat" and reserves the word "flat" for Euclidean space itself. We do not use this terminology here, because it conflicts with standard usage in general relativity that *all* manifolds are locally flat in the sense that they all admit metrics that are flat to first order. (See Appendix D.)

We have already seen that if the metric tensor has constant components then the Riemann tensor vanishes, so one part of the proof of Theorem 8.7 is immediate. The converse, however, is much more involved. We offer two proofs, one here and one in Section 8.8, because each is instructive. The proof given here uses the language of differential forms and relies upon the Frobenius theorem.[16] First we need the following lemma.

Lemma 8.8 *Let $\{e_a\}$ be a frame field on a neighborhood U of a point p, and let ω and Ω be the corresponding matrices of connection and curvature forms, respectively. If $\Omega = 0$ everywhere on U, there exists a unique matrix A of smooth functions on U with $A(p) = I$ (the identity matrix) satisfying*

$$\omega = A^{-1} dA. \tag{8.57}$$

---

[13] If the signature is Lorentzian, replace δ_{ij} by η_{ij}.

[14] Technically, complete flat manifolds are quotients of Euclidean space by the free action of some discrete subgroup of the Euclidean group of rigid motions. (These manifolds are of particular importance in the field of crystallography.) For more information, see e.g. [91].

[15] We said earlier that manifolds "look like" Euclidean space near a point, but that only means they are locally homeomorphic to a piece of $\mathbb{R}^n$; it does not say anything about distances or angles, which are generally distorted by a homeomorphism, nor does it say anything about global topology.

[16] See Appendix F. See also [25], Section 7.4, which we follow here.

Proof Uniqueness is easy. If $\omega = B^{-1} dB$ with $B(p) = I$ then

$$d(AB^{-1}) = (dA)B^{-1} - AB^{-1} dB B^{-1} = (A\omega) B^{-1} - A (\omega B^{-1}) = 0.$$

Hence $AB^{-1} = AB^{-1}(p) = I$, whence $A = B$.

For the existence part, we embed M in an $(n + n^2)$-dimensional manifold E, equipped with n local coordinates $x^1, \ldots, x^n$ from U and n^2 additional coordinates $z^i{}_j$ ($1 \le i, j \le n$) chosen to satisfy the constraint $\det Z \ne 0$, where $Z := (z^i{}_j)$.[17] Define an $n \times n$ matrix of 1-forms on E by

$$\Psi = dZ - Z\omega.$$

Then, as $d\omega = -\omega \wedge \omega$,

$$d\Psi = -dZ \wedge \omega - Z \, d\omega = -(\Psi + Z\omega) \wedge \omega + Z\omega \wedge \omega = -\Psi \wedge \omega.$$

Therefore, by Frobenius' theorem, the distribution $\Psi = 0$ is completely integrable, with integral submanifold $\Sigma \subset E$. This means that we can solve the differential equations $\Psi = 0$ to get Σ. But those equations read

$$dZ = Z\omega,$$

which can be viewed as differential equations for the z_{ij} in terms of the x^i. By integrability, there is a matrix A of functions of x such that $Z = A$ solves this system subject to some initial condition that we may choose to be $A(p) = I$. Substituting A for Z gives

$$dA = A\omega \qquad \Rightarrow \qquad \omega = A^{-1} dA.$$

$\square$

Proof of Theorem 8.7 We want to show that vanishing curvature implies a flat metric. Let $\{e_{\hat{a}}\}$ be an orthonormal frame field on some subset U, with dual frame field $\{\theta^{\hat{a}}\}$. (The dual frame field is automatically orthonormal.) By Lemma 8.8 there exists a matrix A of smooth functions on U with $\omega = A^{-1} dA$, which by virtue of Exercise 8.7 must be orthogonal. Define a new coframe field by

$$v = A\theta. \tag{8.58}$$

As A is orthogonal and θ is orthonormal, v is also orthonormal. By (8.57),

$$dv = dA \wedge \theta + A \, d\theta = A\omega \wedge \theta - A\omega \wedge \theta = 0, \tag{8.59}$$

where we have used the fact that the torsion form (8.29) vanishes. Hence each 1-form v^i is closed, so by the Poincaré lemma there exist functions x^i on U such

---

[17] The *cognoscenti* will recognize a principal bundle lurking about.

that $v^i = dx^i$. It follows immediately that the dual frame field $\partial/\partial x^i$ is also orthonormal. In particular, $g_{ij} = \delta_{ij}$ everywhere on U.[18] $\qquad\square$

8.6 Parallel transport and geodesics

Suppose that you were living in a particular manifold. How could you tell whether it is flat or curved? It turns out there are several techniques one can use, each of which is helpful in different circumstances. They all involve the idea of parallel transport.

Let $I \subseteq \mathbb{R}$ be an interval, and let $\gamma : I \to M$ be a parameterized curve in M with tangent vector field $X = \gamma_*(d/dt)$. Let Y be any other vector field on M. We say that Y is **parallel transported** (or **parallel translated**) along γ if $\nabla_X Y = 0$. Intuitively, Y is parallel transported along γ if it remains "as parallel to itself as possible" along γ, so that the only change in Y along the curve is that induced by the curvature of the manifold in which it is situated.[19]

It is instructive to examine the equation for parallel transport in local coordinates. Suppose that $X = X^i \partial_i$ and $Y = Y^j \partial_j$. Then

$$
\begin{aligned}
\nabla_X Y &= \nabla_X(Y^j \partial_j) \\
&= X(Y^j)\partial_j + Y^j \nabla_X \partial_j \\
&= X(Y^j)\partial_j + Y^j X^i \nabla_{\partial_i} \partial_j \\
&= \left[X(Y^k) + \Gamma^k{}_{ij} X^i Y^j \right] \partial_k.
\end{aligned}
$$

Hence $\nabla_X Y = 0$, provided that

$$
X(Y^k) + \Gamma^k{}_{ij} X^i Y^j = 0. \tag{8.60}
$$

But (cf. (3.95))

$$
X^i = X(x^i) = \gamma_*\left(\frac{d}{dt}\right)(x^i) = \frac{d}{dt}(x^i \circ \gamma) = \frac{d\gamma^i(t)}{dt}, \tag{8.61}
$$

so (8.60) can be written as

$$
\frac{d}{dt}Y^k + \Gamma^k{}_{ij}\frac{d\gamma^i}{dt}Y^j = 0 \tag{8.62}
$$

or, equivalently,

$$
\frac{d\gamma^i}{dt}\left(\frac{\partial Y^k}{\partial x^i} + \Gamma^k{}_{ij} Y^j\right) = 0. \tag{8.63}
$$

---

[18] When the metric is indefinite we need to consider the positive definite and negative definite subspaces separately.

[19] For an extensive discussion of the geometry of parallel transport and its relation to covariant derivatives, see Chapter 10 of [60].

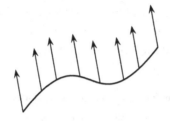

Figure 8.1 Parallel transport in Euclidean space.

An easy special case occurs when the manifold is ordinary Euclidean space. In Cartesian coordinates the Christoffel symbols vanish, and the covariant derivative reduces to the ordinary partial derivative. In that case parallel transport just means that the Cartesian components of Y do not change along the curve, which is the same thing as saying that Y remains parallel along the curve. (See Figure 8.1.)

> **EXERCISE 8.18** Show that parallel translation depends only on the curve γ and not on its parameterization.

> **EXERCISE 8.19** Let Y and Z be two vector fields parallel translated along the same curve. Show that their lengths and the angle between them remain constant along the curve. *Hint:* The length of Y is $g(Y, Y)^{1/2}$, and the cosine of the angle between Y and Z is $g(Y, Z)/[g(Y, Y)g(Z, Z)]^{1/2}$.

Note that (8.62) is *linear* in the components of Y, so in principle it admits a solution for all t. This means that, given a curve $\gamma(t)$ and a fixed tangent vector Y_0 at $\gamma(0)$, solving (8.62) yields a vector field Y (necessarily unique) such that $Y_{\gamma(t)}$ is the parallel transported continuation of Y_0 along the length of the curve. The map $\vartheta_t : T_{\gamma(0)}M \to T_{\gamma(t)}M$ given by $Y_0 \mapsto Y_{\gamma(t)}$ is called the **parallel transport** (or **parallel translation**) **map**. It depends on both the curve γ and the connection ∇.[20]

The curve γ is a **geodesic** (or **autoparallel**) if $\nabla_X X = 0$. Alternatively, a geodesic is a curve whose tangent vector is parallel transported along itself. On a Riemannian manifold, geodesics are the shortest-distance paths between any two points.[21] Substituting X for Y in (8.62) gives the **geodesic equation**,

$$\frac{d^2\gamma^k}{dt^2} + \Gamma^k{}_{ij}\frac{d\gamma^i}{dt}\frac{d\gamma^j}{dt} = 0. \tag{8.64}$$

The geodesic equation is also written as

$$\frac{d^2x^k}{dt^2} + \Gamma^k{}_{ij}\frac{dx^i}{dt}\frac{dx^j}{dt} = 0, \tag{8.65}$$

[20] We investigate the parallel transport map in more detail in Section 8.8. For an instance of the parallel transport map on the sphere, see Exercise 8.41.

[21] The proof of this is not too difficult, but we omit it because it requires variational calculus.

where as usual, $x^i(t) = (\gamma^* x^i)(t) = \gamma^i(t)$. Observe that (8.65) is *nonlinear*. This means that, in general, the geodesic equation is much more difficult to solve than the parallel transport equation, and questions about how far a geodesic can be extended are often quite subtle.

> **EXERCISE 8.20** Show that a geodesic *does* depend on its parameterization, by showing that the only transformation of parameters leaving (8.64) (or (8.65)) form invariant are *affine* parameter changes: $t \rightarrow s = at + b$. In other words, two curves could have the same images but different parameterizations and one could be a geodesic while the other is not. For this reason, the 't' in (8.64) (or (8.65)) is called an **affine parameter**.

> **EXERCISE 8.21** Show that, for any geodesic,
>
> $$\frac{d}{dt}\left(g_{ij}\frac{dx^i}{dt}\frac{dx^j}{dt}\right) = 0, \qquad (8.66)$$
>
> or, in other words, $(ds/dt)^2 = C$ for some constant C which, by rescaling, we may choose to satisfy $|C| = 1$. On a Lorentzian manifold we have three choices: (i) $C = -1$, in which case $t = \tau$ is the **proper time** along the curve: (ii) $C = 0$, in which case the curve is a **null geodesic** (and t is any affine parameter); and (iii) $C = 1$, in which case $t = s$ is the **proper distance** along the curve or, equivalently, the curve is **arc length parameterized**. (The latter is the only possibility if the manifold is Riemannian.) *Hint:* This is essentially a reformulation of an earlier exercise.

In Euclidean space in Cartesian coordinates the Christoffel symbols vanish, and (8.65) reduces to

$$\frac{d^2 x^k}{dt^2} = 0, \qquad (8.67)$$

whose solution is a straight line. Even on nonflat manifolds it turns out that we can always choose coordinates for which the Christoffel symbols vanish at a point,[22] so locally a geodesic is just a straight line.[23] Globally a geodesic may be curved, but locally it is straight. Therefore geodesics are the closest analogues of straight lines on a manifold. Also, we can think of $\nabla_X X$ as the "acceleration vector" of the curve, because that's what it reduces to in Euclidean space. In general relativity, a particle that experiences no net force has no acceleration and so follows a geodesic.[24]

---

[22] See Appendix D.

[23] Technically, this is not quite correct. Even if the Christoffel symbols vanish at a some point they may not vanish at a nearby point, so integrating (8.65) would be problematic. Fortunately, it can be shown that there exist coordinates, called **Fermi normal coordinates**, that have the property that the Christoffel symbols vanish in the neighborhood of any point along a geodesic. For a discussion see [60], §13.6.

[24] Gravity is not a force *per se* in general relativity. Instead, spacetime curvature is the manifestation of gravity. Particles moving under the gravitational influence of other bodies and subject to no other forces follow geodesics in a curved Lorentzian manifold.

8.7 Jacobi fields and geodesic deviation

Thinking physically for a moment, imagine two proximate bodies dropped near the Earth's surface. They both fall straight down, so their trajectories are parallel. But if we place the same two bodies far above the Earth's surface and let them go, each one is pulled towards the center of the Earth along trajectories that would meet at the Earth's center. In Einstein's theory of general relativity, this occurs because the Earth's mass causes spacetime to curve around the planet and the two freely falling bodies move along geodesics in this curved spacetime. Thus, the effect of curvature is to cause **geodesic deviation**, which measures how nearby geodesics converge or diverge.

Consider a family $\gamma(t, s)$ of geodesics. Here t is the parameter distance along a single curve, while s labels the different geodesics, of the family. We therefore have a family of tangent vector fields X_s associated to each geodesic, as well as tangent vector fields J_t connecting corresponding points on nearby geodesics, called **Jacobi fields**.

The key feature of the Jacobi fields J_t is that they commute with the fields X_s, so that

$$[X_s, J_t] = 0. \tag{8.68}$$

This follows because we may think of X_s as d/dt and J_t as d/ds, and mixed partial derivatives commute.[25] Using (7.36) (and dropping the subscripts s and t for typographical ease) we can write

$$\nabla_X \nabla_J X - \nabla_J \nabla_X X = R(X, J)X. \tag{8.69}$$

However, ∇ is torsion free so, by (8.68),

$$\nabla_X J - \nabla_J X = [X, J] = 0. \tag{8.70}$$

Combining (8.69) and (8.70) gives

$$\nabla_X \nabla_X J - \nabla_J \nabla_X X = R(X, J)X; \tag{8.71}$$

but the curves are all geodesics, so $\nabla_X X = 0$ and (8.71) reduces to

$$\nabla_X \nabla_X J = R(X, J)X, \tag{8.72}$$

---

[25] Alternatively, the X_s vectors for nearby geodesics are "Lie dragged" along the integral curves of J_t (and vice versa), so $\mathcal{L}_{X_s} J_t = 0$. Equivalently (cf. Appendix F), $\gamma(t, s)$ is a two-dimensional integral submanifold of M.

which is known as the **geodesic deviation equation** or **Jacobi's equation**. It measures how J changes as one moves along the geodesics. For example, in flat space the curvature tensor vanishes and (8.72) reduces to

$$\frac{d^2 J^i}{dt^2} = 0,$$

whose solution is linear: $J^i = a^i t + b^i$. In flat space, initially parallel lines remain parallel (Euclid's parallel postulate). But, in the presence of nontrivial curvature, Euclid's postulate fails; the resulting geometries are therefore called nonEuclidean.

8.8 Holonomy

Geodesic deviation is a local phenomenon having to do with how nearby geodesics behave. There is a more global way to detect the presence of curvature that is sometimes quite useful; it employs parallel transport around closed curves.[26] We have seen that, given a connection ∇ and a curve γ, we get a map ϑ_t between tangent spaces that maps vectors to vectors by parallel transport. What would happen if we were to carry a vector around a closed curve? The answer is that it may not return to its original state. The amount by which it differs from its original state is related to the curvature of the manifold.

   To illustrate, let's see what happens on the unit 2-sphere S^2. By Exercise 8.42, the geodesics of S^2 are just great circles. Consider a geodesic triangle on the sphere consisting of the three great circle arcs γ_1, γ_2, and γ_3, where γ_1 starts at the north pole and goes down a meridian (longitude line) to the equator, γ_2 goes along the equator to the meridian at ψ, and γ_3 goes along the meridian at ψ back to the north pole. Let A be the tangent vector to γ_1 pointing away from the north pole. (See Figure 8.2.)

   By definition, the tangent vector of a geodesic is parallel transported along itself. Let X_i be the tangent vector to γ_i. Then at the north pole $A = X_1$. It points due south on the zeroth longitude. After parallel transport along γ_1, A is still X_1, so it still points due south. Now we transport A along γ_2. By Exercise 8.19 parallel transport preserves the angle between vectors, so A maintains its angle with X_2 at all times. At the start of its journey it is pointing due south at 90 degrees to X_2, so at the end of its journey it is still pointing due south. Finally, during its journey back along γ_3 it remains pointing due south, so when it returns to the north pole it is pointing at an angle of ψ relative to its original direction. If we had tried this little exercise in flat space the vector A would have returned to its original value, so the

---

[26] In this section we assume that M is connected.

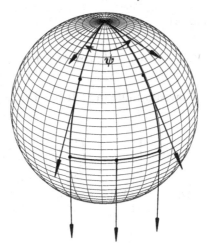

Figure 8.2 Parallel transport of a vector about a geodesic triangle on the sphere.

angular deviation of a vector after parallel transport on a manifold is an indication that we are no longer in flat space.[27]

To discover the relationship between holonomy and curvature, we require an explicit form for the parallel transport operator. The differential equation (8.62) defining parallel transport can be written as

$$\frac{dY}{dt} + AY = 0, \tag{8.73}$$

where Y is a vector and A is a matrix with components $A^k{}_j = \Gamma^k{}_{ij}(d\gamma^i/dt)$. The thing that makes this equation tricky is that A depends on t (through γ). If it did not, the solution of (8.73) would just be

$$Y(t) = e^{-tA}Y(0), \tag{8.74}$$

where for simplicity we write $Y(t)$ instead of $Y(\gamma(t)) = Y_{\gamma(t)}$; the exponential of a matrix is defined by its power series expansion:

$$e^{-tA} = 1 - tA + \frac{(tA)^2}{2!} \pm \cdots . \tag{8.75}$$

But in fact A generally does depend on t, so we need to be more clever.

The way forward is to try an iterative solution. Write (8.73) as

$$\frac{dY}{dt} = -AY,$$

---

[27] Nature abounds with examples of holonomy. For example, one can view the rotation of the plane of oscillation of a Foucault pendulum as being due to parallel translation around a latitude (see [36] or [71]). The relevant computation is performed in Exercise 8.41. Many other examples can be found in the compendium [76].

and integrate both sides to get

$$Y(t) = Y(0) - \int_0^t dt_1 \, A(t_1) Y(t_1),$$

where $A(t) := A(\gamma(t))$. Now substitute the left-hand side into the right-hand side to get

$$Y(t) = Y(0) - \int_0^t dt_1 \, A(t_1) \left[Y(0) - \int_0^{t_1} dt_2 \, A(t_2) Y(t_2) \right]$$

$$= Y(0) - \left[\int_0^t dt_1 \, A(t_1) \right] Y(0) + \int_0^t dt_1 \int_0^{t_1} dt_2 \, A(t_1) A(t_2) Y(t_2).$$

Continuing in this manner we get

$$Y(t) = \left[\sum_{n=0}^{\infty} (-1)^n \int_0^t dt_1 \int_0^{t_1} dt_2 \cdots \int_0^{t_{n-1}} dt_n \, A(t_1) \cdots A(t_n) \right] Y(0). \quad (8.76)$$

Provided A does not blow up anywhere along γ, this infinite sum converges to a solution of (8.73).

Equation (8.76) is often rewritten using a suggestive kind of shorthand notation. You will notice that the matrices are ordered in such a way that "later is on the left", which means $t_1 \leq t_2 \leq \cdots \leq t_n$; this motivates the introduction of a **path ordering operator** $\mathcal{P}$, which reorders any matrix product in such a way that "later is on the left". If $i_1 \leq \cdots \leq i_n$, and if $\sigma \in \mathfrak{S}_n$ is any permutation,

$$\mathcal{P}\left(A\left(t_{i_{\sigma(1)}}\right) \cdots A(t_{i_{\sigma(n)}}) \right) = A(t_{i_1}) \cdots A(t_{i_n}). \quad (8.77)$$

With this notation the multiple integral in (8.76) can be written

$$\frac{1}{n!} \int_0^t dt_1 \int_0^t dt_2 \cdots \int_0^t dt_n \, \mathcal{P}(A(t_1) \cdots A(t_n)), \quad (8.78)$$

or, more simply (and with a shocking abuse of notation),

$$\frac{1}{n!} \mathcal{P} \left(\int_0^t A(t) \, dt \right)^n. \quad (8.79)$$

EXERCISE 8.22 Prove that (8.78) does indeed equal the multiple integral in (8.76).

Finally, then, we define the **path ordered exponential** by

$$\mathcal{P} e^{-\int_0^t A(t) \, dt} := \sum_{n=0}^{\infty} \frac{(-1)^n}{n!} \mathcal{P} \left(\int_0^t A(t) \, dt \right)^n, \quad (8.80)$$

whereupon (8.76) can be written, by analogy with (8.74) (to which it reduces when A is constant),

$$Y(t) = \mathcal{P}e^{-\int_0^t A(t)\,dt}Y(0). \tag{8.81}$$

Thus, we see that the path ordered exponential operator is precisely the parallel transport map ϑ_t (see the text before (8.64)):

$$\vartheta_t = \mathcal{P}e^{-\int_0^t A(t)\,dt}. \tag{8.82}$$

More generally, we write

$$\vartheta_{st}(\gamma, \nabla) := \mathcal{P}e^{-\int_s^t A(t)\,dt} \tag{8.83}$$

for the operator parallel transporting vectors along γ from $T_{\gamma(s)}M$ to $T_{\gamma(t)}M$ relative to the connection ∇. If $\gamma = \gamma_1 \cdots \gamma_m$ is a piecewise smooth path[28] then

$$\vartheta_{0,1}(\gamma, \nabla) = \vartheta_{0,1}(\gamma_m, \nabla) \cdots \vartheta_{0,1}(\gamma_1, \nabla). \tag{8.84}$$

By Exercise 8.18, this is independent of the parameterizations of the individual pieces and therefore depends only on the curve γ (and the connection ∇).

> **EXERCISE 8.23** The inverse of a curve γ is the curve γ^{-1} given by $\gamma^{-1}(t) = \gamma(1-t)$. The trivial curve (at p) is the map $\epsilon_p : [0, 1] \to M$ with $\epsilon_p(t) = p$. Show that $\vartheta(\epsilon_p, \nabla) = id$ and that $\vartheta(\gamma^{-1}, \nabla) = [\vartheta(\gamma, \nabla)]^{-1}$. *Hint:* For the second claim recall that $A(t) = \Gamma^k{}_{ij}(d\gamma^i/dt)$. Note that $\gamma\gamma^{-1} \neq \epsilon_p$.

Now consider the set $\mathcal{C}_p$ of all loops based at p.[29] Although these loops do not form a group (because, for example, for nontrivial $\gamma \in \mathcal{C}_p$ we have $\gamma\gamma^{-1} \neq \epsilon_p$), Exercise 8.23 shows that their holonomies do, because the product of two loops is another loop. The **holonomy group of ∇ based at p** is

$$H(\nabla; p) := \{\vartheta(\gamma, \nabla) : \gamma \in \mathcal{C}_p\}. \tag{8.85}$$

> **EXERCISE 8.24** Show that if γ is a curve with $\gamma(0) = p$ and $\gamma(1) = q$ then
>
> $$H(\nabla; q) = \vartheta(\gamma, \nabla)H(\nabla; p)\vartheta(\gamma^{-1}, \nabla), \tag{8.86}$$
>
> i.e., every $H' \in H(\nabla; q)$ is of the form $\vartheta H \vartheta^{-1}$ for $H \in H(\nabla; p)$. One says that holonomy groups at different basepoints are *conjugate*.

The **restricted holonomy group** $\widetilde{H}(\nabla; p)$ of ∇ based at p is the holonomy group of all the *contractible* loops at p. It is clearly a subgroup of the full holonomy group. It is also geometrically significant.

---

[28] For the definition of a product of paths, see Exercise 4.11.
[29] A loop based at p is just a path that starts and ends at p. See Exercise 4.11.

Theorem 8.9 $\widetilde{H}(\nabla; p)$ *is trivial (i.e., it consists only of the identity element) if and only if the curvature vanishes everywhere.*

Proof In Appendix E we show that the holonomy around an infinitesimal coordinate loop of parameter area α^2 centered at a point p can be written

$$\vartheta = 1 + \alpha^2 \mathcal{R} + \mathcal{O}(\alpha^3), \tag{8.87}$$

where $\mathcal{R}$ is basically the Riemann curvature tensor at p and $\mathcal{O}(\alpha^3)$ represents terms of order α^3 and higher. It then follows that if the holonomy group is trivial then the curvature tensor must vanish at every point.

The converse follows from a famous result of Ambrose and Singer [3] (see also [47]), which is unfortunately beyond the scope of this discussion. Using only (8.87) we can give a "physicist-style" proof.[30] Consider a coordinate square of side length L. Divide it up into N^2 smaller squares of side length L/N. By traversing the perimeter of each square in the same direction, one ends up walking around the individual loops in such a way that the internal edges are traversed in opposite directions. Thus, the holonomy of the larger loop is just the product of the holonomies of the smaller loops (because the holonomies along the internal edges cancel pairwise). If the curvature tensor vanishes then from (8.87) the holonomy of the large square is approximately

$$\vartheta \approx (1 + \mathcal{O}((L/N)^3))^{N^2} \approx 1 + \mathcal{O}(1/N),$$

which vanishes in the limit that $N \to \infty$. (This trick only works if the large loop is contractible, which is why that condition is required.)[31] $\square$

Remark The theorem is not true if we replace the restricted holonomy group by the full holonomy group because, even if the curvature tensor vanishes, a noncontractible loop can still have a nontrivial holonomy if the manifold fails to be simply connected. An example is provided by the Aharanov-Bohm effect. (Look it up!)

We can now take care of some unfinished business.

Second proof of Theorem 8.7 Pick a point $p \in M$ and a vector $X_p \in T_p M$. If the Riemann curvature tensor vanishes then, by Theorem 8.9, parallel transport is path independent.[32] This means that we obtain a *unique* vector field X by parallel transporting X_p from one place to another in M. Moreover, since X is parallel transported along every curve, $\nabla_Y X = 0$ for any vector field Y. In this way we may

---

[30] The following argument is adapted from [86].

[31] Of course, the converse would also follow from Theorem 8.7 but, as we wish to use Theorem 8.9 to prove Theorem 8.7, we cannot very well invoke Theorem 8.7 here.

[32] This is just the usual argument. Let γ_1 and γ_2 be two paths from p to q, say. Then, as parallel transport around a closed loop is trivial, $1 = \vartheta_{pq}(\gamma_1)\vartheta_{qp}(\gamma_2^{-1}) = \vartheta_{pq}(\gamma_1)[\vartheta_{pq}(\gamma_2)]^{-1}$, so $\vartheta_{pq}(\gamma_1) = \vartheta_{pq}(\gamma_2)$.

extend any vector at p to a covariantly constant vector field on M. In particular, given an orthonormal frame $X_1, \ldots, X_p$ we can extend it in the same fashion to a global frame field on M. As parallel transport preserves lengths and angles, this frame field is everywhere orthonormal so, with respect to this frame, $g_{ij} = \delta_{ij}$. As the connection is torsion free we have

$$[X_i, X_j] = \nabla_{X_i} X_j - \nabla_{X_j} X_i = 0, \tag{8.88}$$

so by Theorem 8.2 there exist local coordinates x^i such that $X_i = \partial/\partial x^i$. These are the coordinates we seek. $\qquad \square$

This proof reveals the relationship between holonomic frame fields and holonomy: if the holonomy group is trivial then every frame field is holonomic (on some neighborhood), whereas this is not true if there is nontrivial holonomy.

8.9 Hodge theory

In this section we assume that M is an n-dimensional compact, orientable, Riemannian manifold without boundary. In particular, we assume that the metric is positive definite. It is a deep result of Hodge that, on such a manifold, every k-form admits a canonical decomposition with very nice properties and this decomposition provides information about the cohomology of M. We briefly examine some of these ideas here.

Let g be the metric on M. In terms of local coordinates $x^1, \ldots, x^n$ on a neighborhood U of a point p, we define an n-form

$$\sigma = \sqrt{G}\, dx^1 \wedge \cdots \wedge dx^n, \qquad G := \det g_{ij}. \tag{8.89}$$

As M is orientable, this form extends to a global nowhere-vanishing n-form on M. It is called the (canonical) **volume form** of M.[33]

> **EXERCISE 8.25** Verify this last assertion, namely that σ extends to a global nowhere-vanishing n-form. *Hint:* You may want to complete Exercise 8.52(c) first.

> **EXERCISE 8.26** Let $\theta^1, \ldots, \theta^n$ be an orthonormal basis of 1-forms on a neighborhood U of p. Show that on U we may write
>
> $$\sigma = \theta^1 \wedge \cdots \wedge \theta^n. \tag{8.90}$$

The manifold M is equipped with a metric, so we have an inner product in each tangent space $T_p M$ as well as in each cotangent space $T_p^* M$. Suppose that we have

---

[33] Equation (8.89) also defines the canonical volume form on a pseudo-Riemannian manifold, provided one replaces $\sqrt{G}$ by $\sqrt{|G|}$.

a k-form ω on M. Then, as in Section 2.9, we can define its Hodge dual $\star\omega$ to be the unique $n - k$ form satisfying

$$\omega \wedge \mu = g(\star\omega, \mu)\sigma \tag{8.91}$$

for any $(n - k)$-form μ, where σ is the volume form defined above and g is the induced metric (2.55) applied to $(n - k)$-forms. All the properties of the Hodge star operator then carry through as before.

The set $\Omega^k(M)$ of all differential k-forms on M is a vector space under the operations of pointwise addition and scalar multiplication. Hodge's fruitful idea was to turn this space into an inner product space using his star operator. Thus, given $\eta, \lambda \in \Omega^k(M)$, we define their inner product by

$$(\eta, \lambda) = \int_M \eta \wedge \star\lambda. \tag{8.92}$$

By (2.61) (and the fact that $(-1)^d = 1$ on a Riemannian manifold)

$$(\eta, \lambda) = \int_M g(\eta, \lambda)\sigma. \tag{8.93}$$

It follows that Hodge's inner product is positive definite.

Now define a new operator $\delta : \Omega^k(M) \rightarrow \Omega^{k-1}(M)$ by the rule

$$\delta\eta = (-1)^{nk+n+1} \star d \star \eta. \tag{8.94}$$

The operator δ is sometimes called the **co-differential operator**.

EXERCISE 8.27 Show that $\delta^2 = 0$.

The key fact about the operator δ is that it is precisely the adjoint of the differential operator d with respect to the Hodge inner product:

Theorem 8.10 *If $\eta \in \Omega^k(M)$ and $\lambda \in \Omega^{k+1}(M)$ then*

$$(d\eta, \lambda) = (\eta, \delta\lambda). \tag{8.95}$$

Proof The manifold M is compact without boundary so, by Stokes' theorem and the properties of the exterior derivative operator,

$$0 = \int_M d(\eta \wedge \star\lambda) = \int_M d\eta \wedge \star\lambda + (-1)^k \int_M \eta \wedge d \star \lambda.$$

Note that λ is a $(k+1)$-form, so $\star\lambda$ is an $(n-k-1)$-form and $d\star\lambda$ is an $(n-k)$-form. Thus, from the definition (8.94),

$$\delta\lambda = (-1)^{n(k+1)+n+1}\star d\star\lambda = (-1)^{nk+1}\star d\star\lambda.$$

Employing (2.60) gives

$$\star \delta \lambda = (-1)^{nk+1} \star^2 d \star \lambda = (-1)^{nk+1} (-1)^{k(n-k)} d \star \lambda = (-1)^{k^2+1} d \star \lambda.$$

Combining everything we obtain

$$0 = (d\eta, \lambda) + (-1)^{k(k+1)+1} (\eta, \delta\lambda),$$

from which the conclusion follows. $\qquad \square$

Next we introduce the elegant **Laplace–de Rham operator** (sometimes also called the **Hodge–de Rham Laplacian**),

$$\Delta := d\delta + \delta d, \qquad (8.96)$$

which generalizes the usual Laplacian operator.[34] We say that a k-form ω is closed if $d\omega = 0$, **co-closed** if $\delta\omega = 0$, and **harmonic** if $\Delta\omega = 0$. Obviously, if ω is closed and co-closed then it is harmonic. Interestingly, the converse also holds. If ω is harmonic then, by Theorem 8.10,

$$0 = (\Delta\omega, \omega) = (d\delta\omega, \omega) + (\delta d\omega, \omega) = (\delta\omega, \delta\omega) + (d\omega, d\omega).$$

But the inner product is positive definite, so both terms on the right must vanish separately and this can only happen if $d\omega = \delta\omega = 0$.

Remark It follows that Δ is an elliptic (positive definite) self-adjoint differential operator. That is, $(\Delta\lambda, \eta) = (\lambda, \Delta\eta)$ and $(\Delta\omega, \omega) \geq 0$ with equality holding only if $\Delta\omega = 0$.

> **EXERCISE 8.28** Assume that M is connected and that *dim* $M = n$. Show that any harmonic function on M is necessarily a constant function. Furthermore, show that any harmonic n-form is a constant multiple of the volume form. *Hint for the second part*: Any top-dimensional form is a smooth function times the volume form. Now use the first part together the fact (which you should prove) that $\star\Delta = \Delta\star$.

This brings us to Hodge's powerful result.

Theorem 8.11 (Hodge decomposition theorem) *Let M be a compact, orientable, Riemannian manifold without boundary. Any k-form ω on M can be decomposed uniquely into the sum of a closed form, a co-closed form, and a harmonic form:*

$$\omega = d\alpha + \delta\beta + \gamma. \qquad (8.97)$$

---

[34] See Exercise 8.56.

Proof Uniqueness is easy. By linearity, it suffices to show that

$$d\alpha + \delta\beta + \gamma = 0 \tag{8.98}$$

implies that $d\alpha = \delta\beta = \gamma = 0$. Applying d to (8.98) (and remembering that γ is harmonic and therefore closed) gives

$$d\delta\beta = 0 \quad \Rightarrow \quad (d\delta\beta, \beta) = 0 \quad \Rightarrow \quad (\delta\beta, \delta\beta) = 0 \quad \Rightarrow \quad \delta\beta = 0. \tag{8.99}$$

Applying δ to (8.98) gives $d\alpha = 0$ by a similar argument, so $\gamma = 0$ as well. Existence boils down to proving an existence theorem for the solution of an elliptic second-order partial differential equation, so we merely refer the intrepid reader to ([29], Appendix C, or [88], §6.8). $\qquad\square$

If ω is closed then the same argument as given above shows that we can decompose ω uniquely as $\omega = d\alpha + \gamma$. In other words, ω can be expressed in a canonical way as the sum of a harmonic form and an exact form. In particular, γ is a cohomology class representative for $[\omega]$. Therefore, we have another important theorem of Hodge.

Theorem 8.12 (Hodge) *Let M be a compact, orientable, Riemannian manifold M without boundary. Every closed form ω on M on has a unique harmonic cohomology class representative. In other words, the de Rham cohomology $H_{\mathrm{dR}}^k(M)$ is isomorphic to $Y_k(M)$, the space of harmonic k-forms on M.*

We can use Hodge's theorem to prove an important result in cohomology.

Theorem 8.13 (Poincaré duality) *Let M be a compact, connected, orientable, n-dimensional manifold without boundary. For every k,*

$$H_{\mathrm{dR}}^{n-k}(M) \cong H_{\mathrm{dR}}^k(M)^*. \tag{8.100}$$

In particular, $\beta_{n-k} = \beta_k$, where $\beta_k = \dim H_{\mathrm{dR}}^k(M)$ is the kth Betti number of M.

Proof Let $\eta \in \Omega^k(M)$ and $\lambda \in \Omega^{n-k}(M)$ be closed. There is a natural pairing between $H_{\mathrm{dR}}^k(M)$ and $H_{\mathrm{dR}}^{n-k}(M)$, given by

$$([\eta], [\lambda]) := \int_M \eta \wedge \lambda. \tag{8.101}$$

This bilinear form is well defined, as can be shown in Exercise 8.29 below. It is also nondegenerate. To show this, we must show that for any nonzero class $[\eta]$ there exists a nonzero class $[\lambda]$ such that $([\eta], [\lambda]) \neq 0$.

This can be done as follows. Pick a Riemannian metric on M.[35] Let $[\eta]$ be a nonzero class. By Theorem 8.12 we may assume that η is harmonic. Define $\lambda := \star\eta$. It is closed because η is co-closed and, by construction,

$$([\eta], [\lambda]) = (\eta, \eta) \neq 0 \tag{8.102}$$

so by the Riesz lemma the proof is complete. $\qquad\qquad\qquad\qquad\qquad\square$

EXERCISE 8.29 Show that the pairing (8.101) is well defined (i.e., independent of class representative).

EXERCISE 8.30 Show that a closed (compact, without boundary) connected orientable manifold is never contractible. *Hint:* What does Poincaré duality say about the top-dimensional cohomology?

Additional exercises

8.31 Covariant derivatives of tensor fields and semicolon notation. In a coordinate basis, the covariant derivative of a tensor field T of type (r, s) has components

$$
\begin{aligned}
(\nabla_k T)&^{i_1\ldots i_r}{}_{j_1\ldots j_s} \\
&= T^{i_1\ldots i_r}{}_{j_1\ldots j_s,k} \\
&\quad + \Gamma^{i_1}{}_{k\ell} T^{\ell i_2\ldots i_r}{}_{j_1\ldots j_s} + \Gamma^{i_2}{}_{k\ell} T^{i_1\ell\ldots i_r}{}_{j_1\ldots j_s} + \cdots + \Gamma^{i_r}{}_{k\ell} T^{i_1\ldots i_{r-1}\ell}{}_{j_1\ldots j_s} \\
&\quad - \Gamma^{\ell}{}_{kj_1} T^{i_1\ldots i_r}{}_{\ell j_2\ldots j_s} - \Gamma^{\ell}{}_{kj_2} T^{i_1\ldots i_r}{}_{j_1\ell\ldots j_s} + \cdots - \Gamma^{\ell}{}_{kj_s} T^{i_1\ldots i_r}{}_{j_1\ldots j_{s-1}\ell},
\end{aligned}
\tag{8.103}
$$

where $\nabla_k := \nabla_{\partial_k}$ and $f_{,i} := \partial_i f$. In shorthand notation (which, in fairness, is used by both mathematicians and physicists), the left-hand side of this expression is often written

$$\nabla_k T^{i_1\ldots i_r}{}_{j_1\ldots j_s} \qquad \text{or} \qquad T^{i_1\ldots i_r}{}_{j_1\ldots j_s;k}. \tag{8.104}$$

The latter expression defines the meaning of the **semicolon notation**.

Confirm (8.103) in a special case by verifying that it holds for a tensor field of type $(1, 1)$. In other words, show that

$$T^i{}_{j;k} = T^i{}_{j,k} + \Gamma^i{}_{k\ell} T^{\ell}{}_j - \Gamma^{\ell}{}_{kj} T^i{}_{\ell}. \tag{8.105}$$

Confirm the general formula if you really like indices.

---

[35] We can do this by virtue of Theorem 8.1. Apparently lawyers are useful for something.

8.32 Commas to semicolons   Show that the expression for the Lie derivative given in (3.124) remains valid if the ordinary derivatives are replaced everywhere by covariant derivatives or, equivalently, if the commas in (3.124) are replaced by semicolons (as defined in the previous exercise).

8.33 Covariant constancy of the metric   Show that the metric tensor is **covariantly constant**, namely that

$$\nabla_X g = 0, \tag{8.106}$$

by showing that, in a coordinate basis, $g_{ij;k} = 0$. *Hint:* Use (8.103) and metric compatibility.

8.34 Covariant derivatives commute with contractions   In a coordinate basis, the raising and lowering of indices commutes with semicolons. Confirm this assertion in a special case by verifying the following equation:

$$g_{im} T^i{}_{j;\ell} = T_{mj;\ell}. \tag{8.107}$$

Hint: Use the previous exercise.

8.35 The Bianchi identity revisited   Show that, in a coordinate basis, the Bianchi identity (7.20) becomes

$$R^i{}_{j[k\ell;m]} = 0, \tag{8.108}$$

where, as usual, square brackets denote antisymmetrization. Equivalently, show that the Bianchi identity can also be written as

$$[\nabla_k, [\nabla_\ell, \nabla_m]] + [\nabla_\ell, [\nabla_m, \nabla_k]] + [\nabla_m, [\nabla_k, \nabla_\ell]] = 0.$$

Hint: You may wish to use equations (8.35), (8.45), and (8.103).

8.36 Ricci identity   For any tensor field $\Psi^{i_1 \dots i_r}{}_{j_1 \dots j_s}$ of type (r, s), **Ricci's identity** is

$$[\nabla_k, \nabla_\ell] \Psi^{i_1 \dots i_r}{}_{j_1 \dots j_s} = \sum_{p=1}^{r} \Psi^{i_1 \dots i_{p-1} i i_{p+1} \dots i_r}{}_{j_1 \dots j_s} R^{i_p}{}_{ik\ell}$$

$$- \sum_{q=1}^{s} \Psi^{i_1 \dots i_r}{}_{j_1 \dots j_{q-1} j j_{q+1} \dots j_s} R^j{}_{j_q k\ell}, \tag{8.109}$$

where $[A, B] := AB - BA$. Prove this formula for the special case $r = 2$, $s = 0$. *Warning:* Be careful – partial derivatives and Christoffel symbols are not tensors so you must be clever when applying (8.103). Alternatively, you could do the proof abstractly by writing $\Psi = X \otimes Y$ for two vector fields or 2-forms X and Y and then extend everything to higher-order tensors by linearity.

8.37 Killing fields   The vector field X is called a **Killing field** (after the German mathematician Wilhelm Killing) if $\mathcal{L}_X g = 0$, where g is the metric tensor.

(a) Show that X is a Killing field if and only if

$$g(\nabla_Y X, Z) + g(Y, \nabla_Z X) = 0, \tag{8.110}$$

where ∇ is the Levi-Civita connection. This is called **Killing's equation**. *Hint:* Use (3.121).

(b) If $X = X^i \partial_i$ in a coordinate basis, show that Killing's equation can be written as

$$X_{i;j} + X_{j;i} = 0, \tag{8.111}$$

for all pairs i, j, where $X_i := g_{ik} X^k$.

(c) Let γ be a geodesic with tangent vector Y. Show that $g(X, Y)$ is a constant along γ.

(d) Suppose that we choose local coordinates such that $X = \partial/\partial x^1$. (By the existence and uniqueness theorem for ordinary differential equations we can always do this. See Lemma F.4.) Show that the components of the metric tensor do not depend on x^1, i.e., show that $\partial g_{jk}/\partial x^1 = 0$.

(e) More generally, show that the flow of a Killing field is an isometry of the manifold. *Hint:* Use $\mathcal{L}_X g = 0$ and integrate.

(f) Show that the set of Killing fields forms a Lie algebra under the Lie bracket. *Remark:* The resulting Lie algebra is precisely that of the symmetry group of the manifold.

8.38 Spatial symmetries   In Euclidean space the Christoffel symbols vanish so, in the standard coordinate basis, Killing's equation reduces to

$$\xi_{i,j} + \xi_{j,i} = 0,$$

where $\xi_{i,j} = \partial \xi_i / \partial x^j$.

(a) Show that in three-dimensional Euclidean space the general solution to Killing's equation is

$$\xi_i = \alpha_{ij} x^j + a_i, \tag{8.112}$$

where α_{ij} is a constant antisymmetric tensor and a_i is a constant vector. *Hint:* Show that $\xi_{i,jk} = 0$ by alternating the application of Killing's equation and the fact that mixed partials commute. *Remark:* Observe that there are three independent degrees of freedom from α_{ij} and another three from a_i giving a total of six independent Killing vectors. It turns out that this is the maximal number of Killing fields on any three-dimensional manifold (although, of course, there can only be three linearly independent Killing fields at any given point).

(b) According to Exercise 8.37 the (co-)vector fields (8.112) are the infinitesimal generators of symmetries. With this interpretation the a_i represent the three independent translation symmetries of Euclidean space. We wish to interpret the other Killing fields geometrically. Consider the Killing field with $a_i = 0$, $\alpha_{12} = -\alpha_{21} = 1$, and all other components of α_{ij} equal to zero. Show that the integral curves of this Killing field are circles around the origin in the xy plane, thereby verifying that this Killing field is the infinitesimal generator of rotations about the z axis. (The rotations about the x and y axes are obtained similarly, from $\alpha_{23} = 1$ and $\alpha_{31} = 1$, respectively.) *Remark:* The solutions of Killing's equation give all the *continuous* symmetries of Euclidean space, namely translations and rotations. There are additional *discrete* symmetries, namely the reflections $x^i \rightarrow -x^i$. The set of all these symmetries is called the **Euclidean group** and usually denoted $E(n)$, where n is the dimension of the Euclidean space. It is a noncompact Lie group.

(c) Show that, in spherical polar coordinates, the three (normalized) rotational Killing fields of Euclidean space about the x, y, and z axes are, respectively,

$$\xi^{(1)} = -\sin\phi\,\frac{\partial}{\partial\theta} - \cot\theta\cos\phi\,\frac{\partial}{\partial\phi},$$
$$\xi^{(2)} = \cos\phi\,\frac{\partial}{\partial\theta} - \cot\theta\sin\phi\,\frac{\partial}{\partial\phi},$$
$$\xi^{(3)} = \frac{\partial}{\partial\phi}.$$

8.39 Electric and magnetic monopoles in spherically symmetric spacetimes
A spherically symmetric spacetime (four-dimensional Lorentzian manifold) admits coordinates $(\tau, \rho, \theta, \phi)$ in which the metric takes the form[36]

$$ds^2 = -a^2\,d\tau^2 + b^2\,d\rho^2 + r^2(d\theta^2 + \sin^2\theta\,d\phi^2),$$

where $a = a(\tau, \rho), b = b(\tau, \rho)$, and $r = r(\tau, \rho)$ are undetermined functions.

---

[36] See e.g. [39], Appendix B.

(a) Let F be a 2-form. Show that the condition that F be invariant under rotations implies that it can be written as

$$F = A \, d\tau \wedge d\rho + B \sin\theta \, d\theta \wedge d\phi,$$

for some functions $A = A(\tau, \rho)$ and $B = B(\tau, \rho)$. *Hint:* Set $\mathcal{L}_{\xi^{(j)}} F = 0$ for $j = 1, 2, 3$, where the $\xi^{(j)}$ are the Killing fields from Exercise 8.38(c). The coordinates θ and ϕ in that exercise and in this one may be taken to be the same because they define two-dimensional spherically symmetric integral submanifolds in both spaces. Lastly, by convention, $F = \frac{1}{2} F_{ij} \, dx^i \wedge dx^j$.

(b) Show that

$$\star F = -\frac{Bab}{r^2} \, d\tau \wedge d\rho + \frac{Ar^2 \sin\theta}{ab} \, d\theta \wedge d\phi.$$

Hint: Use (8.91) together with $\sigma = \sqrt{|G|} \, d\tau \wedge d\rho \wedge d\theta \wedge d\phi$. Guess $\star(d\tau \wedge d\rho) = \alpha \, d\theta \wedge d\phi$ and compute α, etc.

(c) Assume now that F is the electromagnetic 2-form. Show that Maxwell's equations $dF = d \star F = 0$ imply that

$$(F_{\hat{i}\hat{j}}) = \begin{pmatrix} 0 & -Q_e/r^2 & 0 & 0 \\ Q_e/r^2 & 0 & 0 & 0 \\ 0 & 0 & 0 & Q_m/r^2 \\ 0 & 0 & -Q_m/r^2 & 0 \end{pmatrix},$$

where $F = \frac{1}{2} F_{\hat{i}\hat{j}} \theta^{\hat{i}} \wedge \theta^{\hat{j}}$, $\{\theta^{\hat{i}}\}$ is an orthonormal coframe, and Q_e and Q_m are constants. *Remark:* F represents the electromagnetic field at $r \neq 0$ due to a point source (i.e., a monopole) with electric charge Q_e and magnetic charge Q_m located at the origin of coordinates.

8.40 Schwarzschild geometry In 1916, shortly after Einstein created his general theory of relativity, Karl Schwarzschild discovered a solution (i.e., a metric) describing an isolated nonrotating black hole. In terms of the coordinates t, r, θ, and ϕ (corresponding roughly to time and the usual three-dimensional spherical polar coordinates), the **Schwarzschild line element** is

$$ds^2 = -\Phi(r) \, dt^2 + \Phi(r)^{-1} \, dr^2 + r^2(d\theta^2 + \sin^2\theta \, d\phi^2),$$

where

$$\Phi(r) := 1 - \frac{2m}{r}.$$

(a) Show that the following 1-forms constitute a dual orthonormal frame
field:

$$\theta^{\hat{0}} = \Phi^{1/2}\, dt, \quad \theta^{\hat{1}} = \Phi^{-1/2}\, dr, \quad \theta^{\hat{2}} = r\, d\theta, \quad \text{and} \quad \theta^{\hat{3}} = r\sin\theta\, d\phi.$$

Hint: In an orthonormal frame the (Minkowski) metric components are
$g_{\hat{0}\hat{0}} = g^{\hat{0}\hat{0}} = -1$, $g_{\hat{i}\hat{i}} = g^{\hat{i}\hat{i}} = 1$, for $i = 1, 2, 3$.

(b) Show that the Levi-Civita connection forms in the orthonormal basis
of part (a) satisfy $\omega^{\hat{0}}{}_{\hat{0}} = 0$, $\omega^{\hat{0}}{}_{\hat{i}} = \omega^{\hat{i}}{}_{\hat{0}}$, and $\omega^{\hat{i}}{}_{\hat{j}} = -\omega^{\hat{j}}{}_{\hat{i}}$, where
$i, j = 1, 2, 3$. *Hint:* Exercise 8.6 implies that $\omega_{\hat{\nu}\hat{\mu}} = -\omega_{\hat{\mu}\hat{\nu}}$, where
$\mu, \nu = 0, 1, 2, 3$.

(c) Because the torsion vanishes, we have

$$d\theta^{\hat{a}} = -\omega^{\hat{a}}{}_{\hat{b}} \wedge \theta^{\hat{b}}.$$

Show that $\omega^{\hat{0}}{}_{\hat{1}} = \omega^{\hat{1}}{}_{\hat{0}} = (m/r^2)\, dt$. *Hint:* Evaluate $d\theta^{\hat{0}}$ and guess the
connection form. Note that there is some ambiguity in this procedure,
which can only be removed by computing all the connection forms.
For reference, we record the remaining nonzero matrix elements here:

$$\omega^{\hat{1}}{}_{\hat{2}} = -\omega^{\hat{2}}{}_{\hat{1}} = -\Phi^{1/2}\, d\theta,$$
$$\omega^{\hat{1}}{}_{\hat{3}} = -\omega^{\hat{3}}{}_{\hat{1}} = -\Phi^{1/2}\sin\theta\, d\phi,$$
$$\omega^{\hat{2}}{}_{\hat{3}} = -\omega^{\hat{3}}{}_{\hat{2}} = -\cos\theta\, d\phi.$$

(d) Use the matrix of connection 1-forms to compute the curvature forms $\Omega^{\hat{0}}{}_{\hat{1}}$
and $\Omega^{\hat{1}}{}_{\hat{2}}$ in the orthonormal frame field basis.

8.41 Parallel transport on the 2-sphere Let ∇ be the Levi-Civita connection
on the 2-sphere, and let $\gamma(t)$ be a curve that traces out a latitude $\theta = \theta_0$.

(a) Show that the map that parallel transports vectors around γ is given
explicitly by

$$\vartheta(\gamma, \nabla) = \begin{pmatrix} \cos(\phi\cos\theta_0) & \sin\theta_0\sin(\phi\cos\theta_0) \\ -\sin(\phi\cos\theta_0)/\sin\theta_0 & \cos(\phi\cos\theta_0) \end{pmatrix}.$$

Hint: Solve the parallel transport equation (8.63). You will need to use
the results of Exercise 8.5. Assume the initial condition $Y(\theta_0, 0) = Y_0$.

(b) Show that a vector parallel transported around γ turns through an angle

$$\psi = 2\pi\cos\theta_0.$$

Hint: Recall that the cosine of the angle between two vectors X and Y is

$$g(X, Y)/\sqrt{g(X, X)g(Y, Y)}.$$

Remark: In flat space, a vector that is parallel transported around a circle turns through an angle of 2π. (Why?) The difference $2\pi(1 - \cos\theta_0)$ is called the *spherical excess* associated with the curve γ, and it is a measure of the total curvature enclosed. In two dimensions the (intrinsic) curvature of a surface at a point can be defined by

$$\text{curvature} = \lim_{A \to 0} \frac{1}{A}(\text{spherical excess}),$$

where A is the area enclosed by the curve. For the sphere, the area enclosed by the latitude is $2\pi R^2(1-\cos\theta_0)$, so the curvature of the sphere is just $1/R^2$.

8.42 Geodesics on the 2-sphere Compute the geodesics on the sphere and show that their images are great circles. *Hint:* A few tricks are needed for this exercise. From the results of Exercise 8.5 one derives the (coupled, second-order) geodesic equations

$$\ddot{\theta} - (\sin\theta\cos\theta)\dot{\phi}^2 = 0, \tag{8.113}$$

$$\ddot{\phi} + 2(\cot\theta)\dot{\phi}\dot{\theta} = 0. \tag{8.114}$$

However, these are difficult to solve directly. Instead we seek so-called "first integrals" of the motion to reduce them to first-order equations. Show that (8.114) can be integrated to yield

$$(\sin^2\theta)\dot{\phi} = J = \text{constant}. \tag{8.115}$$

Substitute this into (8.113) and integrate to get

$$\dot{\theta}^2 + J^2\csc^2\theta = a^2 = \text{constant}. \tag{8.116}$$

Now integrate (8.116) and plug the result back into (8.115); then integrate again to get

$$\cos\theta = -\sqrt{1 - (J/a)^2}\sin(at + c) \tag{8.117}$$

and

$$\tan(\phi - \phi_0) = (J/a)\tan(at + c). \tag{8.118}$$

For the shape of the curve, substitute (8.118) into (8.117) to obtain

$$\cot\theta = C\sin(\phi - \phi_0), \tag{8.119}$$

where $C := -\sqrt{(a/J)^2 - 1}$. Identifying the resulting curves may be easier if you transform the solution back to Cartesian coordinates. Of course, if you are willing to wave your hands a bit you could just observe that if $\phi = $ constant then (8.115) gives $J = 0$, whereupon (8.116) gives $\dot{\theta} = a$, or $\theta = at + c$. This is a great circle (a longitude line), so by symmetry all the geodesics are great circles.

8.43 Constant-curvature manifolds Let M be a Riemannian manifold, and let Π be a two-dimensional subspace of $T_p M$ (a "2-plane" of M). The **sectional curvature** $K(\Pi)$ associated to the 2-plane Π is

$$K(\Pi) := \frac{g(X, R(X, Y)Y)}{g(X, X)g(Y, Y) - g(X, Y)^2}, \qquad (8.120)$$

where X and Y are any two tangent vectors spanning Π.[37] One can think of the sectional curvature as a kind of probe of the curvature at every point. Just as one can reconstruct a three-dimensional object from the two-dimensional sections of a computerized tomographical scan, so too can one recover the full Riemann curvature tensor at any point from its two-dimensional sectional curvatures.

(a) Show that $K(\Pi)$ does not depend on the choice of spanning vectors. *Hint:* Replace X and Y by linear combinations of X and Y. Simplify the problem by invoking some of the symmetry properties (8.47)–(8.50).

(b) Prove a uniqueness theorem: *The Riemann curvature tensor of M at p is determined uniquely if one knows the sectional curvature of every 2-plane of M through p.* *Hint:* The proof is purely algebraic. Suppose that R and R' are two curvature tensors satisfying (8.120). Then $g(X, R(X, Y)Y) = g(X, R'(X, Y)Y)$. Send X to $X + W$ in this equation and use the symmetry properties of the Riemann tensor to show that $g(W, R(X, Y)Y) = g(W, R'(X, Y)Y)$. Use a similar trick (and (8.49)) to obtain $g(W, R(X, Y)Z) = g(W, R'(X, Y)Z)$ for arbitrary vectors X, Y, Z, and W, and thereby conclude that $R = R'$.

(c) If $K(\Pi)$ is independent of the particular 2-plane Π at p then M is **isotropic at** p, and we write K_p for $K(\Pi)$. The manifold M is **isotropic** if it is isotropic for all $p \in M$. Show that M is isotropic if and only if

$$R_{ijk\ell} = K_p(g_{ik}g_{j\ell} - g_{i\ell}g_{jk}) \qquad (8.121)$$

---

[37] The vectors X and Y are linearly independent, so by the Cauchy–Schwarz inequality the denominator is strictly positive.

in local coordinates about any point p. *Hint:* One direction is easy. For the other, define a tensor R' by

$$g(W, R'(X, Y)Z) := g(X, W)g(Y, Z) - g(X, Z)g(Y, W)$$

and use (8.120) together with the result of part (b) to show that $R = K_p R'$.

(d) Choose an orthonormal coframe field $\{\theta^{\hat{a}}\}$ on a neighborhood of p. Show that (8.121) is equivalent to the statement that

$$\Omega^{\hat{a}\hat{b}} = K_p \, \theta^{\hat{a}} \wedge \theta^{\hat{b}}, \tag{8.122}$$

where Ω is the curvature 2-form.

(e) If $K_p =: K$ is constant (i.e., independent of p), M is said to have **constant curvature**. Prove Schur's theorem: *If M is a connected, isotropic Riemannian manifold of dimension at least 3 then M has constant curvature.* In other words, in higher-dimensional manifolds isotropy forces the sectional curvature to be constant. *Hint:* Take the exterior derivative of (8.122) and use the Bianchi identity and the vanishing of the torsion form to show that $dK \wedge \theta^{\hat{a}} \wedge \theta^{\hat{b}} = 0$ for $1 \leq \hat{a}, \hat{b} \leq n$. (You may also want to use the result of Exercise 8.6.)

8.44 Gaussian curvature If M is two dimensional then there is only a single 2-plane through any point, so the isotropy condition of Exercise 8.43 is trivial and Schur's theorem is inapplicable. Instead, the sectional curvature $K_p := K(T_p M)$, now called the **Gaussian curvature** of M, can and usually does vary from point to point. From the definition (8.120) we see that the Gaussian curvature contains all the curvature information of M, because there is only one linearly independent component of the Riemann tensor and K is basically this component. Show that, in fact, the Gaussian curvature can be expressed as one-half the Ricci scalar curvature.

8.45 Hyperbolic geometry It is a classical fact (e.g. [23], Chapter 8) that if M is a complete,[38] connected and simply connected manifold with constant curvature then it can only be one of three types: (i) a sphere, (ii) Euclidean space, or (iii) hyperbolic space. The purpose of this problem is to investigate the last space.

There are several models of **hyperbolic space**; here we use the Poincaré upper half-plane model, which consists of the upper half-plane $\mathbb{H}^2_+ := \{(x, y) \in \mathbb{R}^2 : y > 0\}$ equipped with the line element

---

[38] A (pseudo)-Riemannian manifold is (**geodesically**) **complete** if, for any vector X_p, the exponential map $\mathrm{Exp}\, t X_p$ (defined in Appendix D) exists for all $t \in \mathbb{R}$. Intuitively, you can walk forever along a geodesic and not worry about falling off the manifold.

$$ds^2 = \frac{dx^2 + dy^2}{y^2}. \tag{8.123}$$

(a) Show that all the points on the x axis are at an infinite distance from the points off the axis.

(b) Compute all the Christoffel symbols. (Here's one: $\Gamma^x{}_{xy} = -1/y$.)

(c) Write down and solve the geodesic equations for this metric. What do the curves look like geometrically? *Hint:* The geodesic equations are

$$y\ddot{x} - 2\dot{x}\dot{y} = 0,$$
$$y\ddot{y} + \dot{x}^2 - \dot{y}^2 = 0.$$

The first equation is separable and yields $\dot{x} = cy^2$ for some constant c. Set $f := \dot{y}/y$. If $c = 0$ then the second equation gives $\dot{f} = 0$, which you can easily solve. If $c \neq 0$ then the second equation becomes $\dot{f} = -c\dot{x}$, which can be integrated to give $x\dot{x} + y\dot{y} = a\dot{x}$. Integrate again to get the second class of geodesics.

(d) Show that the Poincaré upper half-plane has constant (Gaussian) curvature $K = -1$.

Remark: Hyperbolic geometry has great historical significance. Two thousand three hundred years ago Euclid set down his axioms of geometry in the form of five postulates. All the postulates seemed self-evident except the fifth, which states that, for every straight line L and point P not on L, there is a unique line L' through P and parallel to L. For many centuries, people tried to prove the fifth postulate from the other four, but failed. The reason for their failure, which surprised almost everyone, was that there are other perfectly consistent geometries in which the first four axioms hold but the fifth does not, a fact discovered around the same time by Bolyai, Gauss, and Lobachevsky. Gauss observed that, in spherical geometry, given a "straight line" (meaning a geodesic or great circle) and a point P not on that "line", every "line" through P meets the original "line". Bolyai and Lobachevsky discovered hyperbolic geometry, in which there are an infinite number of "straight" lines (geodesics) through a fixed point P that do not meet a given "line". Beltrami supplied several models of hyperbolic geometry, including the Poincaré upper half-plane.

8.46 Möbius transformations   Let $A = \begin{pmatrix} a & b \\ c & d \end{pmatrix} \in GL(2, \mathbb{C})$, the Lie group of 2×2 complex matrices with nonzero determinant. The fractional linear map $T_A : \mathbb{C} \to \mathbb{C}$ given by

$$z \mapsto \frac{az+b}{cz+d}$$

is called a **Möbius transformation**. It has all sorts of surprising properties and plays an important role in geometry and number theory.

(a) Show that the set $\{T_A\}$ of all Möbius transformations forms a group under composition called the **Möbius group**, denoted here by $\mathcal{M}$.

(b) Show that the map $A \mapsto T_A$ is a homomorphism of $GL(2, \mathbb{C})$ into $\mathcal{M}$ with kernel consisting of all nonzero complex multiples of the identity matrix. *Remark:* By the first isomorphism theorem of group theory this shows that $\mathcal{M} \cong GL(2, \mathbb{C})/\{\lambda I\} = PGL(2, \mathbb{C})$, the **projective general linear group** of 2×2 complex matrices.

(c) Let $\mathcal{M}'$ denote the subgroup of Möbius transformations for which $A \in SL(2, \mathbb{R})$.[39] Show that $T_A \in \mathcal{M}'$ is a diffeomorphism from the upper half plane $\mathbb{H}^2_+$ to itself (where we make the usual identification of $\mathbb{R}^2$ and $\mathbb{C}$ via $z = x + iy$).

(d) Show that the elements of $\mathcal{M}'$ are isometries of hyperbolic geometry, by showing that $T_A^* ds^2 = ds^2$ where ds^2 is the hyperbolic metric (8.123).

8.47 Induced metrics Let $f : M \to N$ be a smooth immersion of a manifold M into a geometric manifold N equipped with a metric g. Then M is naturally a geometric manifold when equipped with the **induced metric** f^*g. Suppose Σ to be a two-dimensional surface in $\mathbb{R}^3$ given by the parameterization

$$\sigma : (u, v) \to (x(u, v), y(u, v), z(u, v)),$$

and let h be the induced metric on Σ. Define

$$\sigma_u := \left(\frac{\partial x}{\partial u}, \frac{\partial y}{\partial u}, \frac{\partial z}{\partial u}\right) \qquad \text{and} \qquad \sigma_v := \left(\frac{\partial x}{\partial v}, \frac{\partial y}{\partial v}, \frac{\partial z}{\partial v}\right).$$

(a) Show that we can write

$$\begin{aligned}
h &= E\, du \otimes du + F\, du \otimes dv + F\, dv \otimes du + G\, dv \otimes dv \\
&= E\, du^2 + 2F\, du\, dv + G\, dv^2,
\end{aligned}$$

where

$$E := g(\sigma_u, \sigma_u), \qquad F := g(\sigma_u, \sigma_v), \qquad \text{and} \qquad G := g(\sigma_v, \sigma_v).$$

---

[39] The **special linear group** $SL(n, \mathbb{R})$ consists of all $n \times n$ real matrices with unit determinant.

Hint: Identify f with the parameterization map. Write

$$g = dx^2 + dy^2 + dz^2 = dx \otimes dx + dy \otimes dy + dz \otimes dz$$

and use the fact that $f^*(\mu \otimes \nu) = f^*\mu \otimes f^*\nu$. *Remark:* For more fun with immersions, see Appendix H.

(b) The usual parameterization of the sphere in terms of spherical polar coordinates is

$$\sigma(\theta, \phi) = (\sin\theta \cos\phi, \sin\theta \sin\phi, \cos\theta), \quad 0 \leq \theta \leq \pi, 0 \leq \phi < 2\pi.$$

(Yes, the parameterization is not well behaved at the poles – just ignore those points.) Compute the induced metric tensor on S^2 in spherical polar coordinates and verify that it agrees with the metric given in Exercise 8.5.

8.48 Gaussian curvature of the catenoid   A **catenoid** is a two-dimensional surface of revolution in $\mathbb{R}^3$ obtained by rotating a catenary about an appropriate axis. One parameterization is given by

$$\sigma(u, v) = (a \cosh v \cos u, a \cosh v \sin u, av),$$
$$0 < u < 2\pi, \quad -\infty < v < \infty.$$

Show that, in these coordinates, the Gaussian curvature of the catenoid is $K = -a^{-2} \cosh^{-4} v$. *Hint:* Find an orthonormal coframe field $\{\hat{\theta}^1, \hat{\theta}^2\}$ on the surface and compute the curvature 2-form Ω. Use (8.122). For another method, see Appendix H.

8.49 Gaussian curvature of the standard torus   One parameterization for the standard embedding of the torus in $\mathbb{R}^3$ is given by

$$x = (b + a \cos v) \cos u,$$
$$y = (b + a \cos v) \sin u,$$
$$z = a \sin v,$$

where $0 \leq u, v < 2\pi$ and $a < b$. Calculate its Gaussian curvature in the induced metric and verify that it is not identically zero. *Hint:* See the previous hint.

8.50 Geometry of Lie groups   (For readers who completed the problems on Lie groups in Chapter 3 and Exercise 3.52 in particular.) If G is a semisimple Lie group then its Killing form is a nondegenerate inner product on the tangent space to G at the identity. This form can be extended naturally to a metric on the whole of G, defined as follows:

$$(X_g, Y_g) := (L_{g^{-1}*}X_g, L_{g^{-1}*}Y_g)_e, \tag{8.124}$$

for any vectors $X_g, Y_g \in T_g G$. In this way G becomes a natural Riemannian or pseudo-Riemannian manifold, depending on the signature of the Killing form.

(a) In addition to the natural left action of G on itself, given by $L_g h = gh$, there is a natural right action given by $R_g h = hg$. By construction L_g is an isometry, so we say the metric is **left-invariant**.[40] Show that the metric is actually **bi-invariant**, meaning that R_g is an isometry as well. *Hint:* Note that $\text{Ad}\, g = \varphi(g)_{*,e}$, where $\varphi(g) = L_g \circ R_{g^{-1}} = R_{g^{-1}} \circ L_g$. Now use the result of Exercise 3.52(b).

(b) Let ∇ be the Levi-Civita connection on G determined by the Killing form and let X and Y be left-invariant vector fields. Show that

$$\nabla_X Y = \frac{1}{2}[X, Y]. \tag{8.125}$$

Hint: Let Z be a left-invariant vector field. Consider $(\nabla_X Y, Z)$ and apply the Koszul formula. Note that (X, Y), (Y, Z), and (X, Z) are constant. Why? *Remark:* The formula above implies that $\nabla_X X = 0$ for any left-invariant vector field X. This means that the integral curves of X are geodesics. In particular, the geodesics through the identity are precisely the one-parameter subgroups of G.[41]

(c) In the next three parts of the exercise, let X, Y, and Z be left invariant vector fields. Show that

$$R(X, Y)Z = \frac{1}{4}[[X, Y], Z], \tag{8.126}$$

---

[40] This is supposed to be obvious from the definition, but it may be difficult to see when using the parenthetical notation for the metric (8.124). Instead, temporarily denote the metric by the letter m (so as not to confuse it with the group element g), and let B denote the Killing form. The statement that L_g is an isometry is just the statement that $L_g^* m = m$. To prove this, note that for any $h \in G$ and any vectors X_h and Y_h,

$$
\begin{aligned}
(L_g^* m)(X_h, Y_h)_h &= m(L_{g*} X_h, L_{g*} Y_h)_{gh} && \text{(by (3.90))}, \\
&= B(L_{(gh)^{-1}*} L_{g*} X_h, L_{(gh)^{-1}*} L_{g*} Y_h) && \text{(by (8.124))}, \\
&= B(L_{h^{-1}g^{-1}*} L_{g*} X_h, L_{h^{-1}g^{-1}*} L_{g*} Y_h) \\
&= B(L_{h^{-1}*} L_{g^{-1}*} L_{g*} X_h, L_{h^{-1}*} L_{g^{-1}*} L_{g*} Y_h) && \text{(by the chain rule)}, \\
&= B(L_{h^{-1}*} X_h, L_{h^{-1}*} Y_h), \\
&= m(X_h, Y_h) && \text{(by (8.124))}.
\end{aligned}
$$

(It is worth emphasizing that X_h and Y_h are just vectors, not specialized vector fields.)

[41] These geodesics are given by the exponential map, which explains the use in Appendix D of the same terminology for more general Riemannian manifolds.

where R is the Riemann curvature tensor. *Hint:* Use part (b) and the Jacobi identity. *Remark:* This formula completely determines the Riemann curvature tensor, because we can always find a basis of left-invariant vector fields on G.

(d) Show that the sectional curvature of G is given by

$$K(X, Y) = \frac{1}{4} \frac{([X, Y], [X, Y])}{(X, X)(Y, Y) - (X, Y)^2}. \tag{8.127}$$

(e) Show that the Ricci tensor equals the bi-invariant metric (8.124) up to a constant. That is,

$$\text{Ric}(X, Y) = -\frac{1}{4}(X, Y). \tag{8.128}$$

A manifold for which the Ricci tensor is proportional to the metric tensor is called an **Einstein manifold**, so semisimple Lie groups are Einstein manifolds.

8.51 Standard metrics for constant curvature 3-manifolds Let M be an isotropic three-dimensional Riemannian manifold with local coordinates r, θ, ϕ and line element

$$ds^2 = dr^2 + f^2(r)(d\theta^2 + \sin^2 \theta \, d\phi^2),$$

where $f \to 0$ as $r \to 0$.

(a) Starting from the orthonormal coframe

$$\theta^{\hat{1}} = dr, \qquad \theta^{\hat{2}} = f(r) \, d\theta, \qquad \theta^{\hat{3}} = f(r) \sin \theta \, d\phi,$$

compute the connection 1-forms $\omega^{\hat{i}}{}_{\hat{j}}$.

(b) Compute the curvature 2-forms $\Omega^{\hat{i}}{}_{\hat{j}}$.

(c) By Schur's theorem, M must be a constant curvature manifold. By rescaling coordinates if necessary we may take $K = 1$, $K = 0$, or $K = -1$. Show that these cases correspond to the respective choices $f(r) = \sin r$, r, and $\sinh r$. *Remark:* To get these forms you may have to flip the sign of r, which is all right because $r \to -r$ turns out to be an isometry of M.

8.52 Pseudotensor densities and the Levi-Civita alternating symbol (In this exercise a coordinate basis is assumed throughout, so for ease of exposition we make no distinction between a tensor and its components. Also, we assume that dim $M = n$.) The **Levi-Civita alternating symbol** (or, **epsilon symbol**) $\varepsilon_{i_1 \dots i_n}$ is defined by the following two properties: (i) $\varepsilon_{123 \dots n} = +1$ and (ii) $\varepsilon_{i_1 \dots i_n}$ flips sign if any two indices are interchanged (total antisymmetry).

Equivalently, $\varepsilon_{i_1\ldots i_n}$ vanishes if any two indices are the same and equals the sign of the permutation $(i_1 \ldots i_n)$ otherwise.

(a) Show that, if $(A^i{}_j)$ is any $n \times n$ matrix,

$$\varepsilon_{i_1\ldots i_n} A^{i_1}{}_{j_1} \cdots A^{i_n}{}_{j_n} = \varepsilon_{j_1\ldots j_n} \det A. \tag{8.129}$$

(b) A multi-indexed object $T_{i_1,\ldots,i_n}$ is called a **pseudotensor** if, under a change of coordinates, it transforms according to

$$T_{i'_1\ldots i'_n} = (\text{sgn } J)\frac{\partial x^{i_1}}{\partial y^{i'_1}} \cdots \frac{\partial x^{i_n}}{\partial y^{i'_n}} T_{i_1\ldots i_n}, \tag{8.130}$$

where $J := \det(\partial y^{i'}/\partial x^j)$ is the Jacobian determinant.[42] It is called a **tensor density of weight** w if

$$T_{i'_1\ldots i'_n} = |J|^{-w}\frac{\partial x^{i_1}}{\partial y^{i'_1}} \cdots \frac{\partial x^{i_n}}{\partial y^{i'_n}} T_{i_1\ldots i_n}, \tag{8.131}$$

where $|\cdot|$ means the absolute value. We extend the Levi-Civita symbol $\varepsilon_{i_1\ldots i_n}$ to the entire manifold by demanding that it transform under a coordinate change according to

$$\varepsilon_{i'_1\ldots i'_n} = \xi\frac{\partial x^{i_1}}{\partial y^{i'_1}} \cdots \frac{\partial x^{i_n}}{\partial y^{i'_n}}\varepsilon_{i_1\ldots i_n}. \tag{8.132}$$

Show that $\xi = J$ and thereby show that $\varepsilon_{i_1\ldots i_n}$ is a peudotensor density of weight -1.

(c) Let $G := \det g_{ij}$. Show that G is a scalar density of weight 2, i.e., show that

$$G' = J^{-2}G. \tag{8.133}$$

(d) Define

$$\epsilon_{i_1\ldots i_n} := \sqrt{|G|}\varepsilon_{i_1\ldots i_n}. \tag{8.134}$$

---

[42] A pseudotensor is just like a regular tensor except that it suffers an additional sign change under an orientation-reversing coordinate transformation. Note that on an oriented manifold we can choose the Jacobian to be positive everywhere, so for most purposes the distinction between tensors and pseudotensors is immaterial. The distinction becomes important in physics where one wishes to consider the hypothetical possibility of a **parity transformation** (a global coordinate inversion sending $x^i \to -x^i$). The reason is that certain physical quantities such as the position vector r or the velocity vector v (so-called **polar vectors**) flip sign under a parity transformation whereas other quantities such as the angular momentum vector L (a **pseudovector** or **axial vector**) do not (because $L = mr \times v$). The behavior of objects under parity transformations places constraints on the way in which they can appear in the various laws of physics, especially if it is known that the corresponding phenomena are (or are not) parity invariant. For example, if parity were conserved in a certain physical process then a term such as $r \cdot L$ (a **pseudoscalar**) could not appear in the equation of motion as it would violate parity conservation.

Show that this is a pseudotensor. Despite this fact, it is almost always referred to as the **Levi-Civita alternating tensor** or **epsilon tensor**.

(e) Show that the volume element can be written in local coordinates as

$$\sigma = \frac{1}{n!}\epsilon_{i_1\dots i_n}\,dx^{i_1}\wedge\cdots\wedge dx^{i_n}. \tag{8.135}$$

(f) Define $\varepsilon^{i_1\dots i_n}$, the contravariant cousin of $\varepsilon_{i_1\dots i_n}$ by requiring that $\varepsilon^{i_1\dots i_n}$ be the sign of the permutation $(i_1,\dots,i_n)$ and that it transform according to

$$\varepsilon^{i'_1\dots i'_n} = \eta\frac{\partial y^{i'_1}}{\partial x^{i_1}}\cdots\frac{\partial y^{i'_n}}{\partial x^{i_n}}\varepsilon^{i_1\dots i_n}. \tag{8.136}$$

Show that $\eta = J^{-1}$, so that $\varepsilon^{i_1\dots i_n}$ is a pseudotensor density of weight $+1$.

(g) Now define the upstairs version of the Levi-Civita alternating tensor by

$$\epsilon^{i_1\dots i_n} := \pm\frac{1}{\sqrt{|G|}}\varepsilon^{i_1\dots i_n}, \tag{8.137}$$

where the sign is chosen so that

$$\epsilon^{i_1\dots i_n} = g^{i_1 j_1}\cdots g^{i_n j_n}\epsilon_{j_1\dots j_n}. \tag{8.138}$$

Show that the sign is $+1$ if the signature is Euclidean and -1 if the signature is Lorentzian.

(h) Show that the covariant derivative of the Levi-Civita tensor is zero.

8.53 Epsilon tensor identities   Assume that M is an n-dimensional manifold (so that all indices run from 1 to n).

(a) Prove the following so-called **epsilon symbol identities**. For $0 \le k \le n$,

$$\varepsilon^{i_1\dots i_k i_{k+1}\dots i_n}\varepsilon_{j_1\dots j_k i_{k+1}\dots i_n} = (n-k)!\delta^{i_1\dots i_k}_{j_1\dots j_k}, \tag{8.139}$$

where

$$\delta^{i_1\dots i_k}_{j_1\dots j_k} := \det\begin{pmatrix} \delta^{i_1}_{j_1} & \cdots & \delta^{i_1}_{j_k} \\ \vdots & \ddots & \vdots \\ \delta^{i_k}_{j_1} & \cdots & \delta^{i_k}_{j_k} \end{pmatrix}. \tag{8.140}$$

In this formula δ^i_j is the usual Kronecker delta and $\delta^{i_1\dots i_k}_{j_1\dots j_k}$ is the **generalized Kronecker symbol**. (By convention $i_k = j_k = 0$ for $k < 0$ or $k > n$.) *Hint:* First prove the result for $k = n$ by showing that

both sides have the same symmetry properties and that they agree when
$(i_1, \ldots, i_n) = (j_1, \ldots, j_n) = (1, \ldots, n)$, then work down from n to 0.

(b) What is the analogue of (8.139) for the epsilon tensor? Consider both the Riemannian and Lorentzian cases.

8.54 Hodge duals and Levi-Civita tensors Let M be a Riemannian manifold of dimension n, and let

$$\alpha = \frac{1}{k!} a_{i_1 \ldots i_k} \, dx^{i_1} \wedge \cdots \wedge dx^{i_k} \tag{8.141}$$

be a k-form expressed in local coordinates. Verify that the Hodge dual of α can be written as

$$\star \alpha = \frac{1}{(n-k)!} a^*_{i_{k+1} \ldots i_n} \, dx^{i_{k+1}} \wedge \cdots \wedge dx^{i_n}, \tag{8.142}$$

where

$$a^*_{i_{k+1} \ldots i_n} := \frac{1}{k!} a^{i_1 \ldots i_k} \epsilon_{i_1 \ldots i_k i_{k+1} \ldots i_n}. \tag{8.143}$$

Here $\epsilon_{i_1 \ldots i_n}$ is the Levi-Civita alternating tensor, and indices are raised and lowered by the metric. *Remark:* If you prefer, you can rewrite these formulae without the factorials by using the parenthetical notation introduced in Chapter 2. In that case you would write

$$\alpha = a_{(i_1, \ldots, i_k)} \, dx^{i_1} \wedge \cdots \wedge dx^{i_k}, \tag{8.144}$$

$$\star \alpha = a^*_{(i_{k+1}, \ldots, i_n)} \, dx^{i_{k+1}} \wedge \cdots \wedge dx^{i_n}, \tag{8.145}$$

and

$$a^*_{i_{k+1} \ldots i_n} := a^{(i_1, \ldots, i_k)} \epsilon_{i_1 \ldots i_k i_{k+1} \ldots i_n}, \tag{8.146}$$

respectively, where $(i_1, \ldots, i_k)$ denotes the ordered index collection $i_1 \ldots i_k$ if $i_1 < \cdots < i_k$ and zero otherwise.

8.55 More Hodge duals Let α and β be k-forms. In local coordinates we have

$$\alpha = \frac{1}{k!} a_{i_1 \ldots i_k} \, dx^{i_1} \wedge \cdots \wedge dx^{i_k}$$

and

$$\beta = \frac{1}{k!} b_{i_1 \ldots i_k} \, dx^{i_1} \wedge \cdots \wedge dx^{i_k}.$$

Show that

$$\alpha \wedge \star \beta = \frac{1}{k!} a^{i_1 \ldots i_k} b_{i_1 \ldots i_k} \sigma, \tag{8.147}$$

where σ is the volume form. This can also be written, using the parenthetical index notation of Chapter 2, as

$$\alpha \wedge \star\beta = a^{(i_1,\dots,i_k)} b_{(i_1,\dots,i_k)} \sigma. \tag{8.148}$$

8.56 div, grad, curl, and all that   Let f be a smooth function and $X = X^i \partial_i$ a smooth vector field on a Riemannian manifold. By analogy with their flat-space counterparts, we define the **divergence, curl,** and **gradient** as follows:

$$\operatorname{div} X := \nabla_i X^i, \tag{8.149}$$

$$\operatorname{curl} X := \epsilon^{ijk}(\nabla_i X_j)\partial_k, \tag{8.150}$$

$$\operatorname{grad} f := g^{ij}(\nabla_j f)\partial_i = (\nabla^i f)\partial_i, \tag{8.151}$$

where ∇ is the Levi-Civita connection. The **Laplacian** or **Laplace–Beltrami operator** is

$$\operatorname{div} \circ \operatorname{grad} f = g^{ij}\nabla_i \nabla_j f = \nabla^i \nabla_i f = \nabla_i \nabla^i f = \nabla^2 f. \tag{8.152}$$

The divergence, gradient, and Laplacian are defined in any dimension but the curl is defined only in three dimensions.[43]

(a) Set $G := \det g_{ij}$. Show that we can write

$$\operatorname{div} X = G^{-1/2}(G^{1/2}X^j)_{,j} \tag{8.153}$$

so that

$$\nabla^2 f = G^{-1/2}(G^{1/2}g^{ij} f_{,i})_{,j}. \tag{8.154}$$

Hint: Show that $(G^{1/2})_{,j} = G^{1/2}\Gamma^k{}_{kj}$. You may find Exercise 1.36 helpful. Don't forget the chain rule.

(b) Show that

$$\mathcal{L}_X \sigma = (\operatorname{div} X)\sigma, \tag{8.155}$$

where σ is the canonical volume form. *Hint:* You may wish to use the fact that $\mathcal{L}_X$ is a derivation and that it commutes with the exterior derivative on forms. *Remark:* Equation (8.155) is sometimes taken to be the definition of the divergence.

---

[43] In fact the cross product can only be defined in three dimensions. It is perhaps worth observing that these definitions are easy to state in terms of components but awkard in the extreme to formulate otherwise. For example, to define the curl of the vector field X you must first introduce the 1-form $\widetilde{X} = g(\cdot, X)$, then take $d\widetilde{X}$ to get a 2-form, then take the Hodge dual to get another 1-form $\star d\widetilde{X}$, and finally turn that back into a vector field using the inverse metric to get the vector field $\hat{g}(\cdot, \star d\widetilde{X})$. Believe it or not, this is the curl of X. Score one for the index crowd.

(c) In flat three-dimensional space, the divergence of a curl and the curl of a gradient vanish. Show that the same thing is true in a curved three-dimensional space.

8.57 div, curl, grad, and Laplacian in spherical polar coordinates Spherical polar coordinates (r, θ, ϕ) on $\mathbb{R}^3$ are related to Cartesian coordinates by

$$x = r \sin\theta \cos\phi, \qquad y = r \sin\theta \sin\phi, \qquad \text{and} \qquad z = r \cos\theta.$$

(a) Show that the Euclidean metric tensor components in spherical polar coordinates are given by

$$(g_{ij}) = \begin{pmatrix} 1 & 0 & 0 \\ 0 & r^2 & 0 \\ 0 & 0 & r^2 \sin^2\theta \end{pmatrix}. \tag{8.156}$$

(b) Verify that

$$e_{\hat{r}} := \frac{\partial}{\partial r}, \qquad e_{\hat{\theta}} := \frac{1}{r}\frac{\partial}{\partial\theta}, \qquad \text{and} \qquad e_{\hat{\phi}} := \frac{1}{r\sin\theta}\frac{\partial}{\partial\phi} \tag{8.157}$$

constitute an orthonormal frame field.

(c) Using (8.153), (8.150), (8.151), and (8.154), show that, in spherical polar coordinates,

$$\operatorname{div} X = \frac{1}{r^2}\frac{\partial}{\partial r}\left(X^{\hat{r}}r^2\right) + \frac{1}{r\sin\theta}\frac{\partial}{\partial\theta}(X^{\hat{\theta}}\sin\theta) + \frac{1}{r\sin\theta}\frac{\partial}{\partial\phi}X^{\hat{\phi}},$$

$$\operatorname{curl} X = \frac{1}{r\sin\theta}\left[\frac{\partial}{\partial\theta}(\sin\theta X^{\hat{\phi}}) - \frac{\partial X^{\hat{\theta}}}{\partial\phi}\right]e_{\hat{r}}$$

$$+ \frac{1}{r}\left[\frac{1}{\sin\theta}\frac{\partial X^{\hat{r}}}{\partial\phi} - \frac{\partial}{\partial r}(rX^{\hat{\phi}})\right]e_{\hat{\theta}}$$

$$+ \frac{1}{r}\left[\frac{\partial}{\partial r}(rX^{\hat{\theta}}) - \frac{\partial X^{\hat{r}}}{\partial\theta}\right]e_{\hat{\phi}},$$

$$\operatorname{grad} f = \frac{\partial f}{\partial r}e_{\hat{r}} + \frac{1}{r}\frac{\partial f}{\partial\theta}e_{\hat{\theta}} + \frac{1}{r\sin\theta}\frac{\partial f}{\partial\phi}e_{\hat{\phi}},$$

and

$$\nabla^2 f = \frac{1}{r^2}\frac{\partial}{\partial r}\left(r^2\frac{\partial f}{\partial r}\right) + \frac{1}{r^2\sin\theta}\frac{\partial}{\partial\theta}\left(\sin\theta\frac{\partial f}{\partial\theta}\right) + \frac{1}{r^2\sin^2\theta}\frac{\partial^2 f}{\partial\phi^2}.$$

Watch out! The components of X are defined relative to the orthonormal frame field by the expansion $X = X^{\hat{r}}e_{\hat{r}} + X^{\hat{\theta}}e_{\hat{\theta}} + X^{\hat{\phi}}e_{\hat{\phi}}$, but all the formulae assume a coordinate basis.

8.58 The Weitzenböck formula (*Warning:* This problem may cause index blindness and headaches.) Let α be a k-form given by

$$\alpha = \frac{1}{k!} a_{i_1 \ldots i_k} \, dx^{i_1} \wedge \cdots \wedge dx^{i_k}$$

in local coordinates.

(a) Show that

$$d\alpha = \frac{1}{k!} \nabla_{i_1} a_{i_2 \ldots i_{k+1}} \, dx^{i_1} \wedge \cdots \wedge dx^{i_{k+1}} \tag{8.158}$$

$$= \frac{1}{k!} \nabla_{[i_1} a_{i_2 \ldots i_{k+1}]} \, dx^{i_1} \wedge \cdots \wedge dx^{i_{k+1}} \tag{8.159}$$

$$= \frac{1}{(k+1)!} \sum_{r=1}^{k+1} (-1)^{r-1} \nabla_{i_r} a_{i_1 \ldots \hat{i}_r \ldots i_{k+1}} \, dx^{i_1} \wedge \cdots \wedge dx^{i_{k+1}}, \tag{8.160}$$

where we are using the shorthand notation introduced in Exercise 8.31, in which $\nabla_j T_{i_1 \ldots i_k} = T_{i_1 \ldots i_k; j}$, and the bracket in (8.159) means that you are to totally antisymmetrize the indices. *Remark:* The point is that we can replace the ordinary derivatives appearing in the exterior derivative by covariant derivatives, because the components $a_{i_1 \ldots i_k}$ are totally antisymmetric and the Christoffel symbols are symmetric.

(b) Similarly, show that, if δ is the co-differential operator,

$$\delta\alpha = -\frac{1}{(k-1)!} \nabla^{i_1} a_{i_1 \ldots i_k} dx^{i_2} \wedge \cdots \wedge dx^{i_k} \tag{8.161}$$

$$:= -\frac{1}{(k-1)!} g^{i_1 j} \nabla_j a_{i_1 \ldots i_k} dx^{i_2} \wedge \cdots \wedge dx^{i_k}.$$

Hint: You will probably want to use the Leibniz rule and metric compatibility. At one point you may want to use the fact (see Exercise 8.52(h)) that the covariant derivative of the Levi-Civita tensor vanishes. *Remark:* This exercise shows that if $X = X^i \partial_i$ is a vector field and $\widetilde{X} = X_i \, dx^i$ is the corresponding covector field then $\delta \widetilde{X} = -\operatorname{div} X$, the divergence of X. (See Exercise 8.56.)

(c) Let $\Delta = d\delta + \delta d$ be the Laplace–de Rham operator. Writing

$$\Delta\alpha = \frac{1}{k!} (\Delta a)_{i_1 \ldots i_k} \, dx^{i_1} \wedge \cdots \wedge dx^{i_k}, \tag{8.162}$$

prove the **Weitzenböck formula,**

$$(\Delta a)_{i_1\ldots i_k} = -\nabla^2 a_{i_1\ldots i_k} + \sum_{r=1}^{k} a_{i_1\ldots i_{r-1} i i_{r+1}\ldots i_k} R^i{}_{i_r}$$

$$-\frac{1}{2}\sum_{\substack{r=1\ldots k\\ s=1\ldots k\\ r\neq s}} a_{i_1\ldots i_{r-1} i i_{r+1}\ldots i_{s-1} j i_{s+1}\ldots i_k} R^{ij}{}_{i_r i_s}, \qquad (8.163)$$

where

$$R^i{}_j := g^{im} R_{mj} \qquad \text{and} \qquad R^{ij}{}_{k\ell} := g^{jm} R^i{}_{mk\ell} \qquad (8.164)$$

are raised-index versions of the Ricci and Riemann tensors, respectively. *Hint:* You may wish to use (8.109). *Remark:* This is the first in a long line of Weitzenböck-type formulas.

8.59 Bochner's theorem Again let α be a k-form ($0 < k < n$) with local components $a_{(i_1,\ldots,i_k)}$. Assume that M is a compact, connected, orientable Riemannian manifold without boundary.

(a) Assuming that α is harmonic, show that

$$\int_M \left[-(\nabla^2 a_{i_1\ldots i_k})a^{i_1\ldots i_k} + kF(\alpha)\right]\sigma = 0, \qquad (8.165)$$

where

$$F(\alpha) := R_{ij}a^{ii_2\ldots i_k}a^j{}_{i_2\ldots i_k} - \frac{1}{2}(k-1)R_{ijpq}a^{iji_3\ldots i_k}a^{pq}{}_{i_3\ldots i_k}. \qquad (8.166)$$

Hint: Consider $(\Delta\alpha, \alpha)$ and use (8.147) and (8.163).

(b) Prove the following integration by parts formula:

$$\int_M (a^{i_1\ldots i_k}\nabla^2 a_{i_1\ldots i_k})\sigma = -\int_M (\nabla^j a^{i_1\ldots i_k})(\nabla_j a_{i_1\ldots i_k})\sigma. \qquad (8.167)$$

Hint: Show that $\int_M \nabla^2(a_{i_1\ldots i_k}a^{i_1\ldots i_k})\sigma = 0$ (∗). Do this by applying the Weitzenböck formula (8.163) to $\Delta(a_{i_1\ldots i_k}a^{i_1\ldots i_k})$ and using Theorem 8.10. Now expand (∗) using the Leibniz rule.

(c) Prove a celebrated theorem due to Bochner: *If M is a compact, connected, orientable Riemannian manifold without boundary and with constant positive curvature then M is a homology sphere (i.e., all the Betti numbers b_k ($0 < k < n$) vanish. (We already know that $b_0 = 1$, because M is connected, and that $b_n = 1$, by Poincaré duality.)* *Hint:* Combine (a) and (b). Use (8.121) to show that $F(\alpha) = k!(n-k)K\|\alpha\|^2$, where K is the constant positive curvature and $\|\alpha\|^2 := a^{(i_1,\ldots,i_k)}a_{(i_1,\ldots,i_k)}$. Don't forget Theorem 8.12 of Hodge.

8.60 Areas of spheres and volumes of balls. Let

$$\omega = dx^1 \wedge \cdots \wedge dx^{n+1}$$

be the standard volume element on $\mathbb{R}^{n+1}$, and define $r^2 = \sum_{i=1}^{n+1}(x^i)^2$. Denote by V_{n+1} the volume of the unit ball $B^{n+1} = \{x \in \mathbb{R}^{n+1} : r^2 \leq 1\}$, so that

$$V_{n+1} = \int_{r^2 \leq 1} \omega.$$

Also, let A_n denote the n-dimensional "area" of the unit n-sphere $S^n = \partial B^{n+1} = \{x \in \mathbb{R}^{n+1} : r^2 = 1\}$, so that

$$A_n = \int_{S^n} \sigma,$$

where σ is the volume form on S^n. We seek an explicit expression for σ in Cartesian coordinates. One way to do this would be to compute σ in multi dimensional spherical polar coordinates and then convert back. A more elegant approach, which we follow here (with slightly different notation), is that of [25], p. 74.

(a) Show that

$$V_{n+1} = \frac{1}{n+1} A_n.$$

For example, $V_2 = \pi = A_1/2$ and $V_3 = 4\pi/3 = A_2/3$. *Hint:* By a dimensional argument, the area of an n-sphere of radius r is $r^n A_n$. The volume of the ball B^{n+1} may be obtained by integrating spherical shells of thickness dr.

(b) A key fact is that σ must be spherically symmetric (i.e., invariant under rotations). To find such a form, we first differentiate both sides of

$$r^2 = (x^1)^2 + \cdots + (x^{n+1})^2$$

to get

$$r\,dr = x^1\,dx^1 + \cdots + x^{n+1}\,dx^{n+1}.$$

This is a spherically symmetric 1-form on $\mathbb{R}^{n+1}$. Its Hodge dual is the n-form

$$\tau := \star r\,dr = \sum_{i=1}^{n+1}(-1)^{n+1}x^i\,dx^1 \wedge \cdots \wedge \widehat{dx^i} \wedge \cdots \wedge dx^{n+1}.$$

This, too, is spherically symmetric, so restricting it to the n-sphere must yield a multiple of σ. Thus

$$\sigma = c\tau$$

for some constant c.[44] Show that $c = 1$. *Hint:* Compute A_n using Stokes' theorem.

(c) For fun, show that

$$V_n = \frac{2}{n} \frac{\pi^{n/2}}{\Gamma(n/2)},$$

where $\Gamma(x)$ is the gamma function. *Hint:* There are several ways to do this. One pretty approach proceeds from Dirichlet's generalization of the beta integral ([4], p. 32). Another method uses the trick of integrating e^{-r^2} over the whole of $\mathbb{R}^n$ in both Cartesian and spherical polar coordinates ([46], p. 63). A third approach is to slice the n-ball into disks and use recursion ([25], p. 74). A related technique is given in ([5], p. 411).

8.61 Spherical harmonics Let f be a harmonic function on $\mathbb{R}^3$, so that $\Delta f = -\nabla^2 f = 0$. (See Exercise 8.58.) Suppose that, in spherical polar coordinates,

$$f(r, \theta, \phi) = r^n h(\theta, \phi) \tag{8.168}$$

for some n and some function h. Then h is called a **spherical harmonic** (of degree n).

(a) By treating h as a function on the 2-sphere S^2, show that

$$\Delta h = n(n + 1)h, \tag{8.169}$$

where Δ is now the Laplace–de Rham operator on S^2. *Hint:* Use Exercise 8.56 together with the easy-to-prove fact that, acting on functions in spherical polar coordinates, ∇^2 on the 2-sphere is just the spherical part of ∇^2 on $\mathbb{R}^3$ with $r = 1$.

(b) Show that spherical harmonics of different degrees are orthogonal with respect to the Hodge inner product.

8.62 Harmonic forms and the Hodge decomposition Fix a k-form λ. Show that there exists a k-form ω such that

$$\Delta\omega = \lambda \tag{8.170}$$

---

[44] Technically, $\sigma = c\iota^*\tau$ where $\iota : S^n \to \mathbb{R}^{n+1}$ is the inclusion map.

if and only if $(\lambda, \eta) = 0$ for every harmonic form η. *Hint:* One direction is easy. For the other, invoke the Hodge decomposition theorem, Theorem 8.11, to show that $\lambda = d\alpha + \delta\beta$ for some α and β. Write $\alpha = d\alpha_1 + \delta\beta_1 + \gamma_1$ then decompose again to show that $d\alpha$ is harmonic. A similar argument shows that $\delta\beta$ is also harmonic.

9

The degree of a smooth map

All differences in this world are of degree, and not of
kind, because oneness is the secret of everything.
Swami Vivekananda

Suppose you are walking your dog Spot. You put the leash on Spot and take him
to his favorite tree to do his business, but he sees a squirrel on the tree and takes
off after it. The squirrel is really tired of being chased, and decides to teach Spot
a lesson. So, instead of climbing back up the tree, he runs in a counterclockwise
direction around the trunk with Spot not far behind. As Spot follows him around
the tree, the leash gets wound around the tree k times (assuming you stay in place
while he is running). At this point Spot gives up and sits down near your feet to
bark sullenly at the squirrel. We say that the *winding number* of the leash is k.
No matter how you try to move the leash, unless you cut it, it will remain wound
around the tree k times.[1] That is, its winding number is a *homotopy invariant*. This
prosaic example generalizes to higher dimensions and has interesting mathematical
and physical applications.[2]

We start with the **stack of records theorem**, so-called because it reveals that all
smooth maps from a compact manifold to another manifold of the same dimension
look like a smooth covering by a stack of records. (See Figure 9.1.)

Theorem 9.1 (Stack of records theorem) *Let $f : M \to N$ be a smooth map
between manifolds of the same dimension, with M compact. If q is a regular value
of f then $f^{-1}(q)$ is a finite set of points $\{p_1, \ldots, p_r\}$. Moreover, there exists a
neighborhood V_q of q in N such that $f^{-1}(V_q)$ is a disjoint union of neighborhoods
of p_i, each of which is mapped diffeomorphically onto V_q.*

---

[1] Again, we assume that both you and Spot remain fixed in place, and that you do not detach the leash from your
wrist or Spot's collar.

[2] In condensed matter physics, quantum mechanics, and quantum field theory, higher-dimensional winding
numbers are usually called *topological quantum numbers*. For more on this see [66] or [67].

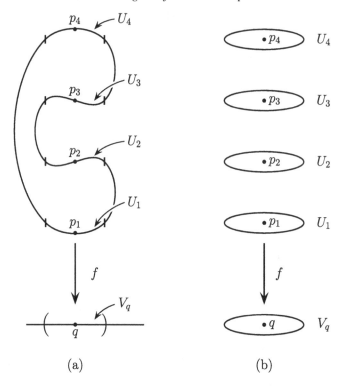

Figure 9.1 A stack of records in (a) one dimension (the domain manifold is shown) and (b) higher dimensions (fragments of the domain manifold are shown).

Proof Let $P := f^{-1}(q) \subseteq M$. By the definition of a regular value together with the inverse function theorem, each point $p \in P$ has an open neighborhood U'_p with the property that f restricted to U'_p is a diffeomorphism. The union $C' := \bigcup_p U'_p$ is an open cover of P. But M is compact, so P is compact (cf. Exercise 3.12). Hence we may choose a finite subcover $C \subseteq C'$. Each $U'_p \in C$ can contain only one preimage of q (otherwise f is not locally bijective), so the number of points $p \in P$ is finite, say $p_1, \ldots, p_r$. By taking smaller neighborhoods if necessary we may also assume that the U'_{p_i} are disjoint. Now define $V'_i := f(U'_{p_i})$ and set $V_q = \bigcap_i V'_i$. Define $U_i := f^{-1}(V_q) \cap U'_{p_i}$. Then $f^{-1}(V_q) = \coprod_i U_i$ (a disjoint union). By construction, f restricted to each U_i is a diffeomorphism. $\qquad\square$

Now let $f : M \to N$ be a smooth map between compact, connected, oriented manifolds without boundary and of the same dimension. Let q be a regular value of f, and let $f^{-1}(q) = \{p_1, \ldots, p_r\}$. Define the **index** $I(f; p_i)$ of the point p_i to be the sign of the Jacobian determinant of f at p_i. (This sign is well defined because both manifolds have a fixed orientation, so f is either orientation preserving, with sign $+1$, or orientation reversing, with sign -1.) The **degree** of f is the sum of the indices of all the points in $f^{-1}(q)$:

$$\deg f := \sum_{i=1}^{r} I(f; p_i). \tag{9.1}$$

As the notation suggests, the degree does not depend on the regular value chosen for its definition. Moreover, it is a homotopy invariant, as $f \sim g$ implies $\deg f = \deg g$. We will prove both these facts by developing an alternate definition of the degree of a map.

By Poincaré duality and the connectivity of M we have $H^n(M) \cong H^0(M) \cong \mathbb{R}$, where the isomorphism $H^n(M) \cong \mathbb{R}$ is given by integration over M. As $f^* : H^n(N) \to H^n(M)$ is an isomorphism of vector spaces we get the following commutative diagram:

$$
\begin{array}{ccc}
H^n(N) & \xrightarrow{f^*} & H^n(M) \\
{\scriptstyle \int_N} \downarrow & & {\scriptstyle \int_M} \downarrow \\
\mathbb{R} & \xrightarrow{\text{Deg } f} & \mathbb{R}
\end{array} \tag{9.2}
$$

The lower map is a linear isomorphism between one-dimensional vector spaces and is therefore simply multiplication by an element $\text{Deg } f \in \mathbb{R}$ that depends only on f. Many authors define this to be the degree of f, which is justified by the following fact.

Theorem 9.2 *Let f, M, and N be as above. Then*

$$\text{Deg } f = \deg f.$$

Proof We apply the stack of records theorem (and use the same notation). Assume q is a regular value of f. Let $\omega \in \Omega^n(N)$ be an n-form whose support is contained in V_q. Define $f_i := f|_{U_i}$. Then, as each f_i is a diffeomorphism, the change of variables theorem gives

$$\int_M f^* \omega = \sum_i \int_{U_i} f_i^* \omega = \sum_i I(f; p_i) \int_{V_q} \omega = (\deg f) \int_N \omega. \qquad \square$$

It follows that $\deg f$ does not depend on the regular value q used for its definition, as the definition of $\text{Deg } f$ makes no mention of regular values. Moreover, homotopic maps induce the same map in cohomology, so $\text{Deg } f$ is a homotopy invariant. Thus, $\deg f$ enjoys this attribute as well. But Theorem 9.2 is a two-way street, for if one were to define the degree of f to be $\mathrm{Deg}(f)$ then there would be no *a priori* reason for it to be an integer. Yet it is, because deg f is always an integer.

EXERCISE 9.1 Given $f : M \to N$ and $G : N \to P$, show that the degree is multiplicative:

$$\mathrm{Deg}(g \circ f) = (\mathrm{Deg}\, g)(\mathrm{Deg}\, f).$$

EXERCISE 9.2 Let $f : S^2 \to T^2$ be a smooth map from the 2-sphere to the 2-torus. Show that deg $f = 0$. *Hint:* The form $dx \wedge dy$ on the torus is a generator of its top-dimensional cohomology. What happens when you pull it back and integrate it over the 2-sphere?

EXERCISE 9.3 Let M be a compact, connected, oriented, smooth manifold without boundary. Fix a point $x \in \mathbb{R}^{n+1}$, and let $f : M \to \mathbb{R}^{n+1} - \{x\}$ be a smooth map. Define a smooth map $g : M \to S^n \in \mathbb{R}^{n+1}$ by

$$g(p) = \frac{f(p) - x}{\|f(p) - x\|}.$$

By definition, the **winding number** of f (about x) is the degree of g. Show that, when $M = S^1$ and $n = 1$, the winding number of f is precisely the winding number we encountered when walking Spot. That is, show that deg g counts the number of times f wraps the circle around the point x. *Hint:* There are many ways to do this but perhaps the quickest is just to observe that, by homotopy invariance, we may assume that f is given by the map $\theta \mapsto \varphi = k\theta$. Choose the closed but not exact 1-form $d\varphi$ on the image circle and pull it back to the domain circle. Otherwise, you have to show that the inverse image of a regular point contains k points and that all the Jacobians are of the same sign.

The above exercise shows that the degree of a map is a generalization of the notion of a winding number to higher dimensions. Intuitively, it counts the number of times M is wrapped around S^n by f.

9.1 The hairy ball theorem and the Hopf fibration

Of particular interest is the degree of a map from one sphere to another. It is immediate from the definitions that the identity map $id : S^n \to S^n$ has degree equal to $+1$. It is also easy to see that the **antipodal map** $-id : S^n \to S^n$ that sends $x \mapsto -x$ (viewing S^n as the unit sphere in $\mathbb{R}^{n+1}$) has degree equal to $(-1)^{n+1}$.

EXERCISE 9.4 Prove that the antipodal map has degree $(-1)^{n+1}$. *Hint:* See Exercise 8.60.

We can use these facts to prove a beautiful little theorem with an ugly name: the **hairy ball theorem**.[3] This theorem says that only odd-dimensional spheres have nowhere-vanishing tangent vector fields. The reason why people call it the hairy ball theorem is that the theorem says that you cannot comb the hair (really, fuzz) on a tennis ball without leaving a cowlick somewhere.[4] Alternatively, and much more interestingly, the theorem implies that there is always at least one point on the Earth where the wind is not blowing.

Theorem 9.3 (The hairy ball theorem) *The sphere S^n admits a nowhere-zero tangent vector field if and only if n is odd.*

Proof Identify $x \in S^n$ with the vector from the origin to x in $\mathbb{R}^{n+1}$. Define the unit normal vector $N = x/\|x\|$. If $n = 2k - 1$ is odd then the vector field $X = (x^2, -x^1, x^4, -x^3, \ldots, x^{2k}, -x^{2k-1}) \in \mathbb{R}^{2k}$ is everywhere orthogonal to N, so it is everywhere tangent to the sphere and nonvanishing.

Conversely, let n be even and suppose X is a nowhere-zero tangent vector field on the sphere. Then there exists a nowhere-zero tangent vector field $Y := X/\|X\|$ on the sphere of unit length. Consider the map

$$h(x, t) : S^n \times [0, 1] \to \mathbb{R}^{n+1}$$

given by

$$h(x, t) = \cos(\pi t)\, N + \sin(\pi t)\, Y.$$

For fixed t, h maps the sphere to itself (because $\|h(t, x)\| = 1$). Also, $h(x, t) = x$ for $t = 0$ and $h(x, t) = -x$ for $t = 1$, so h is a homotopy between id and $-id$. Hence $1 = \deg(id) = \deg(-id) = (-1)^{n+1}$, a contradiction. $\square$

EXERCISE 9.5 Show that, if $f : S^n \to S^n$ has no fixed points, $\deg f = (-1)^{n+1}$. *Hint:* Define

$$F : I \times S^n \to S^n \quad \text{by} \quad F(t, x) = \frac{(1 - t)f(x) + t(-x)}{\|(1 - t)f(x) + t(-x)\|}.$$

---

[3] Many Europeans prefer to call it the **hedgehog theorem**, which is much more civilized.
[4] Note that you *can* comb the hair on your head without leaving a cowlick, because your head (or at least, the place where you have long hair) is more like a hemisphere, so the hair can all be directed to one side.

Show that $F(t, x)$ is a well-defined smooth homotopy between f and the antipodal map.

EXERCISE 9.6 Suppose that $f : S^n \to S^n$ fails to be surjective. Show that deg $f = 0$. *Hint:* See Exercise 4.12.

EXERCISE 9.7 Use the notion of the degree of a map to prove Brouwer's fixed point theorem (cf. Example 4.6). *Hint:* Let $N = \{x \in S^n \cap \mathbb{R}^{n+1} : x^{n+1} \geq 0\}$ be the northern hemisphere of the n-sphere, and define the southern hemisphere S similarly. Let $\varphi_N : N \to B^n$ and $\varphi_S : S \to B^n$ be diffeomorphisms that agree on $N \cap S$. Let $f : B^n \to B^n$ be a smooth map without fixed points, and define a new (smooth) map $g : S^n \to S^n$ by

$$
g(x) = \begin{cases} \varphi_S^{-1} \circ f \circ \varphi_N(x), & \text{if } x \in N, \text{ and} \\ \varphi_S^{-1} \circ f \circ \varphi_S(x) & \text{if } x \in S. \end{cases}
$$

Look at possible fixed points of g and use Exercises 9.5 and 9.6.

So far we have discussed maps between spheres of the same dimension and have discovered that the degree of such a map is always an integer. Put another way, all the maps from one sphere to another of the same dimension fall into homotopy classes labeled by the integers.

Can we say anything about maps from S^m to S^n when $m \neq n$? The answer turns out to be easy when $m < n$. To see this, consider a map from a circle into a 2-sphere. The 2-sphere is simply connected, so we can always shrink the image of the circle to a point while remaining on the 2-sphere. In other words, every such map is null homotopic. This kind of behavior occurs again in higher dimensions, because any map from a lower-dimensional sphere to a higher dimensional sphere can be deformed to a map that misses at least one point, so by Exercise 4.12 it must be homotopically trivial.

The case $m > n$ is quite subtle, and is still not entirely understood despite much work. One of the most striking things is really that such maps exist at all.[5] The classic example in this area is the **Hopf fibration**, which is a certain smooth map $h : S^3 \to S^2$ whose precise definition would take us too far afield. It is called a "fibration" because it expresses the 3-sphere as a smooth collection of fibers (the sets $f^{-1}(x)$), all of which are copies of S^1. It turns out that the Hopf fibration has degree 1, so there is at least one continuous map from S^3 to S^2 that cannot be deformed to a point, a highly nonintuitive result![6]

---

[5] Of course, there exist all sorts of maps from S^m to S^n with $m > n$. The surprising fact is that some of them are continuous and even smooth.

[6] For a discussion of the Hopf map and a computation of its degree, see e.g. [12], Ex. 17.23.

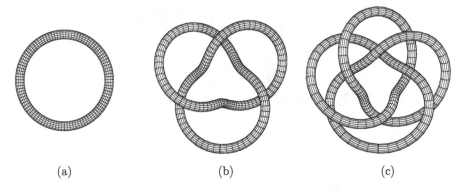

(a)   (b)   (c)

Figure 9.2 Some knots. (a) The unknot. (b) The trefoil. (c) The torus knot $(3, 4)$.

9.2 Linking numbers and magnetostatics

The degree of a map has many other interesting uses. Here's one that is related to knot theory, a fascinating branch of mathematics that studies the topological properties of simple closed curves embedded in three-dimensional manifolds. [7]

For our purposes, we define a **knot** to be a circle embedded in $\mathbb{R}^3$ and a **link** to be a bunch of disjoint knots (called the **components** of the link). The **unknot** is the simplest knot, namely just the circle, and the **unlink** is a bunch of disjoint circles in the plane. (See Figures 9.2 and 9.3.)

Intuitively we can think of a knot as a closed string, and two knots are the same if we can deform one until it looks just like the other (without cutting the string, of course). More precisely, two knots K_1 and K_2 are **isotopic** if there exists a smooth map $f : \mathbb{R}^3 \times [0, 1] \to \mathbb{R}^3$ such that $f_t := f(\cdot, t) : \mathbb{R}^3 \to \mathbb{R}^3$ is a diffeomorphism satisfying $f_0 = id$ and $f_1(K_1) = K_2$. One of the primary objectives of knot theory is to find good tools to distinguish which knots and links are trivial (isotopic to the unknot or the unlink) and which are not.

Given a link, its **linking number** is defined as follows. The **diagram** of a link is just a generic plane projection of the link, with under and overcrossings denoted in the usual way. (See Figure 9.4.) Orient each component of the link arbitrarily. Associate a sign $+1$ to a right-handed crossing and a sign -1 to a left-handed crossing, as shown in Figure 9.5. Then the linking number of the link is one-half the sum of the signs of all the crossings of different components in any diagram of the oriented link. (Self-crossings do not count.) Figure 9.6 provides an example. Using something called "Reidemeister moves" one can show that the linking number is indeed invariant under an isotopy. Unfortunately, the linking number is not really

---

[7] Nice introductions to knot theory are provided by [2] and [52]. Connections between knot theory and physics are discussed in [7] and [45].

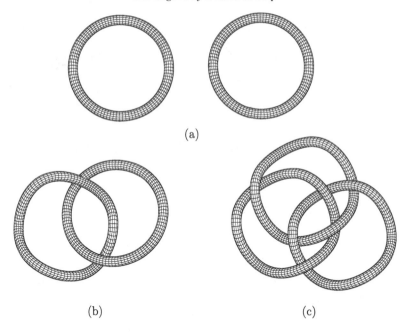

(a)

(b)   (c)

Figure 9.3 Some links. (a) An unlink with two components. (b) A link with two components. (c) The Borromean rings.

Figure 9.4 A diagram of the trefoil.

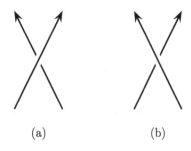

(a)   (b)

Figure 9.5 (a) A right-handed crossing with sign $+1$. (b) A left-handed crossing with sign -1.

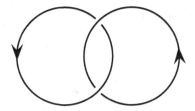

Figure 9.6 The linking number of this link is -1.

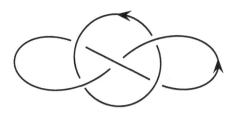

Figure 9.7 The linking number of the Whitehead link is zero.

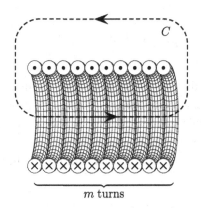

m turns

Figure 9.8 A cross section of a solenoid with m turns carrying a current out of the paper at the top, and into the paper at the bottom.

that useful, as one can see from the example of the Whitehead link, illustrated in Figure 9.7. It has linking number zero, but it is evidently not isotopic to the unlink.

Actually the idea of a linking number is far older than this modern definition. Indeed, it was Gauss who first introduced the notion of linking while pondering the magnetic interaction of two current-carrying wires. Consider the solenoid depicted in Figure 9.8. Assuming a current I in the solenoid, on the one hand Ampère's law

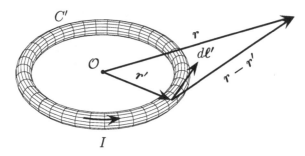

Figure 9.9 A current-carrying wire loop.

tells us that the circulation of the magnetic field around C is given by

$$\oint_C \boldsymbol{B} \cdot d\boldsymbol{\ell} = m\mu_0 I,$$

where m is the number of windings in the solenoid.

On the other hand, consider a wire loop C' carrying a current I, as in Figure 9.9. The magnetic field at a point $\boldsymbol{r}$ produced by this loop is, according to the law of Biot and Savart,

$$\boldsymbol{B}(\boldsymbol{r}) = \frac{\mu_0 I}{4\pi} \oint_{C'} \frac{d\boldsymbol{\ell}' \times (\boldsymbol{r} - \boldsymbol{r}')}{|\boldsymbol{r} - \boldsymbol{r}'|^3},$$

where $\boldsymbol{r}'$ points from the origin to a current element on C' and $d\boldsymbol{\ell}'$ is the infinitesimal displacement vector along C' in the direction of the current.

Now let C' be the solenoid with m turns considered previously. Combining Ampère's law with the expression obtained from Biot and Savart and using the triple product identity yields **Gauss's formula** for the linking number of the two curves C and C':

$$m = \frac{1}{4\pi} \oint_C \oint_{C'} \frac{(\boldsymbol{r} - \boldsymbol{r}')}{|\boldsymbol{r} - \boldsymbol{r}'|^3} \cdot (d\boldsymbol{\ell} \times d\boldsymbol{\ell}'). \tag{9.3}$$

Although it would be difficult (if not impossible) actually to compute the right-hand side of Gauss's formula in any particular case, it does have the virtue of being manifestly symmetric under the interchange of C and C'.

Gauss's expression was generalized to higher dimensions by Călugăreanu and White (see e.g. [61]). Let M and N be two oriented compact boundaryless submanifolds of $\mathbb{R}^n$ with $\dim M + \dim N = n - 1$. Then their **linking number**, $\mathrm{link}(M, N)$, is defined to be the degree of the map $f : M \times N \to \mathbb{R}^n - \{0\}$ given by

$$f(x, y) = y - x.$$

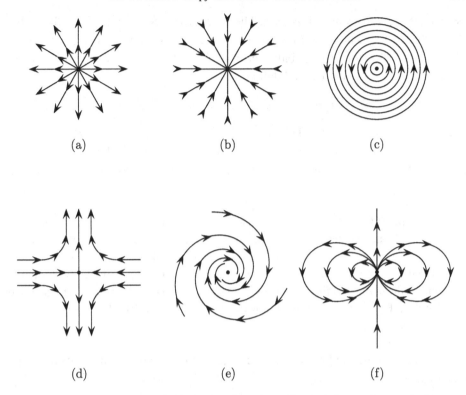

(a)   (b)   (c)

(d)   (e)   (f)

Figure 9.10 Some vector fields in the plane (actually, their integral curves): (a) a
source, (b) a sink, (c) a circulation, (d) a saddle, (e) a spiral, and (f) a dipole.

One can show ([54], Theorem 11.14) that, for the case $M = N = S^1$, this definition
reduces to Gauss's definition and coincides with the knot-theoretic definition of the
linking number given above. As the degree is a homotopy invariant, this provides
another proof that the linking number is invariant under isotopy.

9.3 The Poincaré–Hopf index theorem and the Gauss–Bonnet theorem

We cannot abandon the subject of mapping degrees without at least mentioning
two of the most important results of differential topology. To set the stage, let us
consider a few vector fields X in the plane, as shown in Figure 9.10. The first
observation is that all the interesting behavior occurs near a **zero** (or **singularity**)
p, where $X_p = 0$, because away from a zero the smoothness of the vector field
ensures that X does not change much in magnitude or direction.

The notion of the degree of a map allows us to construct a measure of the behavior of a vector field near a zero. Although the definition appears a bit contrived at
first glance, it is remarkably powerful. A zero p of X is **isolated** if it is possible
to find a neighborhood U of p that contains no other zeros of X. Henceforth we

assume all our zeros are of this type. Then we may choose a closed neighborhood B of p that contains no zeros of X other than p and a diffeomorphism $\varphi : B \to B^n$, where $B^n \in \mathbb{R}^n$ is the usual n-ball. Let $\pi : \mathbb{R}^n - \{0\} \to S^{n-1}$ denote the standard projection map $x \to x/\|x\|$, and define

$$\psi := \pi \circ \varphi_* \circ X \circ \varphi^{-1} : \partial B^n \cong S^{n-1} \to S^{n-1}.$$

In other words, let $x \in \partial B^n$. Then $q := \varphi^{-1}(x) \in \partial B$ and $X_q \neq 0$, so by linearity $\varphi_* X_q \in \mathbb{R}^n - \{0\}$. Hence $\psi(x) \in S^{n-1}$. Now we define the **index** ind(X, p) of X to be the degree of ψ. Although ind(X, p) appears to depend on φ, in fact it does not (see e.g. [58], p. 34).[8]

> **Example 9.1** Let X be a vector field in $\mathbb{R}^2$. Pick a zero of X and choose a small circle C_1 surrounding it. Associate to each point x of C_1 a point y on a second circle C_2, where $y = X(x)/\|X(x)\|$. Then as you walk around C_1, y moves on C_2. The degree deg ψ counts the number of times y moves counterclockwise around C_2 minus the number of times y moves clockwise around C_2. Applying this to the vector fields in Figure 9.10 gives the following indices: (a) $+1$, (b) $+1$, (c) -1, (d) -1, (e) $+1$, and (f) $+2$.

We can now state the first result, a striking connection between vector fields and topology.

Theorem 9.4 (Poincaré–Hopf index theorem) *Let M be a compact manifold without boundary, and let X be a vector field on M with only isolated zeros. Then*

$$\sum_p \text{ind}(X, p) = \chi(M),$$

where p ranges over all zeros of X and $\chi(M)$ is the Euler characteristic of M.

We refer the reader to [33] or [58] for a proof. As a simple application, observe that the theorem implies the difficult part of the hairy ball theorem, because all even-dimensional spheres have Euler characteristic equal to 2 (cf. Example 5.6), so every vector field on such a sphere must vanish somewhere. Incidentally, the Poincaré–Hopf theorem is itself a special case of a very far-reaching theorem, called the **Lefschetz–Hopf theorem**, which relates the Lefschetz number of a smooth map f (defined in Exercise 5.12) to something called the total fixed point index of f. For more on this, see [30].

---

[8] Note that the index of a vector field X, as we have just defined it, is the degree of a certain map ψ. But the degree of a map is defined as the sum of the indices of the map. In this context, the index is the sign of the Jacobian at a preimage of a regular value of the map. If there is only one preimage of the chosen regular value, then the index of the vector field X is just the index of the map ψ. This explains the use of the word index for both concepts.

The second, absolutely fundamental, result reveals a deep connection between the curvature of a manifold and its topology. In its simplest form, it relates the Gaussian curvature K of a two-dimensional manifold to its Euler characteristic $\chi(M)$.

Theorem 9.5 (Gauss–Bonnet) *Let M be a compact, oriented, two-dimensional Riemannian manifold without boundary. Then*

$$\int_M K\, dA = 2\pi\,\chi(M),$$

where dA is the area element of M.

The theorem was known to Gauss and was published by Bonnet in 1848. Although it was first proved using more elementary methods, one can prove it, as well as a generalization to $2n$-dimensional submanifolds of $\mathbb{R}^{2n+1}$, using some of the fancy machinery at our disposal. Very roughly, one embeds M in Euclidean space and shows that the left-hand side of the above equation is the degree of a certain map, called the **Gauss map** (see Appendix H). The degree of the Gauss map is then related to the index of a certain vector field, to which the Poincaré–Hopf theorem is then applied. (For the details see [33].)

An intrinsic proof of the generalized Gauss–Bonnet theorem was discovered by Chern in the 1940s. To state that generalization we need one last definition. Let Q be a skew symmetric $2n \times 2n$ matrix. The skew symmetry of Q ensures that the determinant of Q is a perfect square, and pf Q, the **Pfaffian** of Q, is defined to be its positive square root:

$$\mathrm{pf}\ Q = \sqrt{\det Q}. \tag{9.4}$$

Explicitly,

$$\mathrm{pf}\ Q = \frac{1}{2^n n!} \sum_{\sigma \in \mathfrak{S}_{2n}} (-1)^\sigma Q_{\sigma(1)\sigma(2)} \cdots Q_{\sigma(2n-1)\sigma(2n)}. \tag{9.5}$$

It can be shown that, for any matrix A, $\mathrm{pf}(A^T Q A) = (\det A)(\mathrm{pf}\ Q)$, so the Pfaffian is invariant under (special) orthogonal transformations.

EXERCISE 9.8 Define a 2-form $\lambda := \frac{1}{2} Q_{ij}\, \theta^i \wedge \theta^j$ in $\mathbb{R}^{2n}$. Show that

$$\lambda^n = \underbrace{\lambda \wedge \cdots \wedge \lambda}_{n \text{ times}} = n!(\mathrm{pf}\ Q)\,\sigma$$

where $\sigma = \theta^1 \wedge \cdots \wedge \theta^{2n}$ is the volume form. Use this expression to prove that $\mathrm{pf}(A^T Q A) = (\det A)(\mathrm{pf}\ Q)$. *Hint for the second part*: Consider what happens to λ when $\theta \mapsto \theta' = A^{-1}\theta$.

Let Ω be the matrix of curvature 2-forms associated to the Levi-Civita connection of a Riemannian manifold M of dimension $2n$. As observed in Exercise 8.6, the matrix $\Omega = (\Omega_{\hat{a}\hat{b}})$ is skew symmetric in an orthonormal basis, so we may plug it into the above definition to get a $2n$-form pf Ω (where the ordinary product is replaced by the wedge product). Under a change of frame, Ω changes by a special orthogonal transformation (see (7.32)), so pf Ω is a well-defined (global) $2n$-form on M.

Example 9.2   Although it was constructed using an orthonormal basis, the Pfaffian pf Ω is globally defined so it cannot depend on the basis. This is made evident by rewriting (9.5) in terms of the Levi-Civita tensor:

$$\text{pf } \Omega = \frac{1}{2^n n!} \epsilon^{i_1 i_2 \ldots i_{2n-1} i_{2n}} \Omega_{i_1 i_2} \wedge \cdots \wedge \Omega_{i_{2n-1} i_{2n}}. \tag{9.6}$$

This expression, in turn, can be written in terms of the Riemann tensor using $\Omega_{ab} = \frac{1}{2} R_{abcd} \theta^c \wedge \theta^c$. In particular, in dimensions 2 and 4 we have

$$\text{pf } \Omega = \begin{cases} \frac{1}{2}\epsilon^{ab}\Omega_{ab} = \Omega_{\hat{1}\hat{2}} = R_{\hat{1}\hat{2}\hat{1}\hat{2}}\,\sigma = K\,\sigma & (n=1), \\ \frac{1}{8}\epsilon^{abcd}\Omega_{ab} \wedge \Omega_{cd} = \frac{1}{8}\epsilon^{abcd}\epsilon^{pqrs}R_{abpq}R_{cdrs}\,\sigma & (n=2), \end{cases}$$

where, as always, σ is the volume form and the carets indicate an orthonormal frame.

An argument similar to the one given in Exercise 7.9 shows that pf Ω, and therefore eu $\Omega := (\text{pf } \Omega)/(2\pi)^n$, defines a cohomology class in $H^{2n}(M)$. The class eu Ω is called the **Euler class** of E because of the following theorem.[9]

Theorem 9.6 (Chern–Gauss–Bonnet [16])   *Let M be a compact, oriented, $2n$-dimensional Riemannian manifold without boundary, and let Ω be the curvature 2-form of its Levi-Civita connection. Then*

$$\int_M \text{eu } \Omega = \chi(M),$$

where $\chi(M)$ is the Euler characteristic of M.

This seems a fitting place to end our brief tour of the geometry and topology of manifolds.

---

[9] The theorem applies without change (although with a different proof) to pseudo-Riemannian manifolds [17].

Appendix A

Mathematical background

A.1 Sets and maps

A **set** X is a collection of objects, which we call **elements**. We write $x \in X$ if x is an element of X. We also write $X = \{x, y, \ldots\}$ to denote the elements of X. The **empty set** $\emptyset = \{\}$ is the unique set containing no elements. A set U is a **subset** of X, written $U \subseteq X$, if $x \in U$ implies $x \in X$. A set U is a **proper** subset of X, written $U \subset X$, if $U \subseteq X$ and $U \neq X$. The **union** $X \cup Y$ of two sets X and Y is the set of all elements in X or in Y (or in both), whereas the **intersection** $X \cap Y$ is the set of all elements that are in both X and Y. If $X \cap Y \neq \emptyset$ then X **meets** Y (and Y meets X). The **(set-theoretic) difference** of two sets X and Y is the set $X \backslash Y = X - Y = \{x \in X : x \notin Y\}$. (This definition does not require that Y be contained in X.) The **complement** of $U \subseteq X$ is $\overline{U} := X - U$. For any collection $\{U_i\}$ of subsets of X, **de Morgan's laws** hold:

$$\overline{\bigcup_i U_i} = \bigcap_i \overline{U_i}, \tag{A.1}$$

$$\overline{\bigcap_i U_i} = \bigcup_i \overline{U_i}. \tag{A.2}$$

A **map** ϕ from a set X to a set Y, written $\phi : X \to Y$, assigns a unique point $y \in Y$ to each point $x \in X$. We write $\phi(x) = y$ or $x \mapsto y$ to denote this assignment, and we call X the **domain** of ϕ and Y the **target** of ϕ. If $\phi(x) = y$ for some $x \in X$ and $y \in Y$ then y is the **image** of x and x is a **preimage** or **inverse image** of y. The set of all preimages of y is denoted $\phi^{-1}(y)$. The **image** or **range** of ϕ is the set of all points in the target that are the image of some point in the domain.

A map $\phi : X \to Y$ between two sets X and Y is said to be **injective** or **one-to-one** or **into** if $\phi(x_1) = \phi(x_2)$ implies $x_1 = x_2$. A map ϕ is **surjective** or **onto** if, for every $y \in Y$, there exists an $x \in X$ such that $\phi(x) = y$. A map ϕ is **bijective** if it is both injective and surjective. (See Figure A.1.)

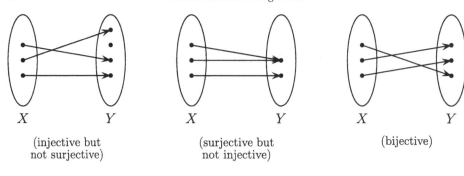

Figure A.1 Types of map.

If there is a bijection between X and Y then the two sets have the same number of elements. We say they have the same **cardinality**. The cardinality of a set X is denoted $|X|$. Sets can have finite or infinite cardinality. The **natural numbers** are the elements of $\mathbb{N} = \{1, 2, \ldots\}$. If an infinite set has the same cardinality as the natural numbers it is called **countably infinite** (or simply **countable** or **denumerable**); otherwise, it is **uncountably infinite** (or simply **uncountable**).

Given a map $\phi : X \to Y$ and a map $\psi : Y \to Z$, the **composition** of ψ and ϕ, written $\psi \circ \phi$, is a map from X to Z given by $(\psi \circ \phi)(x) := \psi(\phi(x))$. (Note that the composition $\phi \circ \psi$ is not defined.) The **inverse** of a map $\phi : X \to Y$ is the map $\phi^{-1} : Y \to X$ satisfying $\phi \circ \phi^{-1} = \phi^{-1} \circ \phi = id$, where id denotes the **identity** map, given by $id(x) = x$ for all x. Only bijective maps have inverses.[1]

An **equivalence relation** on a set X is a binary relation, denoted $\sim$, satisfying the following three properties. For all $x, y, z \in X$ we have

(1) (reflexivity) $x \sim x$
(2) (symmetry) $x \sim y$ implies $y \sim x$
(3) (transitivity) $x \sim y$ and $y \sim z$ implies $x \sim z$.

Given $x \in X$, the **equivalence class** of x, often denoted $\bar{x}$ or $[x]$, is the set of all elements of X equivalent to x.

> **Example A.1** The **integers** are the elements of the set $\{\ldots, -2, -1, 0, 1, 2, \ldots\}$. Two integers a and b are **equivalent modulo (or mod)** n if the integer n divides $a - b$ without remainder. In that case we write $a = b \mod n$. Equivalence modulo n is easily seen be an equivalence relation on the integers. For example, if $n = 5$ then $12 = 7 \mod 5$. Modulo 5, the equivalence class $[3] = \{\ldots, -7, -2, 3, 8, 13, \ldots\}$.

---

[1] The notation is somewhat ambiguous, as we have also denoted the preimage of a point y under a set map $\phi : X \to Y$ by $\phi^{-1}(y)$, and this makes sense even if ϕ is not bijective. The meaning should always be clear from the context.

A **partition** of X is a collection of subsets $\{X_1, X_2, \ldots, X_r\}$ of X, called the **parts** of the partition, such that every element of X is in one and only one part. A partition of X is the same thing as an equivalence relation on X in which the parts of the partition are the equivalence classes of the equivalence relation.

> **Example A.2** The integers modulo n, denoted $\mathbb{Z}_n$, are partitioned into n equivalence classes $\{[0], [1], \ldots, [n-1]\}$.

An **ordered set** is a set together with a choice of order for the elements. Parentheses are used to distinguish ordered sets from unordered sets. For example, (z, x, y) denotes an ordered set whose underlying set is $\{x, y, z\}$. The **Cartesian product** of two sets X and Y, denoted $X \times Y$, is the set of all ordered pairs (x, y). We can iterate this construction to form the Cartesian product of as many sets as we like. For finite sets, $|X \times Y| = |X||Y|$.

A.2 Groups

A nonempty set G together with a binary operation (often denoted by "$+$" or "$\cdot$" or simply by juxtaposition and called the **group operation**) is said to be a **group** provided the following axioms hold:

(1) $a, b \in G$ implies that $a \cdot b \in G$;
(2) $a, b, c \in G$ implies that $a \cdot (b \cdot c) = (a \cdot b) \cdot c$;
(3) there exists an element $e \in G$ such that $a \cdot e = e \cdot a = a$ for all $a \in G$;
(4) for every $a \in G$ there exists an element $a^{-1} \in G$ such that $a \cdot a^{-1} = a^{-1} \cdot a = e$.

Axiom (1) says that G is closed under the group operation. Axiom (2) says that G is associative. Axiom (3) is the axiom of the identity (e is called the identity element of the group). Axiom (4) is the axiom of inverses, and says that every element in the group has an inverse.

> **Example A.3** Let $G = \mathbb{Z}$, where the group operation is given by ordinary addition. To verify that $\mathbb{Z}$ is a group we must verify that the axioms are satisfied. Well, certainly the integers are closed under addition: the sum of any two integers is again an integer. Ordinary addition is associative, so axiom (2) is satisfied. The identity element is 0, because $a + 0 = 0 + a = a$ for any integer a. Finally, the inverse element of a is just $-a$, because $a + (-a) = (-a) + a = 0$.

> **Example A.4** Let $G = \{-1, 1\}$, where the group operation is now ordinary multiplication. Clearly G is closed under multiplication. Ordinary multiplication is associative, so axiom (2) is satisfied. The identity element is 1, because $a \cdot 1 = 1 \cdot a = a$ for any $a \in G$. Finally, each element is its own inverse. This group is usually denoted $\mathbb{Z}_2$.

Example A.5   Let $G = \{e, a, a^2, a^3, \ldots, a^{n-1}\}$ with the group operation denoted by multiplication. The formal symbol a satisfies the identity $a^n = e$. The group axioms are easily verified. This group is called the **cyclic group of order** n and is usually denoted C_n.

Example A.6   Let $X = \{1, 2, \ldots, n\}$ be a set of n objects. A **permutation** of X is a bijection $\sigma : X \to X$. Permutations can be represented in many different ways, but the simplest is just to write down the images of the numbers 1 to n in a linear order. So, for example, if $\sigma(1) = 2$, $\sigma(2) = 4$, $\sigma(3) = 3$, and $\sigma(4) = 1$ then we write $\sigma = 2431$. Let G be the set of all permutations of X, where the group operation is composition. If σ and τ are permutations we write $\sigma\tau$ for $\sigma \circ \tau$. That is, $(\sigma\tau)(i) = \sigma(\tau(i))$. For example, if $\tau(1) = 4$, $\tau(2) = 2$, $\tau(3) = 3$, and $\tau(4) = 1$ then $\tau = 4231$ and $\sigma\tau = 1432$. Evidently, G is closed under composition (which is also associative). The identity element of G is just the identity map, given by $\sigma(i) = i$ for all i. For example, the identity permutation of $\mathfrak{S}_4$ is $e = 1234$. Finally, the **inverse** of a permutation σ is just the inverse map σ^{-1}, which satisfies $\sigma\sigma^{-1} = \sigma^{-1}\sigma = e$. For example, if $\sigma = 2431$ then $\sigma^{-1} = 4132$. The collection of all permutations of X is called the **symmetric** (or **permutation**) **group** on n elements and is denoted $\mathfrak{S}_n$. It contains $n!$ elements.

A **transposition** is a permutation that switches two numbers and leaves the rest fixed. For example, the permutation 4231 is a transposition because it flips 1 and 4 and leaves 2 and 3 alone. For simplicity one usually denotes this transposition by $1 \leftrightarrow 4$. It is not too difficult to see that $\mathfrak{S}_n$ is generated by transpositions. This means that any permutation σ can be written as the product of transpositions.

A permutation σ is **even** if it can be expressed as the product of an even number of transpositions; otherwise it is **odd**. The **sign** of a permutation σ, written $(-1)^\sigma$, is $+1$ if it is even and -1 if it is odd. One can show that the sign of a permutation is the number of transpositions required to transform it back to the identity permutation. So 2431 is an even permutation (sign $+1$) because we can get back to the identity permutation in two steps: $2431 \xrightarrow{1 \leftrightarrow 2} 1432 \xrightarrow{2 \leftrightarrow 4} 1234$. Although a given permutation σ can be written in many different ways as a product of transpositions, it turns out that the sign of σ is always the same. Furthermore, as the notation is meant to suggest, $(-1)^{\sigma\tau} = (-1)^\sigma(-1)^\tau$. Both these claims require proof, which we omit.

Example A.7   The set $SO(n)$ of all rotations in $\mathbb{R}^n$ (endomorphisms of $\mathbb{R}^n$ that preserve the Euclidean inner product and have unit determinant) is a subgroup (under composition) of the orthogonal group $O(n)$. The set $SO(n)$ is called the special orthogonal group.

Remark   Examples A.3–A.6 are finite groups, while Example A.7 is an infinite group.

A group G is said to be **abelian** or **commutative** if $a \cdot b = b \cdot a$ for every pair of elements a, b in G. The groups $\mathbb{Z}$, $\mathbb{Z}_2$, and C_n are all abelian, whereas $SO(3)$ is nonabelian.

> **Example A.8** For $n > 2$, $\mathfrak{S}_n$ is nonabelian. For example, if $n = 3$ and $\sigma = 132$ and $\tau = 213$ then $\sigma\tau = 231$ whereas $\tau\sigma = 312$. This implies the claim, as $\mathfrak{S}_n$ is naturally a subgroup (a subset that is also a group in its own right) of $\mathfrak{S}_m$ when $n \leq m$.

Next we introduce a crucial concept in the theory of algebraic structures. A map $\phi : G \to H$ from a group G to a group H satisfying $\phi(gh) = \phi(g)\phi(h)$ is called a **(group) homomorphism**.

Remark In the above definition, the first product gh is the product in G, while the second product $\phi(g)\phi(h)$ is the product in H.

Remark The word "homomorphism" is from the Greek roots ὁμóζ, meaning similar or akin, and μορπή, meaning shape or form. Two groups are **homomorphic** if there is a homomorphism between them. In that case they are indeed similar in form, because the homomorphism establishes a link between the two group structures.

> **Example A.9** The map $\phi : C_3 \to \mathfrak{S}_3$ taking a to 231 and extended homomorphically to the rest of C_3 is a homomorphism. ("Extended homomorphically" means that we require the map act homomorphically on the entire group, subject to the initial constraint given for the image of a; in our case this means that we require ϕ to satisfy the condition $\phi(a^m) = \phi(a)^m$ for all m.)

> **Example A.10** The map $\phi : \mathbb{Z} \to \mathbb{Z}_2$ taking every even integer to $+1$ and every odd integer to -1 is a homomorphism, because $\phi(a + b) = \phi(a)\phi(b)$ for all a and b. (Recall that the group "product" in $\mathbb{Z}$ is just addition.)

Let G and H be groups. A homomorphism $\phi : G \to H$ is called an **isomorphism** if ϕ is a bijection. Groups G and H are said to be **isomorphic** if there is an isomorphism between them. Isomorphic groups are essentially the same (the Greek word ἴσοζ means "equal"). A homomorphism $\phi : G \to G$ from a group to itself is called an **endomorphism**. A bijective endomorphism $\phi : G \to G$ is called an **automorphism**. An endomorphism is not necessarily one-to-one or onto, which means it could map G to part of itself in a funny way. An automorphism is just about the most innocuous sort of map around – it basically permutes the group elements.

Example A.11 Let G be C_3, and consider the map $\phi : C_3 \rightarrow C_3$ given by $\phi(a) = a^2$ and extended homomorphically. The map ϕ is an automorphism.

Given a group G and a space X, a (**left**) **action** of G on X is a map $G \times X \rightarrow X$ given by $(g, x) \mapsto gx$ and satisfying two axioms:

(1) $ex = x$ for all $x \in X$;
(2) $(gh)x = g(hx)$ for all $g, h \in G$ and $x \in X$.

If $gx = x$ implies $g = e$ then the action of G on X is **free**.

A.3 Rings and fields

Next we need the idea of a **field**. The intuitive definition of a field is a set of elements (called **scalars**) that possess the same kinds of properties as do the real numbers. So, for example, any three elements a, b, and c in a field satisfy the distributive law: $a \cdot (b + c) = a \cdot b + a \cdot c$.

More precisely, a nonempty set R together with two operations, addition and multiplication, denoted $+$ and $\cdot$ respectively, is said to be an **associative ring** if, for all a, b, and c in R, the following hold:

(1) $a + b \in R$;
(2) $a + b = b + a$;
(3) $(a + b) + c = a + (b + c)$;
(4) there is an element $0 \in R$ such that for every $a \in R$ we have $a + 0 = a$;
(5) for every $a \in R$ there is an element $-a \in R$ such that $a + (-a) = 0$;
(6) $a \cdot b$ is in R;
(7) $a \cdot (b \cdot c) = (a \cdot b) \cdot c$;
(8) $a \cdot (b + c) = a \cdot b + a \cdot c$ and $(b + c) \cdot a = b \cdot a + c \cdot a$;
(9) there exists an element $1 \in R$ such that for every $a \in R$ we have $a \cdot 1 = 1 \cdot a = a$.

Axioms (1) through (5) say that R is an abelian group under addition. Axiom (6) says that R is also closed under multiplication. Axiom (7) says the multiplication is associative.[2] Axiom (8) just says that addition and multiplication are "compatible" – i.e., they satisfy the usual distributive axioms. If Axiom (9) is satisfied, we say that R is a **ring with unit**.

Now, you may have noticed that axioms (6), (7), and (9) are just one axiom shy of making R into a group under multiplication as well. What is missing are the multiplicative inverses. If it should be the case that every nonzero element of the ring has a multiplicative inverse then the ring is called a **division ring** or **skew**

---

[2] There are examples of nonassociative rings (for example, the octonions) but we shall not consider them here.

field. If the multiplication of the division ring is commutative, so that R is an *abelian* group under multiplication (in addition to being an abelian group under addition), then the structure is called a **field**.

Example A.12 The integers $\mathbb{Z}$ with the usual operations of addition and multiplication form a ring with unit.

Example A.13 The even integers with the usual operations of addition and multiplication form a ring without unit.

Example A.14 The rational numbers $\mathbb{Q}$ with the usual operations of addition and multiplication form a field, as do the real numbers $\mathbb{R}$.

Example A.15 Recall how the complex numbers work. We define $i = \sqrt{-1}$. Then any complex number is of the form $z = a + ib$ where a and b are real numbers (called the real and imaginary parts of z, respectively). We define addition and multiplication as follows. Let $z = a + ib$ and $w = c + id$. Then $z + w = (a + c) + i(b + d)$ and $z \cdot w = (a + ib)(c + id) = (ac - bd) + i(ad + bc)$. The set of complex numbers $\mathbb{C}$ with these operations is a field.

The field of complex numbers admits a special linear involution (often denoted by an asterisk, but here denoted by a bar), namely complex conjugation, which leaves real numbers invariant and sends i to $-i$. Thus, for example, $\overline{a + ib} = a - ib$. The modulus of $z \in \mathbb{C}$ is $|z| = (z\bar{z})^{1/2}$.

Example A.16 In this example we define the **quaternions** Q, which were first discovered by Sir William Rowan Hamilton (the same person who gave you the Hamiltonian). Elements of Q are of the form $q = a + ib + jc + kd$, where a, b, c, and d are real numbers and i, j, and k are formal symbols that generalize the i of the complex numbers. We define the addition of quaternions as we did for complex numbers, namely the symbols i, j, and k act as place holders. So, for example, if $q = a + ib + jc + kd$ and $p = e + if + jg + kh$ then $q + p = (a + e) + i(b + f) + j(c + g) + k(d + h)$. We define the multiplicative properties of these formal symbols as follows: $i^2 = j^2 = k^2 = -1$, $ij = k = -ji$, $jk = i = -kj$, and $ki = j = -ik$. (You may notice a certain resemblance to the cross product.) With these rules, the quaternions form a classic example of a skew field (for which multiplication is not commutative).

Example A.17 Let $\mathbb{Z}_p$ be the set of all integers mod p. When p is prime, $\mathbb{Z}_p$ is a field. First notice that $\mathbb{Z}_p$ only has p elements, namely $\{[0], [1], [2], \ldots, [\text{p-1}]\}$. (For this reason, $\mathbb{Z}_p$ is called a **finite field**.) The addition and multiplication of equivalence classes are all defined modulo p. So, for example, let $p = 7$. Then $[4] + [6] = [3]$, because $(4 + 7x) + (6 + 7y) = 3 + 7(x + y + 1)$. Similarly, $[3] \cdot [4] = [5]$ because

$(3 + 7x)(4 + 7y) = 5 + 7(4x + 3y + 1)$. One can show that every class has both an additive and multiplicative inverse, so $\mathbb{Z}_p$ forms a field.

A.4 Vector spaces

At last we come to the formal definition of a vector space. A nonempty set V is said to be a **vector space** over a field $\mathbb{F}$ if V is an abelian group under addition, and for every $a \in \mathbb{F}$ and $v \in V$ we have:

(1) $av \in V$;
(2) $a(v + w) = av + aw$;
(3) $(a + b)v = av + bv$;
(4) $a(bv) = (ab)v$;
(5) $1v = v$.

Here 1 is the multiplicative identity in $\mathbb{F}$. The elements of V are called **vectors** while the elements of $\mathbb{F}$ are called **scalars**. The quantity av is called a **scalar multiple** of v. In general, the quantity $av + bw$ for scalars $a, b \in \mathbb{F}$ and vectors $v, w \in V$ is called a $\mathbb{F}$-**linear combination** (or simply a linear combination if the field $\mathbb{F}$ is obvious) of v and w.

It is worth observing some simple consequences of the axioms. For example, by axiom (3) we have $0v = (0 + 0)v = 0v + 0v$, so $0v = 0$. Similarly one can show that $a0 = 0$ and $(-a)v = -(av)$.

Appendix B

The spectral theorem

In this appendix we provide a quick proof of the spectral theorem of linear algebra. To do so at the right level of generality requires that we work over the complex numbers. Thus, let V be a vector space over the complex numbers equipped with the usual sesquilinear inner product $(\cdot, \cdot)$ (for brevity we drop explicit mention of the map g). An operator A (respectively, matrix A) is called **Hermitian** or **self-adjoint** if $A = A^\dagger$ (respectively, $A = A^\dagger$).[1] An operator U (respectively, matrix U) is **unitary** if $U^\dagger = U^{-1}$ (respectively, $U^\dagger = U^{-1}$). A matrix A is **diagonalized** by a matrix U if $U^{-1}AU = D$ for some diagonal matrix D.

Theorem B.1 (Spectral theorem) *Every Hermitian matrix can be diagonalized by a unitary matrix.*

Proof Let $A : V \to V$ be any linear operator. Then A has at least one (nontrivial) eigenvector v, since

$$Av = \lambda v \quad \Leftrightarrow \quad \det(A - \lambda I) = 0.$$

But the latter is a polynomial equation in λ of degree $n := \dim V$, which always has a solution by virtue of the fundamental theorem of algebra.[2]

Now assume A is Hermitian, and let $W := (\text{span}\, v)^\perp$ be the orthogonal complement of the linear span of the eigenvector v, with $w \in W$. Then

$$(Aw, v) = (w, Av) = (w, \lambda v) = \lambda(w, v) = 0.$$

In particular, A maps any element of W to another element of W (W is called an **invariant subspace** of A). Therefore we may repeat the process to find an

---

[1] Warning: "Hermitian" and "self-adjoint" are distinct notions in infinite dimensions. See e.g. [74].

[2] This is why we need to work over the complex numbers. One can develop a proof of the spectral theorem for real symmetric matrices without going through the complex numbers first, but it takes a little longer and proves less.

eigenvector of A lying in W. By induction, we obtain a basis of V consisting of eigenvectors of A. By construction these eigenvectors are orthogonal, and by normalizing they can be made orthonormal.[3]

Let $\{e_i\}$ be any orthonormal basis for V, and let $\mathbf{A}$ be the matrix representing A in this basis. Let $\{v_i\}$ be the basis of eigenvectors of A just constructed, with corresponding eigenvalues λ_i. Define $U : V \to V$ by $U e_i = v_i$, $1 \le i \le n$. Then U is unitary, because

$$(e_i, U^\dagger U e_j) = (U e_i, U e_j) = (v_i, v_j) = \delta_{ij} = (e_i, e_j),$$

and therefore $U^\dagger U = I$. Also,

$$(e_i, U^{-1} A U e_j) = (U e_i, A U e_j) = (v_i, A v_j) = \lambda_j (v_i, v_j) = \lambda_j \delta_{ij}.$$

It follows that if $\mathbf{U}$ is the matrix representing U in the basis $\{e_i\}$ then $\mathbf{U}^{-1} \mathbf{A} \mathbf{U} = \mathbf{D}$, where $\mathbf{D}$ is a diagonal matrix with the eigenvalues of A along the diagonal. □

Remark As $U e_j = \sum_i U_{ij} e_i$, the columns of the matrix $\mathbf{U}$ are just the components of the vectors $\{v_i\}$ in the basis $\{e_i\}$.

Remark Note that the entries of the diagonal matrix $\mathbf{D}$ in the theorem are necessarily real because A is Hermitian: if v is a nontrivial eigenvector with eigenvalue λ then

$$\overline{\lambda}(v, v) = (Av, v) = (v, Av) = \lambda(v, v),$$

so $\overline{\lambda} = \lambda$ because $(v, v) > 0$.

Corollary B.2 *Every real symmetric matrix can be diagonalized by an orthogonal matrix.*

Proof There are two ways to proceed. Naively, we just restrict everything in the previous theorem to real numbers. A real Hermitian matrix is just a real symmetric matrix, and a real unitary matrix is just an orthogonal matrix, so the result more or less follows. The problem is this: the theorem says that if we start with a real symmetric matrix, namely a special kind of Hermitian matrix, there is a unitary matrix that diagonalizes it. How do we know that the entries of that unitary matrix are real? We need another argument.

Here is one that works. Start with a real symmetric matrix $\mathbf{A}$. By the first part of the proof of the theorem, it has a nontrivial eigenvector v whose entries may be complex. So write $v = x + iy$ where x and y are real. Now

$$\lambda(x + iy) = \lambda v = Av = Ax + iAy.$$

---

[3] In practice, one generally needs Gram–Schmidt orthonormalization. See e.g. [43]. (For the real case, see Exercise 1.52.)

But A and λ are both real, so equating real and imaginary parts gives

$$Ax = \lambda x \qquad \text{and} \qquad Ay = \lambda y.$$

We must have $x \neq 0$ or $y \neq 0$, so we have found a nonzero *real* eigenvector of A. Now proceed as in the proof of the theorem. $\qquad\qquad\qquad\square$

Appendix C

Orientations and top-dimensional forms

Recall that a manifold M with atlas $\{(U_i, \varphi_i)\}$ (see Section 3.4) is oriented if all the transition functions $\varphi_j \circ \varphi_i^{-1}$ have positive Jacobian determinant, and it is orientable if it can be oriented. This is not a very useful characterization because there are too many transition functions to check. Fortunately, we can use the pullback map to devise a nicer criterion for ascertaining orientability.

Theorem C.1 *An n-dimensional manifold M is orientable if and only if it has a global nowhere-vanishing n-form.*

Remark A global nowhere-vanishing n-form on an n-dimensional manifold is called a **volume form**.

Proof Suppose that ω is a global nowhere-vanishing n-form. Any other n-form on M is a multiple of this one, so $\varphi_i^*(dx^1 \wedge \cdots \wedge dx^n) = f_i \omega$ for some smooth nowhere-vanishing function f_i on U_i. We cannot have $f_i > 0$ and $f_j < 0$ for two overlapping coordinate patches, or else by continuity f_i or f_j would have to be zero somewhere. Thus, without loss of generality we may choose $f_i > 0$ everywhere. It follows that, for every pair of patches, the transition functions $\varphi_j \circ \varphi_i^{-1}$ pull back $dx^1 \wedge \cdots \wedge dx^n$ to a positive multiple of itself. This, in turn, implies that the Jacobian determinants of the transition functions are all positive.

Conversely, suppose that M has an oriented atlas $\{(U_i, \varphi_i)\}$. Then

$$(\varphi_j \circ \varphi_i^{-1})^*(dx^1 \wedge \cdots \wedge dx^n) = h\, dx^1 \wedge \cdots \wedge dx^n$$

for some positive function h. Hence,

$$\varphi_j^*(dx^1 \wedge \cdots \wedge dx^n) = (\varphi_i^* h)\varphi_i^*(dx^1 \wedge \cdots \wedge dx^n).$$

Writing ω_i for $\varphi_i^*(dx^1 \wedge \cdots \wedge dx^n)$ we get $\omega_j = f\omega_i$ for some positive function $f = \varphi_i^* h$ on $U_i \cap U_j$.

Let $\{\rho_i\}$ be a partition of unity subordinate to the open cover $\{U_i\}$, and define $\omega = \sum_i \rho_i \omega_i$. At each point in M, the forms $\{\omega_i\}$ are positive multiples of each other whenever they are defined. As $\rho_i \geq 0$ and not all the ρ_i can vanish at a point, ω is nowhere vanishing. $\qquad \square$

As the proof shows, all global nowhere-vanishing n-forms are multiples of each other by nowhere-vanishing functions. Thus, on a connected manifold, the set of all such forms can be divided into two equivalence classes, depending on whether this multiple is positive or negative. Either class determines an **orientation class** (or, more simply, **orientation**) of M.

Appendix D

Riemann normal coordinates

Let $x^1, \ldots, x^n$ be the usual coordinates on $\mathbb{R}^n$. Technically, the corresponding vector fields $\partial/\partial x^1, \ldots, \partial/\partial x^n$ live in $T\mathbb{R}^n$, the tangent space to Euclidean space. But we always identify $\mathbb{R}^n$ and $T\mathbb{R}^n$, because both are vector spaces. Under the natural identification, a step in the coordinate direction x^i is the same thing as moving from the tail to the tip of $\partial/\partial x^i$: this works because $\partial/\partial x^i$ is a unit vector.

There is a natural analogue of this correspondence in the vicinity of a point p on any Riemannian manifold M. Given a tangent vector X_p we define Exp X_p to be the point on M obtained by moving a unit parameter distance along the (unique) geodesic γ whose tangent vector at p is X_p. The map Exp $: TM \to M$ is called the **exponential map**. (See Figure D.1.)

Prima facie this construction seems to be totally arbitrary, as it appears to depend on the parameterization. But appearances can be deceiving. According to (8.14), the distance along the curve corresponding to a unit parameter distance is

$$\int_0^1 \sqrt{g_{ij}(t) \frac{dx^i}{dt} \frac{dx^j}{dt}}\, dt = \int_0^1 \sqrt{g(X, X)}\, dt = \sqrt{g(X_p, X_p)}, \qquad (D.1)$$

where $X = (dx^i/dt)(\partial/\partial x^i)$ denotes the tangent vector field to γ. The second equality holds because the curve is a geodesic and so $g(X, X)$ is constant along the curve. Hence, moving a unit parameter distance along γ corresponds to moving a distance equal to the length of the vector X_p along γ, and the latter quantity is parameterization independent. It follows that we can express the geodesic γ itself in terms of the exponential map:

$$\gamma(t) = \mathrm{Exp}\, t X_p. \qquad (D.2)$$

We digress briefly to discuss why Exp is called the exponential map. Let $a \in \mathbb{R}$ and consider the differential operator on smooth functions given by

$$\mathcal{D} = e^{a(d/dx)} = 1 + a\frac{d}{dx} + \frac{a^2}{2!}\frac{d^2}{dx^2} + \cdots. \qquad (D.3)$$

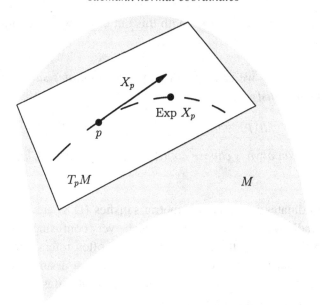

Figure D.1 The exponential map.

By Taylor's theorem

$$\mathcal{D}f(x) = f(x + a) \tag{D.4}$$

for any smooth function f. In this sense, $\mathcal{D}$ can be thought of as a shift operator that translates every point of $\mathbb{R}$ by a. This "exponential map" moves things a distance a along the "geodesic" whose tangent vector is d/dx. In general, if $a \in \mathbb{R}^n$ and

$$\mathcal{D} = e^{a^i(\partial/\partial x^i)} \tag{D.5}$$

then

$$\mathcal{D}f(x) = f(x + a). \tag{D.6}$$

The map $\mathrm{Exp}\, X_p$ is a kind of analogue on a general manifold of e^{X_p} on Euclidean space, whence the nomenclature.[1]

Using the exponential map we can set up a coordinate system in the neighborhood of a point p of M as follows. Let $X_1, \ldots, X_n$ be a basis of $T_p M$, and let U be a neighborhood of the origin in $\mathbb{R}^n$. Define a smooth map $f : U \subset \mathbb{R}^n \to M$ by

$$f(u^1, \ldots, u^n) = \mathrm{Exp}(u^1 X_1 + \cdots + u^n X_n), \tag{D.7}$$

and define coordinate functions x^i on M by $x^i(f(u^1, \ldots, u^n)) = u^i$. These coordinates are called **geodesic coordinates** or **Riemann normal coordinates** in the

---

[1] The difference is that e^{X_p} is a differential operator whereas $\mathrm{Exp}\, X_p$ is a point of M.

neighborhood of $p = f(0, \dots, 0)$.[2] With this out of the way, we turn to the main theorem.

Theorem D.1 *Near p we can always choose coordinates such that the metric tensor components satisfy*

$$g_{ij}(p) = \delta_{ij} \quad and \quad g_{ij,k}(p) = 0. \tag{D.8}$$

In particular, we can always choose coordinates such that the Christoffel symbols vanish at p.

Remark Coordinates in which the metric satisfies (D.8) are sometimes called **locally flat coordinates**. This terminology can be very confusing, though, because, as observed in Section 8.5, the term "locally flat" often refers a manifold whose Riemann tensor vanishes everywhere, which we do not assume here. Perhaps a better term would therefore be **locally Euclidean coordinates**, although this does not appear to be much of an improvement.

When the signature is Lorentzian instead of Euclidean, the theorem holds with δ_{ij} replaced by η_{ij} (the Minkowski metric tensor components), and the coordinates are called **local Lorentz coordinates** or **local inertial coordinates**. The latter terminology comes from the fact that, in special relativity, constant-velocity observers are inertial observers – for them, Newton's second law holds. In general relativity, Theorem D.1 is one form of the equivalence principle, which says that, in a small enough region, the effects of gravity (as embodied in the Christoffel symbols) can be transformed away.

> EXERCISE **D.1** Show that, in a coordinate system satisfying the conditions of Theorem D.1, the components of the Riemann tensor can be written
>
> $$R_{ijk\ell} = \frac{1}{2}(g_{i\ell,jk} - g_{j\ell,ik} - g_{ik,j\ell} + g_{jk,i\ell}). \tag{D.9}$$

We offer two proofs of Theorem D.1. The first uses Riemann normal coordinates while the second is a less rigorous but nonetheless insightful "counting" proof.

Proof 1 Fix a constant vector a in $\mathbb{R}^n$ and consider the (radial) line in $\mathbb{R}^n$ given by $u^i(t) = a^i t$. The image of this line under f is the geodesic in M given (cf. (D.2)) by

$$\gamma(t) = \mathrm{Exp}\, ta^i X_i, \tag{D.10}$$

---

[2] Generally these coordinates only exist in a small neighborhood of p, because the curvature of the manifold will typically cause the geodesic lines to cross at some point q different from p. For more about this see [11], Section VII.6.

whose tangent vector at $t = 0$ is $a^i X_i$. In terms of our local coordinates this is the same thing as saying that the curve on M given by $x^i(\gamma(t)) = u^i = a^i t$ is a geodesic. Plugging this into the geodesic equation (8.65) gives $\Gamma^k{}_{ij} a^i a^j = 0$. Choosing n linearly independent vectors $a_1, \ldots, a_n$ allows us to conclude that $\Gamma^k{}_{ij} = 0$.[3] From (8.39) we obtain

$$g_{ij,k} = \Gamma_{jki} + \Gamma_{ikj}, \tag{D.11}$$

so $g_{ij,k} = 0$ as advertised.

If $g_{ij} := g(X_i, X_j) \neq \delta_{ij}$ we can always choose a new basis $\{X'_i\}$ for the tangent space such that $g_{i'j'} := g(X'_i, X'_j) = \delta_{i'j'}$ at p. If we define $X'_i := X_j b^j{}_{i'}$ for some nonsingular constant matrix $\boldsymbol{b}$ then

$$g_{i'j'} = g(X'_i, X'_j) = b^k{}_{i'} b^\ell{}_{j'} g_{k\ell}. \tag{D.12}$$

By the same argument as that given in the second proof of Theorem 1.5, we can choose $\boldsymbol{b}$ such that $g_{i'j'} = \delta_{i'j'}$. Moreover, because $\boldsymbol{b}$ is constant, $g_{i'j',k'} = b^m{}_{i'} b^n{}_{j'} g_{mn,k} = 0$. $\quad\square$

Proof 2 Because the metric is a tensor, its components must transform according to (3.53):

$$g_{i'j'}(y) = g_{ij}(x(y)) \frac{\partial x^i}{\partial y^{i'}} \frac{\partial x^j}{\partial y^{j'}}. \tag{D.13}$$

A general coordinate transformation can be Taylor expanded about p:

$$\begin{aligned}
x^i(y) = x^i(y_p) &+ \left(\frac{\partial x^i}{\partial y^{j'}}\right)_p (y^{j'} - y^{j'}_p) \\
&+ \frac{1}{2} \left(\frac{\partial^2 x^i}{\partial y^{j'} \partial y^{k'}}\right)_p (y^{j'} - y^{j'}_p)(y^{k'} - y^{k'}_p) \\
&+ \frac{1}{6} \left(\frac{\partial^3 x^i}{\partial y^{j'} \partial y^{k'} \partial y^{\ell'}}\right)_p (y^{j'} - y^{j'}_p)(y^{k'} - y^{k'}_p)(y^{\ell'} - y^{\ell'}_p) + \cdots.
\end{aligned}$$

At p there are n^2 freely adjustable numbers

$$\left(\frac{\partial x^i}{\partial y^{j'}}\right)_p,$$

which we can use in (D.13) to put the $\binom{n}{2}$ metric components into Euclidean form. The remaining $n(n+1)/2$ parameters are exactly the number of parameters in

---

[3] Technically, we can only conclude that $\Gamma^k{}_{(ij)} = 0$, where $A_{(ij)} := \frac{1}{2}(A_{ij} + A_{ji})$, but as the Christoffel symbol is symmetric in its lower indices we need not worry about the symmetrization issue.

the orthogonal group $O(n)$, every element of which leaves the Euclidean metric invariant.

Differentiating both sides of (D.13) yields $n\binom{n+1}{2}$ equations for the first derivatives of the metric tensor components. Those equations involve second derivatives of the x^i of the form

$$\left(\frac{\partial^2 x^i}{\partial y^{j'} \partial y^{k'}}\right)_p.$$

As there are precisely $n\binom{n+1}{2}$ freely specifiable second derivative terms of this form, we can choose them in such a way that all the first derivatives of the metric vanish.   □

EXERCISE D.2 Show that there are $a := n\binom{n+2}{3}$ choices for the numbers

$$\left(\frac{\partial^3 x^i}{\partial y^{j'} \partial y^{k'} \partial y^{\ell'}}\right)_p,$$

but $b := \binom{n+1}{2}^2$ second derivatives of the metric. Interpret the significance of $b - a$ in the light of (8.51).

Appendix E

Holonomy of an infinitesimal loop

In this appendix we derive the precise form of (8.87). Pick local coordinates $x^1, \ldots, x^n$ in a neighborhood of p on M, and let α be a small number. Fix two indices r and s. Holding all the coordinates constant except x^r and x^s, choose four points p_1 to p_4 as follows: $(x^r(p_1), x^s(p_1)) = (0, 0)$; $(x^r(p_2), x^s(p_2)) = (\alpha, 0)$; $(x^r(p_3), x^s(p_3)) = (\alpha, \alpha)$; and $(x^r(p_4), x^s(p_4)) = (0, \alpha)$. Let $\gamma_a(t)$ be the path following the coordinate curves from p_a to p_{a+1} with $p_5 \equiv p_1$, and set $\gamma = \gamma_1\gamma_2\gamma_3\gamma_4$. (See Figure E.1.)

Recall that $A^k{}_j = \Gamma^k{}_{ij}(d\gamma^i/dt) = \Gamma^k{}_{ij}(dx^i/dt)$. To simplify the notation, define a matrix Γ_i such that $(\Gamma_i)^k{}_j = \Gamma^k{}_{ij}$. Then $A = \Gamma_i(dx^i/dt)$, so that $A\,dt = \Gamma_i dx^i$. Let us write $A^{(a)}$ to denote A restricted to path γ_a. Thus, for example, $A^{(1)}\,dt = \Gamma_r dx^r$ (no sum on r), because γ_1 is just the rth coordinate curve. Hence, using a Taylor expansion about p_1 (and setting $x^k = 0$ for all k except $k = r, s$),

$$
\int_0^\alpha A^{(1)}\,dt = \int_{(0,0)}^{(\alpha,0)} \Gamma_r\,dx^r
$$

$$
\approx \int_{(0,0)}^{(\alpha,0)} \left(\Gamma_r(0, 0) + \frac{\partial \Gamma_r}{\partial x^r}(0, 0)\, x^r \right) dx^r
$$

$$
\approx \alpha\Gamma_r(0, 0) + \frac{\alpha^2}{2}\frac{\partial \Gamma_r}{\partial x^r}(0, 0),
$$

where we have retained only terms of order α^2 or smaller. The argument for $A^{(2)}$ is a little different, because the base point is shifted, so we must use Taylor's theorem again:

$$
\int_0^\alpha A^{(2)}\,dt = \int_{(\alpha,0)}^{(\alpha,\alpha)} \Gamma_s\,dx^s
$$

$$
\approx \int_{(\alpha,0)}^{(\alpha,\alpha)} \left(\Gamma_s(\alpha, 0) + \frac{\partial \Gamma_s}{\partial x^s}(\alpha, 0)\, x^s \right) dx^s
$$

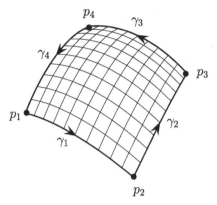

Figure E.1 A small closed curve around a coordinate patch.

$$\approx \int_{(\alpha,0)}^{(\alpha,\alpha)} \left(\Gamma_s(0,0) + \alpha \frac{\partial \Gamma_s}{\partial x^r}(0,0) + \frac{\partial \Gamma_s}{\partial x^s}(0,0)\, x^s \right) dx^s$$

$$\approx \alpha \Gamma_s(0,0) + \alpha^2 \frac{\partial \Gamma_s}{\partial x^r}(0,0) + \frac{\alpha^2}{2} \frac{\partial \Gamma_s}{\partial x^s}(0,0).$$

Also,

$$\int_0^\alpha A^{(3)}\, dt = \int_{(\alpha,\alpha)}^{(0,\alpha)} \Gamma_r\, dx^r$$

$$\approx \int_{(\alpha,\alpha)}^{(0,\alpha)} \left(\Gamma_r(\alpha,\alpha) + \frac{\partial \Gamma_r}{\partial x^r}(\alpha,\alpha)\,(x^r - \alpha) \right) dx^r$$

$$\approx \int_{(\alpha,\alpha)}^{(0,\alpha)} \left(\Gamma_r(0,0) + \alpha \frac{\partial \Gamma_r}{\partial x^r}(0,0) + \alpha \frac{\partial \Gamma_r}{\partial x^s}(0,0) \right.$$

$$\left. + \frac{\partial \Gamma_r}{\partial x^r}(0,0)\,(x^r - \alpha) \right) dx^r$$

$$\approx -\alpha \Gamma_r(0,0) - \alpha^2 \frac{\partial \Gamma_r}{\partial x^s}(0,0) - \frac{\alpha^2}{2} \frac{\partial \Gamma_r}{\partial x^r}(0,0).$$

Lastly,

$$\int_0^\alpha A^{(4)}\, dt = \int_{(0,\alpha)}^{(0,0)} \Gamma_s\, dx^s$$

$$\approx \int_{(0,\alpha)}^{(0,0)} \left(\Gamma_s(0,\alpha) + \frac{\partial \Gamma_s}{\partial x^s}(0,\alpha)\,(x^s - \alpha) \right) dx^s$$

$$\approx \int_{(0,\alpha)}^{(0,0)} \left(\Gamma_s(0,0) + \alpha \frac{\partial \Gamma_s}{\partial x^s}(0,0) + \frac{\partial \Gamma_s}{\partial x^s}(0,0)\,(x^s - \alpha) \right) dx^s$$

$$\approx -\alpha \Gamma_s(0,0) - \frac{\alpha^2}{2} \frac{\partial \Gamma_s}{\partial x^s}(0,0).$$

Keeping only terms up to order α^2 (and dropping $(0,0)$ from the notation), we get, for example,

$$\vartheta_{0,\delta}(\gamma_1) \approx 1 - \int_0^\delta dt_1\, A^{(1)}(t_1) + \int_0^\delta dt_1 \int_0^{t_1} dt_2\, A^{(1)}(t_1)A^{(1)}(t_2)$$

$$= 1 - \alpha\Gamma_r - \frac{\alpha^2}{2}\frac{\partial\Gamma_r}{\partial x^r} + \frac{\alpha^2}{2}\Gamma_r^2.$$

(For the last term we need to retain only the constant term in the Taylor expansion, because this term is already of order α^2.) Analogous formulae hold for all the other segments. Multiplying them all together according to (8.84) gives

$$\vartheta(\gamma) \approx \left(1 + \alpha\Gamma_s + \frac{\alpha^2}{2}\frac{\partial\Gamma_s}{\partial x^s} + \frac{\alpha^2}{2}\Gamma_s^2\right)$$

$$\times \left(1 + \alpha\Gamma_r + \alpha^2\frac{\partial\Gamma_r}{\partial x^s} + \frac{\alpha^2}{2}\frac{\partial\Gamma_r}{\partial x^r} + \frac{\alpha^2}{2}\Gamma_r^2\right)$$

$$\times \left(1 - \alpha\Gamma_s - \alpha^2\frac{\partial\Gamma_s}{\partial x^r} - \frac{\alpha^2}{2}\frac{\partial\Gamma_s}{\partial x^s} + \frac{\alpha^2}{2}\Gamma_s^2\right)$$

$$\times \left(1 - \alpha\Gamma_r - \frac{\alpha^2}{2}\frac{\partial\Gamma_r}{\partial x^r} + \frac{\alpha^2}{2}\Gamma_r^2\right)$$

$$\approx 1 + \alpha^2 \mathcal{R}_{rs},$$

where

$$\mathcal{R}_{rs} := \frac{\partial\Gamma_r}{\partial x^s} - \frac{\partial\Gamma_s}{\partial x^r} + \Gamma_s\Gamma_r - \Gamma_r\Gamma_s \tag{E.1}$$

is a matrix whose entries are precisely the components of the Riemann curvature tensor (cf. (8.46)):

$$(\mathcal{R}_{rs})^i{}_j = \Gamma^i{}_{rj,s} - \Gamma^i{}_{sj,r} + \Gamma^i{}_{sk}\Gamma^k{}_{rj} - \Gamma^i{}_{rk}\Gamma^k{}_{sj} = R^i{}_{jsr}. \tag{E.2}$$

Appendix F
Frobenius' theorem

Frobenius' theorem is a cornerstone of modern differential geometry. It appears in many different guises and is applied in many different circumstances. At its base it has to do with the question of whether certain systems of partial differential equations can be solved (i.e., integrated), but it has a much more elegant and versatile formulation in terms of vector fields (or forms) on manifolds.

F.1 Integrable distributions

To motivate the discussion, we start with a smooth vector field X on a manifold M. By the standard theory of ordinary differential equations, given an initial point we can find an integral curve of X through that point. This integral curve is a one-dimensional submanifold of M. We want to generalize this to higher-dimensions, so first we need a higher-dimensional analogue of a vector field.

A k-dimensional **distribution** Δ of a manifold M is a smooth assignment of a k-dimensional subspace Δ_p of T_pM to each point $p \in M$.[1] The smoothness condition is equivalent to saying that, at every point q in some neighborhood of p, Δ_q is generated by k linearly independent vector fields:

$$\Delta_q = \text{span}\{X_1(q), \ldots, X_k(q)\}. \tag{F.1}$$

The distribution Δ is **integrable** if for each $p \in M$ there exists a submanifold Σ containing p such that $T_q\Sigma = \Delta_q$ for all $q \in \Sigma$. In other words, the tangent spaces to Σ all lie in the distribution Δ. The submanifold Σ is called an **integral submanifold** of Δ. (See Figure F.1.)

We have seen that integral curves exist, so every one-dimensional distribution is integrable. The same is not true of higher-dimensional distributions. To see this,

---

[1] In fancy terminology, a distribution is a **subbundle** of the tangent bundle TM.

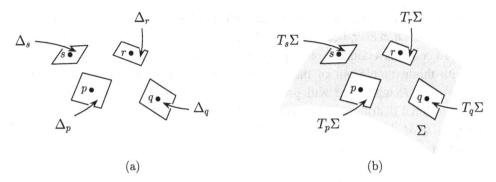

Figure F.1 (a) A distribution and (b) its integral submanifold.

we will obtain a necessary condition for a distribution to be integrable. We say that Δ is **involutive** if, for all $p \in M$ and $1 \le i, j \le k$,

$$[X_i(q), X_j(q)] \in \Delta_q \tag{F.2}$$

for all q in a neighborhood of p. This condition is often written symbolically as

$$[\Delta, \Delta] \subset \Delta, \tag{F.3}$$

because if (F.2) is satisfied then the Lie bracket of any element of Δ with any other element of Δ must again lie in Δ (and vice versa).

If Δ is integrable then it is clearly involutive. This follows because if Σ is an integral submanifold of Δ then the Lie bracket of any vector fields in the tangent space to Σ is again in the tangent space to Σ (*cf.* Exercise 3.20). Surprisingly, the converse is also true. This is the content of Frobenius' theorem.[2]

Before stating the theorem, let's see what can go wrong if Δ is not involutive. Consider the two-dimensional distribution Δ on $\mathbb{R}^3$ spanned by the vector fields

$$X = \frac{\partial}{\partial x} + z\frac{\partial}{\partial y}, \qquad Y = \frac{\partial}{\partial z}. \tag{F.4}$$

Note that $[X, Y] = -\partial/\partial y \notin \Delta$. Suppose that Δ is integrable, and let Σ be the integral surface through the origin. The integral curves of X (which must lie in Σ) are obtained from the equations (*cf.* (3.95))

$$\frac{dx}{dt} = 1, \qquad \frac{dy}{dt} = z(t), \qquad \text{and} \qquad \frac{dz}{dt} = 0, \tag{F.5}$$

whose solution is $(x(t), y(t), z(t)) = (t + a, ct + b, c)$ for some constants a, b, and c. The surface Σ goes through the origin, so $a = b = c = 0$; this means that Σ contains an open subset of the x axis. But the integral curves of Y all contain

---

[2] Frobenius' theorem is not actually due to Frobenius, but rather to A. Clebsch and F. Deahna. Unfortunately, this is how it is remembered by most people nowadays. See e.g. [49].

a segment of the z axis, so Σ must contain a small area of the xz plane. This is a contradiction, however, because if $z \neq 0$ then the vector X is not tangent to the xz plane (it contains a nonzero piece proportional to $\partial/\partial y$).

With this example out of the way, we can now turn to a proof of the theorem itself. Actually, we will prove something a bit stronger. By definition, a k-dimensional distribution Δ is **completely integrable** if it is integrable and for every $p \in M$ there is a neighborhood U of p with local coordinates $x^1, \ldots, x^n$ such that, everywhere on U,

$$\Delta = \text{span} \left\{ \frac{\partial}{\partial x^1}, \ldots, \frac{\partial}{\partial x^k} \right\}. \tag{F.6}$$

If Δ is completely integrable then, in the neighborhood U, the integral submanifold Σ looks like $x^{k+1} = c_{k+1}, x^{k+2} = c_{k+2}, \ldots, x^n = c_n$ for some constants $c_{k+1}, \ldots, c_n$. We say that Σ looks **locally straight** (because we have "straightened out" the integral submanifold so that it can be made to look like $\mathbb{R}^k$ in a neighborhood).[3]

Theorem F.1 (Clebsch–Deahna–Frobenius) *The distribution Δ is completely integrable if and only if it is involutive.*

Proof We follow the proof given in [53]. We have already shown one direction, so we may assume that Δ is involutive. First we show that we can find a coordinate system in which Σ is almost straight and then we show that we can straighten it out completely. $\qquad \square$

Lemma F.2 *There is a neighborhood V of p with coordinates $y^1, \ldots, y^n$ such that Δ is spanned by vector fields of the form*

$$X_i = \frac{\partial}{\partial y^i} + \sum_{j=k+1}^n b_{ij} \frac{\partial}{\partial y^j}, \qquad 1 \leq i \leq k, \tag{F.7}$$

for some $k \times n$ matrix (b_{ij}) of smooth functions.

Proof Choose local coordinates $y^1, \ldots, y^n$ in a neighborhood V' of p and suppose that $\Delta = \text{span}\{Y_1, \ldots, Y_k\}$ on V'. Then

$$Y_i = \sum_{j=1}^n a_{ij} \frac{\partial}{\partial y^j}, \qquad 1 \leq i \leq k,$$

for some $k \times n$ matrix (a_{ij}) of smooth functions on V'. The Y_i are linearly independent, so (a_{ij}) has rank k. We may therefore assume that, in a (possibly smaller)

---

[3] Do not confuse "straight" with "flat". A flat manifold has locally straight neighborhoods, but not necessarily vice versa.

neighborhood $V \subseteq V'$ of p, the $k \times k$ submatrix $A' = (a_{ij})$ with $1 \leq i, j \leq k$ is nonsingular. Now set $X_i := \sum_{j=1}^{k}(A'^{-1})_{ij}Y_j$. □

Corollary F.3 *If the k-dimensional distribution Δ is involutive then the basis* (F.7) *satisfies* $[X_i, X_j] = 0$ *for* $1 \leq i, j \leq k$.

Proof By computing brackets, we note that

$$[X_i, X_j] \in \text{span}\left\{\frac{\partial}{\partial y^{k+1}}, \ldots, \frac{\partial}{\partial y^n}\right\}, \qquad 1 \leq i, j \leq k.$$

But, by the involutivity of Δ,

$$[X_i, X_j] \in \text{span}\{X_1, \ldots, X_k\}.$$

These two vector spaces meet only in the zero vector (since $\text{span}\{X_1, \ldots, X_k\}$ always contains vectors proportional to $\partial/\partial y^i$ for $1 \leq i \leq k$). □

It only remains to show that if $\{X_1, \ldots, X_k\}$ are pairwise commuting linearly independent vector fields on a neighborhood U of a point p then there exist coordinates $x^1, \ldots, x^n$ on U such that $X_i = \partial/\partial x^i$ for $1 \leq i \leq k$. When $k = n$ this is just the content of Theorem 8.2 and, when $k < n$, a similar but more involved argument can be made. Instead, we present a simple proof by induction on k that uses only the fundamental theorem of ordinary differential equations. The base case $k = 1$ is dealt with by the next lemma.

Lemma F.4 *Let X be a vector field nonvanishing at p. Then in a neighborhood of p we can choose coordinates such that $X = \partial/\partial x^1$.*

Proof The basic idea is simple enough. Choose local coordinates $y^1, \ldots, y^n$ in a neighborhood U of p, and consider the integral curves of X through $q \in U$. Suppose the curves look schematically like those in Figure F.2, where the horizontal axis is y^1 and the vertical axis represents all the other coordinates. We then make a coordinate transformation

$$(y^1, \ldots, y^n) \rightarrow (x^1, y^2, \ldots, y^n), \tag{F.8}$$

where x^1 is the parameter along the integral curves. The only technical part of the proof lies in showing that this is a good coordinate transformation everywhere on the neighborhood U. For this, we refer the reader to ([88], Proposition 1.53). □

We can now complete the proof of Frobenius' theorem. By Lemma F.4 the result is true for $k = 1$, so assume it holds for $k - 1$. By Corollary F.3 we may assume

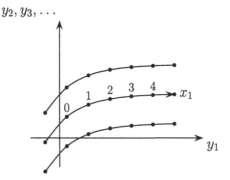

Figure F.2 Coordinates based on integral curves.

that $\Delta = \text{span}\{X_1, \ldots, X_k\}$ with $[X_i, X_j] = 0$. By the inductive hypothesis, there exist coordinates $y^1, \ldots, y^{k-1}$ such that

$$\text{span}\{X_1, \ldots, X_{k-1}\} = \text{span}\left\{\frac{\partial}{\partial y^1}, \ldots, \frac{\partial}{\partial y^{k-1}}\right\}. \tag{F.9}$$

In other words $\partial/\partial y^i = c_{ji} X_j$ for some nonsingular matrix (c_{ij}) with $1 \le i, j \le k - 1$. Hence, by (F.9) again,

$$\left[\frac{\partial}{\partial y^i}, X_k\right] = [c_{ji} X_j, X_k] = -X_k(c_{ji}) X_j \in \text{span}\left\{\frac{\partial}{\partial y^1}, \ldots, \frac{\partial}{\partial y^{k-1}}\right\}. \tag{F.10}$$

Writing $X_k = \sum_{j=1}^n a^j \partial/\partial y^j$ and using (F.10) gives

$$\frac{\partial a^j}{\partial y^i} = 0 \qquad \text{for } 1 \le i \le k - 1 \text{ and } k \le j \le n. \tag{F.11}$$

Now define

$$Y := X_k - \sum_{j=1}^{k-1} a^j \frac{\partial}{\partial y^j} = \sum_{j=k}^n a^j \frac{\partial}{\partial y^j}. \tag{F.12}$$

Then

$$\text{span}\{X_1, \ldots, X_k\} = \text{span}\left\{\frac{\partial}{\partial y^1}, \ldots, \frac{\partial}{\partial y^{k-1}}, Y\right\}. \tag{F.13}$$

Because Y only depends on $y^k, \ldots, y^n$, Lemma F.4 guarantees that we can choose new coordinates $x^\ell = x^\ell(y^k, \ldots, y^n)$, $k \le \ell \le n$, such that $Y = \partial/\partial x^k$. Lastly, set $x^\ell = y^\ell$ for $1 \le \ell \le k - 1$. Then

$$\Delta = \text{span}\left\{\frac{\partial}{\partial x^1}, \ldots, \frac{\partial}{\partial x^k}\right\}, \tag{F.14}$$

as desired. $\square$

Remark To preempt possible confusion, it is worth observing that Frobenius' theorem says something a bit more general but a bit less strong than Theorem 8.2. Given an involutive set of linearly independent vector fields $X_1, \ldots, X_n$, Frobenius' theorem says that we can find local coordinates $x^1, \ldots, x^n$ in which the span of these vector fields coincides with the span of $\partial/\partial x^1, \ldots, \partial/\partial x^n$. It does *not* say that we can find local coordinates such that $X^i = \partial/\partial x^i$ for all i. That is only true if the vector fields satisfy the stronger condition of mutual commutativity, namely $[X_i, X_j] = 0$ for all i, j.

Recall that in Section 3.16 we discussed how vector fields yield integral curves: given a vector field X on a manifold M, through each point of M there is a unique integral curve of X. The fact that these integral curves are unique means that they do not cross anywhere. Hence the collection of all these curves fills up M (cf. Figure 3.13). Similarly, if we have an involutive distribution, Frobenius' theorem guarantees that there is a unique integral submanifold through every point of M, so that these submanifolds fill up M. One says that the integral submanifolds are the **leaves** of a **foliation** of M. The arborial metaphor is intended to suggest a tree (the manifold) filled by its leaves (the integral submanifolds). A better metaphor would be that of an onion, whose various layers are the integral submanifolds.[4] In any event, we have the following equivalent formulation of Frobenius' theorem.[5]

Theorem F.5 *If Δ is an involutive distribution on M then the integral submanifolds of Δ foliate M.*

F.2 Pfaffian systems

Frobenius' theorem can also be rephrased in the language of differential forms. Let $\Theta := \{\theta^1, \ldots, \theta^m\}$ be a linearly independent set of 1-forms on a neighborhood U of $p \in M$ (so that $\theta^1 \wedge \cdots \wedge \theta^m \neq 0$ on U). The set Θ is called a **Pfaffian system** of 1-forms. The **zero set** or **kernel** of Θ is the set of all vector fields X annihilated by Θ: $\{X : \theta^i(X) = 0, \ 1 \leq i \leq m\}$. Each condition $\theta^i(X) = 0$ imposes one constraint on X, so the zero set of Θ is an $(n - m)$-dimensional distribution. Thus, statements about distributions can be translated into statements about forms.[6]

Theorem F.6 *Let $\Theta := \{\theta^1, \ldots, \theta^m\}$ be a Pfaffian system whose kernel coincides (locally) with the distribution Δ. Then the following are equivalent.*

(1) *The distribution Δ is involutive.*
(2) *We have $d\theta^i \wedge (\theta^1 \wedge \cdots \wedge \theta^m) = 0$ for $i = 1, \ldots, m$.*

---

[4] For example, the set of all 2-spheres of different radii foliate $\mathbb{R}^3$.
[5] A foliation has a technical definition but, as we do not need it here, we refer the reader instead to, e.g., [14].
[6] The form versions are due to Frobenius, which is why his name is attached to the theorem.

(3) *We have $d\theta^i = 0$ mod θ^j for $1 \leq i, j \leq m$ (i.e., $d\theta^i = \sum_{j=1}^{m} A^i{}_j \wedge \theta^j$ for some 1-forms $A^i{}_j$). We say that Θ is a* **differential ideal**.

Any of these conditions is sufficient to guarantee that Δ is completely integrable, but sometimes the form versions are more convenient.

EXERCISE F.1 Prove Theorem F.6. *Hint:* You may wish to use (3.123).

The language of forms allows us to restate Frobenius' theorem in a very useful way. We begin by clarifying why the submanifolds in that theorem are called "integral". The basic idea is that the Pfaffian system Θ can be viewed as a system of partial differential equations whose solution sets (with varying initial conditions) are precisely the integral submanifolds guaranteed to exist by Frobenius. This, in turn, ensures that Θ can be expressed in a particularly convenient manner.

As a very simple example, consider the form $\theta = (2xy + z^2)\,dx + x^2\,dy + 2xz\,dz$ on $\mathbb{R}^3$. It is easy to check that $d\theta = 0$, so θ is a trivial differential ideal. Hence, by Frobenius' theorem, θ is completely integrable. But what is the integral submanifold? A hint is provided by the Poincaré lemma, which tells us that θ is exact so that there is a function f such that $\theta = df$. This is equivalent to a set of differential equations for f,

$$\frac{\partial f}{\partial x} = 2xy + z^2, \qquad \frac{\partial f}{\partial y} = x^2, \qquad \frac{\partial f}{\partial z} = 2xz,$$

whose simultaneous solution is $f = x^2y + xz^2$. The equation $\theta = 0$ just says that $df = 0$, or $f = $ constant. We have therefore found our integral submanifolds, namely the surfaces $x^2y + xz^2 = $ constant. The tangent spaces of these integral submanifolds constitute the distribution determined by θ, as they are spanned by vectors that are all annihilated by θ. Moreover, Frobenius' theorem also says that we can find new coordinates x', y', z' such that the integral submanifold looks locally like $z' = $ constant. (In this example we can choose coordinates $x' = x$, $y' = y$, and $z' = f$ whenever $(x, z) \neq (0, 0)$.)

Of course, most 1-forms θ will not satisfy $d\theta = 0$, so in general there is no f such that $\theta = df$. However, as long as $d\theta$ is proportional to θ, the claim is that there are now *two* functions, f and g, such that $\theta = g\,df$. If θ is not identically zero at p then g cannot vanish at p then so again $\theta = 0$ implies that the integral surface is given by $f = $ constant.[7]

This idea generalizes, and the result is sufficiently useful for us to present it as yet another version of Frobenius' theorem.

---

[7] Observe that the condition that θ does not vanish identically at p is part of the hypothesis of a Pfaffian system.

Theorem F.7 *Let $\Theta := \{\theta^1, \ldots, \theta^m\}$ be a Pfaffian system on some neighborhood U. If Θ is a differential ideal then on U there exist m^2 functions $g^i{}_j$ and m functions f^j such that*

$$\theta^i = \sum_{j=1}^{m} g^i{}_j \, df^j, \qquad i = 1, \ldots, m. \tag{F.15}$$

Equivalently, writing $\Theta = (\theta^i)$ (a column matrix), $G := (g^i{}_j)$, and $F = (f^j)$ (another column matrix),

$$\Theta = G \, dF. \tag{F.16}$$

Proof By Frobenius' theorem the distribution Δ annihilated by Θ is completely integrable, with integral submanifold Σ. Hence there exist local coordinates $x^1, \ldots, x^n$ such that Σ is given by $x^{k+1} = c_{k+1}, \ldots, x^n = c_n$, where $k = n - m$. Equivalently, Σ is the solution set to the differential system $df^1 = \cdots = df^m = 0$, where $f^j := x^{k+j}$. It follows that any annihilator of Δ must be a linear combination of the 1-forms df^j for $j = 1, \ldots, m$. $\qquad\qquad\square$

F.3 Holonomicity and constrained systems

There is one more thing that needs to be addressed, namely the relationship between Frobenius' theorem and holonomicity. The story begins with classical mechanics. Consider the spherical pendulum, which consists of a bob of mass m suspended from a single point by a massless rigid rod of length a. To specify the state of the system at any instant, we use (inverted) polar coordinates r, θ, ϕ, as shown in Figure F.3. In effect, the configuration space of the system is a three-dimensional manifold M coordinatized by $q = (r, \theta, \phi)$. The (generalized) velocity of the bob at any instant is $\dot{q} = (\dot{r}, \dot{\theta}, \dot{\phi})$, where the overdot means differentiation with respect to time.

But not all velocities are allowed, because we are assuming that the rod that is rigid. Instead, we have a **constraint**, namely $\dot{r} = 0$. Now, it is easy to see the effect of this constraint: the bob is constrained to move along a sphere of radius a (hence the name "spherical pendulum"). Constraints of physical systems that can be written directly in terms of coordinates (such as $r = a$) or in terms of derivatives that can be integrated (such as $\dot{r} = 0$), are called **holonomic constraints**.

To connect this with Frobenius' theorem we must change our point of view slightly. First, we view $\dot{q}$ as a tangent vector field on M:

$$\dot{q} \to X = \dot{r}\frac{\partial}{\partial r} + \dot{\theta}\frac{\partial}{\partial \theta} + \dot{\phi}\frac{\partial}{\partial \phi}.$$

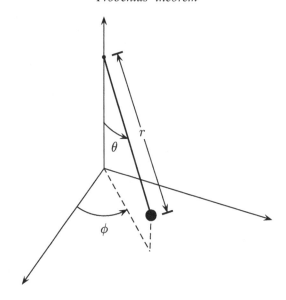

Figure F.3 A spherical pendulum.

The velocity constraint can then be written $\omega(X) = 0$, where $\omega := dr$. The vector fields X annihilated by ω are precisely the allowed velocity vectors, and they define a two-dimensional distribution Δ on M. Frobenius' theorem guarantees that this distribution is integrable, because $d\omega = 0$ (ω itself constitutes a differential ideal). Of course, we already knew that, because we can directly integrate the constraint $\omega = 0$ to get the answer $r = $ constant. In particular, the integral submanifolds of Δ are spheres of different radii, and they foliate M as expected. The motion of the bob is constrained to lie on one of these spheres, namely $r = a$. This is true in general: holonomic constraints lead to motions that are constrained to lie within integral submanifolds; such constraints are called "integrable" because they can be integrated (in principle).

Nonholonomic systems, however, behave quite differently. Consider a disk rolling without slipping on a plane, as shown in Figure F.4. To specify the state of the system at any instant, we need four coordinates: x and y tell you where the disk touches the plane, θ gives its direction relative to the x axis, and ϕ gives the angular location of a point on the rim of the disk relative to the vertical. The configuration space of the system is a four-dimensional manifold M globally coordinatized by $q := (x, y, \theta, \phi)$.

The (generalized) velocity of the disk at any instant is $\dot{q} = (\dot{x}, \dot{y}, \dot{\theta}, \dot{\phi})$, and the condition of rolling without slipping can be written

$$\dot{x} = a \cos \theta \, \dot{\phi},$$
$$\dot{y} = a \sin \theta \, \dot{\phi},$$

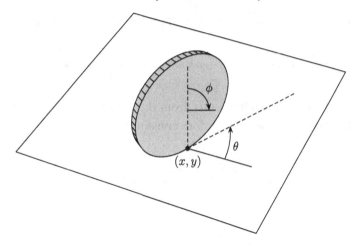

Figure F.4 A disk rolling on a plane.

where a is the radius of the disk. Again we interpret $\dot{q}$ as a vector field X on M, so that

$$\dot{q} \to X := \dot{x}\frac{\partial}{\partial x} + \dot{y}\frac{\partial}{\partial y} + \dot{\theta}\frac{\partial}{\partial \theta} + \dot{\phi}\frac{\partial}{\partial \phi}.$$

The constraints can then be written as $\omega_1(X) = \omega_2(X) = 0$, where

$$\omega_1 := dx - a\cos\theta\, d\phi,$$
$$\omega_2 := dy - a\sin\theta\, d\phi.$$

Thus ω_1 and ω_2 constitute a Pfaffian system whose zero set Δ defines a two-dimensional distribution on M, but the difference now is that Δ is not integrable, because, for example,

$$d\omega_1 = a\sin\theta\, d\theta \wedge d\phi$$

fails to lie in the ideal generated by ω_1 and ω_2.

It is instructive to see what all this looks like in terms of vector fields instead of forms. Denote an arbitrary vector field by

$$X = f\frac{\partial}{\partial x} + g\frac{\partial}{\partial y} + h\frac{\partial}{\partial \theta} + k\frac{\partial}{\partial \phi},$$

where f, g, h, and k are arbitrary functions of x, y, θ, and ϕ. Then the two constraint conditions become conditions on the allowed vector fields:

$$\omega_1(X) = 0 \quad \Rightarrow \quad X = ak\cos\theta\frac{\partial}{\partial x} + g\frac{\partial}{\partial y} + h\frac{\partial}{\partial \theta} + k\frac{\partial}{\partial \phi},$$

$$\omega_2(X) = 0 \quad \Rightarrow \quad X = f\frac{\partial}{\partial x} + ak\sin\theta\frac{\partial}{\partial y} + h\frac{\partial}{\partial \theta} + k\frac{\partial}{\partial \phi}.$$

The two conditions must be satisfied simultaneously, so we must have

$$X = ak \cos\theta \frac{\partial}{\partial x} + ak \sin\theta \frac{\partial}{\partial y} + h\frac{\partial}{\partial \theta} + k\frac{\partial}{\partial \phi}.$$

(We can also obtain this result directly from the velocity constraints themselves.) The following two vector fields therefore constitute a basis for Δ:

$$X_1 := a \cos\theta \frac{\partial}{\partial x} + a \sin\theta \frac{\partial}{\partial y} + \frac{\partial}{\partial \phi},$$

$$X_2 := \frac{\partial}{\partial \theta}.$$

A simple calculation shows that

$$[X_1, X_2] = a \sin\theta \frac{\partial}{\partial x} - a \cos\theta \frac{\partial}{\partial y} \notin \operatorname{span}\{X_1, X_2\}, \tag{F.17}$$

so again we see that Δ is not integrable.

What is the physical significance of the fact that Δ fails to be integrable? One answer is given by the theorem of Chow and Rashevskii, which says, in effect, that if one has a bunch of vector fields that are "maximally noninvolutive" then by following along the integral curves of those vector fields you can travel from any point of M to any other point of M. More precisely, a distribution Δ is called **bracket-generating** if, for any point $p \in M$ and any local basis $\{X_1, \ldots, X_k\}$ for Δ_p, the vector space generated by the basis vectors and all their iterated brackets $[X_i, X_j], [X_i, [X_j, X_k]], \ldots$ coincides with $T_p M$.

Theorem F.8 (Chow–Rashevskii) *If Δ is a bracket-generating distribution on a connected manifold M then any two points of M can be connected by a path whose tangent vectors lie in Δ.*

Proof The idea of the proof is simple enough. If all the iterated brackets span only a proper subspace of the tangent space then they determine a lower-dimensional involutive distribution; so, by Frobenius' theorem, the motion is constrained to lie within a proper integral submanifold. For the details see [62]. □

One can check that the constraint vector fields for the disk system are bracket-generating. Therefore, by the Chow–Rashevskii theorem, legal motions of the system are not constrained to lie on any submanifold. In other words, we can roll the disk without its slipping around the plane and can visit every possible point of the configuration space.[8] The Chow–Rashevskii theorem has applications to various disciplines such as thermodynamics and control theory. For example, in designing a robot arm or a car steering mechanism one would like to know that one has enough degrees of freedom to reach any point in the configuration space, within the constraints inherent to the system. For more details, see e.g. [44].

At last we are able to connect the notion of a holonomic constraint to a holonomic frame field. As should be clear by now, holonomic constraints can be integrated to give integral submanifolds of the distribution spanned by the holonomic vector fields arising from the constraints whereas nonholonomic constraints give rise to nonholonomic vector fields that determine nonintegrable distributions.

---

[8] At first sight this would seem to be impossible. For example, if the disk starts at the origin with $\theta = \phi = 0$, how could we get the disk back to the origin with the same value of θ and a different value of ϕ? One way would be to roll the disk around circles on the plane of different radii. Similarly, one might ask how could we get from the origin to some other point p while leaving θ and ϕ fixed. One answer is to change θ until the disk lies along the line from the origin to p, run it to p, change θ back, then run it around a little circle if necessary to get ϕ back to its original value. (Rotating in place satisfies the constraint of rolling without slipping.)

Appendix G

The topology of electrical circuits

G.1 Kirchhoff's rules

Electrical circuits provide an example of a simple physical system in which homology and cohomology play a small but important part.[1] We consider only the simplest case of a resistive network comprising batteries and resistors obeying Ohm's law (Figure G.1), although by employing complex impedances the theory can be extended to steady state circuits that also contain capacitors and inductors. To solve the circuit (that is, to find the currents through, and voltages across, each element) we may appeal to Kirchhoff's rules:[2]

(1) (junction condition)   the sum of the currents into a node is zero;
(2) (loop law)   the sum of the voltage gains around any closed loop is zero.

The junction condition embodies the law of charge conservation, which says that you cannot create or destroy net charge. The loop law embodies the law of energy conservation (because the work W done on a charge q by a potential difference ΔV is $W = q\Delta V$, so if you the law did not hold, a charge could flow around a closed loop and gain energy *ad infinitum*).

G.2 A graph model

We can model such a circuit by means of a **graph**, which is just a bunch of points (or **nodes** or **vertices**) joined by a bunch of lines (or **edges**).[3] Topologically, a

---

[1] For more details, see [8], Chapter 12, [27], Appendix B, [32], or [78].

[2] We use the standard convention that a current into a node is positive, while a current out of a node is negative. Similarly, a voltage difference between two points is positive if we gain voltage in going from one to the other, and negative if we lose voltage.

[3] The word "line" does not imply a rectilinear edge, although in fact every graph can be drawn in $\mathbb{R}^3$ with straight lines for edges in such a way that edges only meet at their endpoints. One can also view (simple) graphs abstractly as a set of vertices together with a subset of unordered pairs of distinct vertices. For more about graphs, see [35] or [89].

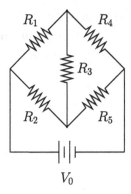

Figure G.1 A simple resistive circuit.

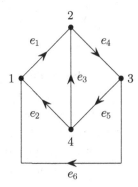

Figure G.2 An oriented graph.

graph is the same thing as a one-dimensional simplicial complex. For simplicity, we will assume that the graph has only one connected component. So, start with a connected graph G with n nodes and m edges that models the topology of the circuit. Label the nodes, then orient each edge arbitrarily.[4] (Figure G.2.) The vector space of i-chains on G is denoted C_i. The nodes are a basis for C_0 and the edges are a basis for C_1, so $\dim C_0 = n$ and $\dim C_1 = m$. The boundary operator $\partial :$ $C_1 \to C_0$ is defined by $\partial \langle p, q \rangle = q - p$, extended by linearity. With respect to the natural basis, ∂ is represented by an $n \times m$ matrix (called the **vertex–edge incidence matrix**). For the graph in Figure G.2,

$$\partial = \begin{pmatrix} -1 & 1 & 0 & 0 & 0 & 1 \\ 1 & 0 & 1 & -1 & 0 & 0 \\ 0 & 0 & 0 & 1 & -1 & -1 \\ 0 & -1 & -1 & 0 & 1 & 0 \end{pmatrix}.$$

---

[4] As with simplices, it should be clear from the context whether one is considering G to be oriented or not, so we do not distinguish between the two cases notationally.

We assign a current i_j to each (oriented) edge e_j and represent the currents altogether by the 1-chain $I = \sum_j i_j e_j$. Kirchhoff's junction condition is then elegantly expressed by the requirement that I be a 1-cycle:

$$\partial I = 0. \tag{G.1}$$

For our example this equation reads

$$\begin{pmatrix} -1 & 1 & 0 & 0 & 0 & 1 \\ 1 & 0 & 1 & -1 & 0 & 0 \\ 0 & 0 & 0 & 1 & -1 & -1 \\ 0 & -1 & -1 & 0 & 1 & 0 \end{pmatrix} \begin{pmatrix} i_1 \\ i_2 \\ i_3 \\ i_4 \\ i_5 \\ i_6 \end{pmatrix} = 0.$$

Next, we seek a similar statement about voltages. Here we are guided by the physics. It makes sense to assign an electric potential Φ to each vertex, so Φ is naturally a map from the vertices to the real numbers. We extend it by linearity, so that Φ becomes a 0-cochain. As before, we denote the space of i-cochains by C^i.

By analogy with the continuous case we denote the natural pairing between chains and cochains by the integral symbol. Thus, if $c = \sum_i c_i e_i$ and $f = \sum_j a_j \tilde{e}_j$, where $\{\tilde{e}_j\}$ is the dual basis to $\{e_j\}$, then

$$\int_c f := \sum_j c_j a_j. \tag{G.2}$$

The notation looks even more natural if we recall the coboundary map $\partial_i^* : C^i \to C^{i+1}$. If c is an i-chain and g is an $(i-1)$-cochain then, in our notation,

$$\int_{\partial c} g = \int_c \partial^* g, \tag{G.3}$$

a formula that ought to look moderately familiar.

Indeed, if you think of ∂^* as a discrete analogue of the differential operator d, and if you recall that the electric field E is the gradient of the electric potential Φ, then you ought to be able to write something like $E = -\partial^* \Phi$. But, in our discrete world, ∂^* is a difference operator rather than a differential operator so instead it makes more sense to write

$$V = -\partial^* \Phi, \tag{G.4}$$

where V is a 1-cochain representing the voltage differences across each edge. To see that this choice is indeed the right one, we "integrate". If $e = \langle pq \rangle$ is an edge then

$$\int_e V = -\int_e \partial^* \Phi = -\int_{\partial e} \Phi = -[\Phi(q) - \Phi(p)] = \Phi(p) - \Phi(q), \qquad (G.5)$$

which is exactly the voltage difference across the edge e, consistently with the idea that current runs downhill in the direction of the arrow, from high potential to low potential.

At this point we must confess that we have pulled a fast one, for (G.4) is nothing other than Kirchhoff's loop law in disguise. The reason is that this law reads, in our notation,

$$\int_z V = 0 \qquad (G.6)$$

for every cycle z. If (G.4) holds then (G.6) holds, because in that case

$$\int_z V = -\int_z \partial^* \Phi = -\int_{\partial z} \Phi = 0$$

as $\partial z = 0$. But the converse holds as well. In essence this is the analogue of the de Rham theorem, which says that a closed form is exact if and only if all its periods vanish.[5]

Of course, we would prefer a more direct proof, and in this simple case such a proof is not difficult. First, we introduce a bit of slightly nonstandard terminology. Let p and q be two nodes of G. A **path from p to q** is an alternating sequence

$$(p_0, \{p_0, p_1\}, p_1, \{p_1, p_2\}, p_2 \ldots, \{p_{k-1}, p_k\}, p_k)$$

of vertices and edges, with $p_0 = p$ and $p_k = q$, where successive vertices are distinct.[6] To every such path in G we assign the 1-chain (element of C_1) $\sum_i \langle p_i, p_{i+1} \rangle$ (which by abuse of terminology we also call a path). The boundary of this path is just $q - p$. A **cycle** of G is just a path that starts and ends at the same point; as an element of C_1 it has no boundary, so it is a 1-cycle.[7]

---

[5] The 1-cochain V is "closed", meaning that $\partial^* V = 0$, simply because there are no 2-chains in G and so there are no 2-cochains either.

[6] In particular, although it is not really necessary for anything that follows, we allow our paths to backtrack, to cover the same edge many times, etc. We do this simply to agree with the definition of a path in a manifold, which need not be an embedded curve. Graph theorists use the word "path" to mean a path with no repeated vertices; this forces all the edges to be distinct as well.

[7] Some graph theorists would call our cycle a circuit and would reserve the word "cycle" for a circuit with no repeated vertices.

Let V be an arbitrary 1-cochain. Fix a point p on G, and let c be a path in G starting at p and ending at some arbitrary point q. Define a 0-cochain Φ by setting $\Phi(p) := 0$ and

$$\Phi(q) := -\int_c V. \tag{G.7}$$

The claim is that Φ is well defined, in the sense that it does not depend on the path chosen. If c' were some other path from p to q then we would have

$$\int_{c'} V - \int_c V = \int_{c'-c} V = 0,$$

because $c' - c$ is a cycle. It follows that Φ is a well-defined 0-cochain. Moreover,

$$\int_c V = \Phi(p) - \Phi(q) = -\int_c \partial^* \Phi,$$

so that

$$\int_c (V + \partial^* \Phi) = 0.$$

By linearity it follows that the integral over every edge vanishes, and hence that $V = -\partial^* \Phi$, as promised.

We close this section with an important observation, namely that the zero-dimensional homology of any path connected space is one-dimensional. In particular, as any connected graph is path connected we have $\dim H_0(G) = 1$. Once again, it is worth proving this directly for graphs, as the general case is similar.

We claim that

$$B_0(G) = \{\textstyle\sum_i a_i p_i : \sum_i a_i = 0\}.$$

Suppose that on the one hand $b \in B_0$. Then there exists a 1-chain $c = \sum_{ij} c_{ij} \langle p_i, p_j \rangle$ with $c_{ij} = -c_{ji}$ and

$$\begin{aligned}
b = \partial c &= \textstyle\sum_{ij} c_{ij}(p_j + p_i) = \sum_i (\sum_j c_{ij} p_j) + \sum_j (\sum_i c_{ij} p_i) \\
&= \textstyle\sum_i (\sum_j c_{ij} p_j) + \sum_i (\sum_j c_{ji} p_j) = \sum_{ij} (c_{ij} + c_{ji}) p_j \\
&= \textstyle\sum_j a_j p_j,
\end{aligned}$$

where $a_j := \sum_i (c_{ij} + c_{ji})$. Thus

$$\textstyle\sum_j a_j = \sum_{ij} (c_{ij} + c_{ji}) = 0.$$

On the other hand, suppose that $b = \sum_i a_i p_i$ is a 0-chain with $\sum_i a_i = 0$. Choose a fiducial point $p \in G$. By connectivity we may choose a family $\{\gamma_i\}$ of paths, where γ_i is a path from p to p_i. Define the 1-chain $c = \sum_i a_i \gamma_i$. Then

$$\partial c = \textstyle\sum_i a_i (p_i - p) = \sum_i a_i p_i - p \sum_i a_i = \sum_i a_i p_i = b,$$

and the claim is proved. $\qquad\square$

Now, every zero chain is a cycle (because there are no lower-dimensional chains) and so $C_0 = Z_0$. Consider the surjective map $\psi : C_0 \to \mathbb{R}$ given by $\sum_i a_i p_i \mapsto \sum_i a_i$. This gives an exact sequence,

$$0 \longrightarrow B_0 \overset{\iota}{\longrightarrow} Z_0 \overset{\psi}{\longrightarrow} \mathbb{R} \longrightarrow 0,$$

where ι is the inclusion map. We conclude that $\dim(Z_0/B_0) = 1$.

G.3 Tellegen's theorem

We observe here that a simple consequence of Kirchhoff's laws is **Tellegen's theorem**, which is just another expression of the law of energy conservation. If the voltages $V = \sum_i V_i \tilde{e}_i$ and currents $I = \sum_j I_j e_j$ obey Kirchhoff's laws $V = -\partial^* \Phi$ and $\partial I = 0$ then the total power dissipated by the circuit is

$$\sum_j i_j V_j = \int_I V = -\int_I \partial^* \Phi = -\int_{\partial I} \Phi = 0, \tag{G.8}$$

so we obtain the result that an isolated circuit conserves energy. Of course we ought to have expected this, considering that we derived Kirchhoff's laws assuming energy conservation, but it is comforting nonetheless.

G.4 Ohm's law

We are finally in a position to write down the "equations of motion" for our system, which in our case is just Ohm's law $V = IR$. But we have defined V as a 1-cochain and I as a 1-chain, so R cannot just be a real number. Once again we draw inspiration from physics. In a bulk medium Ohm's law is $j = \sigma E$, where j is the current density vector, σ is the conductivity tensor, and E is the electric field vector. Equivalently, $E = \rho j$, where $\rho = (\sigma)^{-1}$ is the resistivity tensor. By analogy, then, we write

$$V = ZI + v, \tag{G.9}$$

where V is the vector of voltage differences, Z is the **impedance matrix**, I is the vector of currents, and v is a vector representing the voltages contributed by any batteries on the edges. We may think of the impedance matrix as the matrix representation of a linear transformation $Z : C_1 \to C^1$. In our case the impedance matrix is diagonal, the diagonal elements being the resistances across each edge, but, as mentioned above, it could just as well contain complex impedances to account for capacitors and inductors.

We must now somehow combine Kirchhoff's rules and Ohm's law to solve the circuit. It is not clear at this stage that it can actually be done. But it can, and this is the content of Kirchhoff's theorem. Actually, there are many ways to solve for the currents and voltages, although will discuss just one.

By way of motivation, we may recall that, when we were using Kirchhoff's laws to solve for the currents in a circuit, we only needed to use a few loops rather than all of them. The key point is that most of the information in (G.1) and (G.6) is redundant. The general rule of thumb, taught in elementary physics classes, is to choose just enough loops that every branch of the circuit lies in some loop and then add in as many independent junction conditions as are needed to get an equal number of equations and unknowns. Formalizing these intuitions leads to some really fun graph theory, which we can only discuss in the briefest possible fashion.

G.5 Spanning trees and cycle bases

If a graph contains no cycles then it is called a **forest**, and if it only has one connected component then it is a **tree** (See Figure G.3) A **bridge** is an edge whose removal disconnects G. A **spanning tree** of a connected graph G is a subset of G that contains all the vertices of G and some edges of G and is a tree (See Figure G.4.).

> **EXERCISE G.1** Show that a connected graph is a tree if and only if every edge is a bridge.

Theorem G.1 *A connected graph G with n vertices is a tree if and only if it has $n - 1$ edges.*

Figure G.3 A forest of trees.

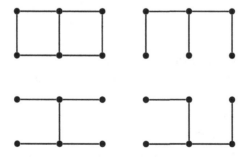

Figure G.4 A graph (upper left) and three of its spanning trees.

Proof The assertion is immediate for a single edge with $n = 2$. So, assume that G is connected, with n vertices. If G is a tree then removing an edge gives two disconnected pieces. Let n_i, $i = 1, 2$, be the number of vertices in each piece, so that $n = n_1 + n_2$. Evidently $n_i < n$, $i = 1, 2$, so by induction on the number of vertices each piece has $n_i - 1$ edges. Accounting for the edge that was removed, the number of edges in the original graph is therefore $(n_1 - 1) + (n_2 - 1) + 1 = n - 1$.

Conversely, suppose G has n vertices but is not a tree. By successively removing edges that are not bridges, we must eventually arrive at a spanning tree, which, as we have shown, must have $n - 1$ edges. Clearly, G must have had more than $n - 1$ edges to begin with. $\qquad\square$

Remark There is a really wonderful theorem, called **Kirchhoff's matrix tree theorem**, which gives a simple formula for the **tree number**, namely the number of spanning trees in a connected graph. The **Laplacian matrix** corresponding to the graph G is defined to be the $n \times n$ matrix $L = \partial\partial^*$, where we identify C_i with C^i, $i = 1, 2$, so that ∂^* is just the transpose of ∂. (This is just a discrete version of the Laplace–de Rham operator (8.96).) It is not difficult to show that L is independent of the choice of orientation of the edges. More remarkable is the fact that all the cofactors of L are equal, and they are all equal to the tree number! (For the proofs see [9] or [89].) Try it for the graph in Figure G.4. Answer: 15. (The matrix tree theorem can be generalized to higher-dimensional simplicial complexes. See *e.g.* [24].)

If G is a connected graph and T is a spanning tree not equal to G, an edge in G but not in T is called a **chord**. If we adjoin a chord to a spanning tree then we get a cycle of G, called an **elementary cycle**. All the cycles obtained in this fashion are linearly independent as elements of C_1, and therefore of Z_1, because each contains an edge not contained in the others, namely, its chord. (See Figure G.5.)

The claim is that the elementary cycles form a basis for Z_1. First, we count the number of chords of a spanning tree. But that's easy. The graph has n vertices and m edges, while the tree has n vertices and $n - 1$ edges, so there must be $m - n + 1$ chords.

Next, consider the chain complex

$$C_1 \xrightarrow{\ \partial_1\ } C_0 \xrightarrow{\ \partial_0\ } 0.$$

The **cycle rank** of a graph G, rk G, is the dimension of the cycle space $Z_1 = \ker \partial_1$. By the rank–nullity theorem, $\dim C_1 = \operatorname{rk} \partial_1 + \dim \ker \partial_1$, which gives $m = \dim B_0 + \operatorname{rk} G$, where m is the number of edges of G and B_0 is the space of 0-boundaries. But the graph is connected, so $1 = \dim H_0 = \dim Z_0 - \dim B_0$.

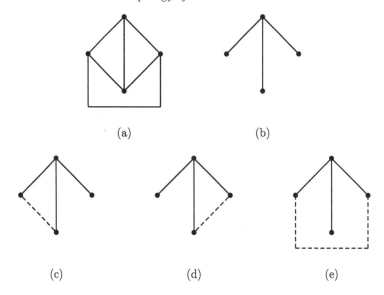

Figure G.5 (a) A graph, (b) a spanning tree, and (c), (d), and (e) its elementary cycles. The broken lines are the chords associated to the tree.

As everything in C_0 is killed by ∂_0, it follows that $Z_0 = C_0$ so we get $1 = n -$ dim B_0. Combining these formulae gives

$$\operatorname{rk} G = m - n + 1. \tag{G.10}$$

This equals the number of chords of G (relative to any spanning tree), so the elementary cycles really do form a basis for Z_1.

G.6 The mesh current method

At this stage it should come as no surprise that the number of independent loops that we must use when applying Kirchhoff's laws is precisely the cycle rank of G. First, pick a spanning tree of G and then write down the corresponding elementary cycles. For example, for the circuit modeled by Figure G.2 we obtain the cycles shown in Figure G.5. In the order (c), (d), and (e) they are

$$z_1 = e_1 + e_2 - e_3,$$
$$z_2 = e_3 + e_4 + e_5,$$
$$z_3 = e_1 + e_4 + e_6.$$

Next, one imagines there are currents j_i flowing around the elementary cycles z_i, called **mesh currents**. These are packaged into a 1-cycle $J = \sum_i j_i z_i$. The inclusion map $\rho : Z_1 \to C_1$ relates the mesh currents to the edge currents. For our circuit, the matrix representation of ρ is

$$\rho = \begin{pmatrix} 1 & 0 & 1 \\ 1 & 0 & 0 \\ -1 & 1 & 0 \\ 0 & 1 & 1 \\ 0 & 1 & 0 \\ 0 & 0 & 1 \end{pmatrix},$$

giving

$$\rho \begin{pmatrix} j_1 \\ j_2 \\ j_3 \end{pmatrix} = \begin{pmatrix} j_1 + j_3 \\ j_1 \\ -j_1 + j_2 \\ j_2 + j_3 \\ j_2 \\ j_3 \end{pmatrix}.$$

Observe that Kirchhoff's junction condition is automatically satisfied for this vector of currents, because ρ takes cycles to cycles, so $\partial \rho J = 0$. The reader may wish to check this for our circuit, using matrix multiplication to verify that $\partial \rho = 0$.

As there are no 2-chains in G, no 1-cycle can be a boundary of something. This means that Z_1 is the same thing as H_1, the first homology group. In particular the inclusion map may be viewed as a map $\rho : H_1 \to C_1$, and it induces a dual map $\pi : C^1 \to H^1$. In terms of the canonical dual bases the matrix representation of π is just the transpose of the matrix representation of ρ, so, in our example,

$$\pi = \begin{pmatrix} 1 & 1 & -1 & 0 & 0 & 0 \\ 0 & 0 & 1 & 1 & 1 & 0 \\ 1 & 0 & 0 & 1 & 0 & 1 \end{pmatrix}.$$

The dual of $\partial \rho$ is $\pi \partial^*$ and, as the former vanishes, so must the latter.

At last we are in a position to write down an algorithm for solving circuits. Write the currents I in terms of the unknown mesh currents J according to $I = \rho J$, so that Kirchhoff's junction condition is satisfied, write $V = -\partial^* \Phi$ so that Kirchhoff's loop law is satisfied, and substitute into Ohm's law (G.9) to get

$$Z\rho J = -\partial^* \Phi - v.$$

Apply π to both sides and use $\pi \partial^* = 0$ to obtain

$$\pi Z \rho J = -\pi v. \tag{G.11}$$

If $\pi Z\rho$ is invertible[8] then we get

$$J = -(\pi Z\rho)^{-1}\pi v. \tag{G.12}$$

Once one has J, one can obtain I from $I = \rho J$ and V from $V = ZI + v$.
    What does all this look like for the circuit in Figure G.1? We compute

$$\pi Z\rho = \begin{pmatrix} 1 & 1 & -1 & 0 & 0 & 0 \\ 0 & 0 & 1 & 1 & 1 & 0 \\ 1 & 0 & 0 & 1 & 0 & 1 \end{pmatrix} \begin{pmatrix} R_1 & 0 & 0 & 0 & 0 & 0 \\ 0 & R_2 & 0 & 0 & 0 & 0 \\ 0 & 0 & R_3 & 0 & 0 & 0 \\ 0 & 0 & 0 & R_4 & 0 & 0 \\ 0 & 0 & 0 & 0 & R_5 & 0 \\ 0 & 0 & 0 & 0 & 0 & 0 \end{pmatrix} \begin{pmatrix} 1 & 0 & 1 \\ 1 & 0 & 0 \\ -1 & 1 & 0 \\ 0 & 1 & 1 \\ 0 & 1 & 0 \\ 0 & 0 & 1 \end{pmatrix}$$

$$= \begin{pmatrix} R_1 + R_2 + R_3 & -R_3 & R_1 \\ -R_3 & R_3 + R_4 + R_5 & R_4 \\ R_1 & R_4 & R_1 + R_4 \end{pmatrix}.$$

For concreteness, let's take $R_1 = R_3 = R_5 = 5\,\Omega$ and $R_2 = R_4 = 10\,\Omega$. Then

$$\pi Z\rho = \begin{pmatrix} 20 & -5 & 5 \\ -5 & 20 & 10 \\ 5 & 10 & 15 \end{pmatrix}$$

and

$$(\pi Z\rho)^{-1} = \begin{pmatrix} 8/105 & 1/21 & -2/35 \\ 1/21 & 11/105 & -3/35 \\ -2/35 & -3/35 & 1/7 \end{pmatrix}.$$

Also, setting $V_0 = 10$ volts, say, we get[9]

$$\pi v = \begin{pmatrix} 1 & 1 & -1 & 0 & 0 & 0 \\ 0 & 0 & 1 & 1 & 1 & 0 \\ 1 & 0 & 0 & 1 & 0 & 1 \end{pmatrix} \begin{pmatrix} 0 \\ 0 \\ 0 \\ 0 \\ 0 \\ -10 \end{pmatrix} = \begin{pmatrix} 0 \\ 0 \\ -10 \end{pmatrix}.$$

[8] There are some subtleties when Z contains complex impedances, but one can show that $\pi Z\rho$ is always invertible for purely resistive circuits. See [8].

[9] We must set $v_6 = -10$ volts because, by our convention, voltage differences are positive if we go from high potential to low potential when following the arrows.

and therefore

$$J = -(\pi Z \rho)^{-1} \pi v = \begin{pmatrix} 8/105 & 1/21 & -2/35 \\ 1/21 & 11/105 & -3/35 \\ -2/35 & -3/35 & 1/7 \end{pmatrix} \begin{pmatrix} 0 \\ 0 \\ -10 \end{pmatrix} = \frac{1}{7} \begin{pmatrix} -4 \\ -6 \\ 10 \end{pmatrix}.$$

Finally,

$$I = \rho J = \frac{1}{7} \begin{pmatrix} 1 & 0 & 1 \\ 1 & 0 & 0 \\ -1 & 1 & 0 \\ 0 & 1 & 1 \\ 0 & 1 & 0 \\ 0 & 0 & 1 \end{pmatrix} \begin{pmatrix} -4 \\ -6 \\ 10 \end{pmatrix} = \frac{1}{7} \begin{pmatrix} 6 \\ -4 \\ -2 \\ 4 \\ -6 \\ 10 \end{pmatrix}$$

and

$$V = ZI + v = \frac{1}{7} \begin{pmatrix} 5 & 0 & 0 & 0 & 0 & 0 \\ 0 & 10 & 0 & 0 & 0 & 0 \\ 0 & 0 & 5 & 0 & 0 & 0 \\ 0 & 0 & 0 & 10 & 0 & 0 \\ 0 & 0 & 0 & 0 & 5 & 0 \\ 0 & 0 & 0 & 0 & 0 & 0 \end{pmatrix} \begin{pmatrix} 6 \\ -4 \\ -2 \\ 4 \\ -6 \\ 10 \end{pmatrix} + \begin{pmatrix} 0 \\ 0 \\ 0 \\ 0 \\ 0 \\ -10 \end{pmatrix} = \frac{1}{7} \begin{pmatrix} 30 \\ -40 \\ -10 \\ 40 \\ -30 \\ -70 \end{pmatrix},$$

where currents are in amperes and voltages are in volts.

Appendix H

Intrinsic and extrinsic curvature

In Section 8.5 we noted that the cylinder is a flat manifold, in the sense that its curvature vanishes. But this conflicts with our intuition, which tells us that a cylinder is obviously curved. These two apparently contradictory facts are reconciled by distinguishing two types of curvature: intrinsic and extrinsic. All the curvatures we have talked about elsewhere in this book are intrinsic – they depend only on the manifold itself. In contrast, extrinsic curvature has to do with how a manifold is embedded in a higher-dimensional space. The cylinder has zero intrinsic curvature but nonzero extrinsic curvature when viewed as a surface in $\mathbb{R}^3$. In this appendix we investigate how these two types of curvature are related.

Let M and N be smooth geometric manifolds of dimensions m and n, respectively, with $m < n$, and let $f : M \to N$ be an immersion (so that M is a submanifold of N). Let g denote the metric on N and $h = f^*g$ the induced metric on M. At every point p the tangent space to N naturally breaks up into a direct sum $T_pN = T_pM \oplus T_p^\perp M$, where $T_p^\perp M$ is the orthogonal complement to T_pM (relative to the metric g); $T_p^\perp M$ is called the **normal space** to M at p and its elements are **normal vectors**. The union of all the normal spaces to M is $T^\perp M$, the **normal bundle** of M. A **normal vector field** is a smooth section of the normal bundle, namely, a smooth vector field of N that is everywhere normal to M.

Let X and Y be tangent vector fields on M. Then, by definition, $h(X, Y)_p = g(f_*X, f_*Y)_{f(p)}$. But the immersion f restricts to the identity on M, so we can identify p and $f(p)$, in which case $f_* : T_pM \to T_pM$ is just the identity map. It follows that $h(X, Y)_p = g(X, Y)_p$. If either X or Y had a normal component it would make no sense to write $h(X, Y)$. Therefore we may safely use g to denote the metric on both M and N, which is the standard convention anyway.

Now let ∇ denote and $\tilde{\nabla}$ denote the Levi-Civita connections of M and N, respectively. The Gauss formula relates these two connections.

Theorem H.1 *Let X and Y be vector fields tangent to M. Then*

$$\tilde{\nabla}_X Y = \nabla_X Y + \alpha(X, Y), \tag{H.1}$$

where $\alpha(X, Y) \in T^\perp M$.

Proof The covariant derivative operator $\tilde{\nabla}$ maps vector fields on N to vector fields on N so, technically, to define $\tilde{\nabla}_X Y$ we must first *extend* X and Y to vector fields $\tilde{X}$ and $\tilde{Y}$ on N. But the whole point of the covariant derivative is that it depends only on what vector fields do at a point and so is independent of any local extension. In other words,

$$\tilde{\nabla}_{\tilde{X}} \tilde{Y} = \tilde{\nabla}_X Y. \tag{H.2}$$

Note also for future use that, on M, $[\tilde{X}, \tilde{Y}] = [X, Y]$ because the Lie bracket preserves tangency.

Now, $\tilde{\nabla}_X Y$ lies in the tangent space to N so we can break it up into two pieces, one tangent to M and the other normal to M. Let $P : T_p N \rightarrow T_p M$ be the linear projection onto $T_p M$, and define $\beta(X, Y)_p := P(\tilde{\nabla}_X Y)_p$ and $\alpha(X, Y)_p := (1 - P)(\tilde{\nabla}_X Y)_p$. We must show that $\beta(X, Y) = \nabla_X Y$, where ∇ is the Levi-Civita connection on M obtained from the induced metric. To do this, we must show that axioms (C1)–(C3), (C7), and (C8) of Sections 7.2 and 8.2 are satisfied.

Axioms (C1)–(C3) follow simply from the corresponding properties of $\tilde{\nabla}$ because P is linear. For example, axiom (C2) applied to $\tilde{\nabla}$ gives, for any vector fields X, Y and function f on M,

$$\beta(X, fY) = P(\tilde{\nabla}_X fY)$$
$$= P(f\tilde{\nabla}_X Y) + P((Xf)Y)$$
$$= f\beta(X, Y) + (Xf)Y.$$

(The term $(Xf)Y$ is already tangent to M.) To show axiom (C7) we must show that, for any vector fields X, Y on M,

$$\beta(X, Y) - \beta(Y, X) = [X, Y]. \tag{H.3}$$

But $\tilde{\nabla}$ is torsion free, so

$$0 = \tilde{\nabla}_{\tilde{X}} \tilde{Y} - \tilde{\nabla}_{\tilde{Y}} \tilde{X} - [\tilde{X}, \tilde{Y}]$$
$$= \tilde{\nabla}_X Y - \tilde{\nabla}_Y X - [X, Y]$$
$$= (\beta(X, Y) + \alpha(X, Y)) - (\beta(Y, X) + \alpha(Y, X)) - [X, Y]. \tag{H.4}$$

The tangential component of (H.4) is just (H.3), while the normal component of (H.4) yields

$$\alpha(X, Y) = \alpha(Y, X). \tag{H.5}$$

Again, let X, Y, and Z be vector fields on M, with corresponding (arbitrary) extensions $\widetilde{X}$, $\widetilde{Y}$, and $\widetilde{Z}$. By the metric compatibility of $\widetilde{\nabla}$,

$$\widetilde{X}g(\widetilde{Y}, \widetilde{Z}) = g(\widetilde{\nabla}_{\widetilde{X}}\widetilde{Y}, \widetilde{Z}) + g(\widetilde{Y}, \widetilde{\nabla}_{\widetilde{X}}\widetilde{Z}).$$

Using (H.2) and the fact that, by construction, $g(\widetilde{X}, \widetilde{Y})_p = g(X, Y)_p$ for all $p \in M$, we get

$$\begin{aligned}
Xg(Y, Z) &= g(\widetilde{\nabla}_X Y, Z) + g(Y, \widetilde{\nabla}_X Z) \\
&= g(\beta(X, Y) + \alpha(X, Y), Z) + g(Y, \beta(X, Z) + \alpha(X, Z)) \\
&= g(\beta(X, Y), Z) + g(Y, \beta(X, Z)),
\end{aligned}$$

because the normal space of M is orthogonal to the tangent space of M. Hence axiom (C8) holds for $\beta(X, Y)$. By the uniqueness of the Levi-Civita connection, we conclude that $\beta(X, Y) = \nabla_X Y$, as promised. $\square$

EXERCISE H.1 Show that α is function linear.

By virtue of equation (H.5) and Exercise H.1, α defines a symmetric tensor field on M with values in the normal space of M. It is called the **second fundamental tensor** of M and depends on the chosen embedding of M into N. If the codimension of M (namely, $n - m$) is unity, M is called a **hypersurface**. If ξ is a field of unit normals to the hypersurface then

$$\alpha(X, Y) = \mathrm{II}_\xi(X, Y)\xi, \tag{H.6}$$

where $\mathrm{II}_\xi(X, Y)$ is sometimes called the **extrinsic curvature tensor** or the **second fundamental form** of the hypersurface M.[1]

We can use the Gauss formula (H.1) together with the definition of the curvature tensor (7.36) to relate the intrinsic and extrinsic Riemann curvatures of M and N. Let $\widetilde{R}$ (respectively, R) denote the Riemann curvature tensor obtained from $\widetilde{\nabla}$ (respectively, ∇). If X, Y, and Z are tangent vector fields on M then, by considering tangential and normal projections, we obtain the **equations of Gauss and Codazzi-Mainardi**:

$$P(\widetilde{R}(X, Y)Z) = R(X, Y)Z + P(\widetilde{\nabla}_X \alpha(Y, Z) - \widetilde{\nabla}_Y \alpha(X, Z)) \tag{H.7}$$

and

$$\begin{aligned}
(1 - P)(\widetilde{R}(X, Y)Z) &= \alpha(X, \nabla_Y Z) - \alpha(Y, \nabla_X Z) - \alpha([X, Y], Z) \\
&\quad + (1 - P)(\widetilde{\nabla}_X \alpha(Y, Z) - \widetilde{\nabla}_Y \alpha(X, Z)). \tag{H.8}
\end{aligned}$$

EXERCISE H.2 Derive (H.7) and (H.8).

---

[1] The **first fundamental form** of M is just the metric (viewed as a quadratic form).

These equations assume their more traditional forms when M is a hypersurface of N with unit normal vector field ξ. The **Weingarten map** or **shape operator** $S_\xi : T_p M \to T_p M$ is given by

$$S_\xi(X) = -P(\widetilde{\nabla}_X \xi), \tag{H.9}$$

where $X \in \Gamma(TM)$ is a vector field tangent to M.[2] Actually, the projection operator P is redundant here because ξ is a unit vector field so $\widetilde{\nabla}_X \xi$ is already tangent to M, as the following computation demonstrates:

$$1 = g(\xi, \xi) \quad \Rightarrow \quad 0 = X g(\xi, \xi) = 2g(\widetilde{\nabla}_X \xi, \xi).$$

EXERCISE H.3 Prove **Weingarten's formula**

$$g(S_\xi(X), Y) = g(\alpha(X, Y), \xi) = \mathrm{II}_\xi(X, Y). \tag{H.10}$$

Hint: Use the metric compatibility of $\widetilde{\nabla}$ and the orthonormality of the tangent and normal spaces.

Remark The Weingarten map has a useful geometrical interpretation when M is a hypersurface in Euclidean space $N = \mathbb{R}^n$. Let g be the ordinary Euclidean metric (so that $\widetilde{\nabla}$ is just the ordinary gradient), and let $\xi = \xi^j \partial_j$ be the hypersurface unit normal. Then $g(\xi, \xi) = \sum_j (\xi^j)^2 = 1$, so we get a natural map $\Xi : M \to S^{n-1}$, given by $p \mapsto (\xi^1(p), \ldots, \xi^n(p))$, called the **Gauss map** or **sphere map**. We claim that, under these circumstances, the Weingarten map can be viewed as the negative of Ξ_*, the differential of the Gauss map.

To see this, first note that Ξ_* maps $T_p M$ to $T_{\Xi(p)} S^{n-1}$. But $T_{\Xi(p)} S^{n-1}$ and $T_p M$ are parallel affine hyperplanes (because each consists of the set of all vectors orthogonal to $\xi(p)$), so we may identify them (up to translation). Let $X = X^k \partial_k$ be a vector field on M and f a smooth function on $\mathbb{R}^n$. Then

$$\Xi_*(X)_{\Xi(p)}(f) = X_p(\Xi^* f) = X_p(f \circ \Xi) = X_p(f(\xi^1, \ldots, \xi^n))$$
$$= X^k(p) \left.\frac{\partial f}{\partial \xi^j}\right|_{\Xi(p)} \left.\frac{\partial \xi^j}{\partial x^k}\right|_p = \left.\left(\widetilde{\nabla}_X(\xi)\right)^j\right|_p \left.\frac{\partial}{\partial \xi^j} f\right|_{\Xi(p)}.$$

Stripping f from both sides and using the identification of tangent spaces allows us to write

$$\Xi_*(X) = \widetilde{\nabla}_X(\xi). \tag{H.11}$$

---

[2] Technically, we should define $S_\xi(X) = -P(\widetilde{\nabla}_{\widetilde{X}} \widetilde{\xi})$ where $\widetilde{X}$ and $\widetilde{\xi}$ are local extensions of X and ξ to N but, as before, the properties of covariant derivatives ensure that S is independent of the manner in which the vector fields are extended. The minus sign is inserted to make Weingarten's formula (H.10) look nice.

Looking at (H.10) we see that, because the right-hand side is symmetric in X and Y, so is the left-hand side. That is, the shape operator S_ξ is a self-adjoint linear operator so it can be diagonalized by an orthogonal transformation. Let $\{e_1, \ldots, e_m\}$ be a basis of T_pM consisting of orthonormal eigenvectors of S_ξ, with corresponding eigenvalues $\{\lambda_1, \ldots, \lambda_m\}$. The eigenvectors e_i are called the **principal directions** (of the embedding) and the eigenvalues λ_i are called the **principal curvatures**. The product of the principal curvatures (i.e., the determinant of the shape operator) is called the **Gauss–Kronecker curvature** of M, and the average of the principal curvatures (i.e., the trace of the shape operator divided by m) is called the **mean curvature** of M. Gauss's *theorema egregium* is the statement that, in the case of a two-dimensional surface embedded in $\mathbb{R}^3$, the Gauss–Kronecker curvature (the product of the principal curvatures) is precisely the Gaussian curvature (the sectional curvature) and is therefore a purely intrinsic quantity, independent of the particular embedding. (See Exercise H.5 below.)

EXERCISE H.4 Let W, X, Y, and Z be tangent vector fields on M, a hypersurface in N with unit normal field ξ. Show that the Gauss and Codazzi–Mainardi equations become

$$g(\widetilde{R}(X,Y)Z, W) = g(R(X,Y)Z, W)$$
$$+ \mathrm{II}_\xi(X,Z)\mathrm{II}_\xi(Y,W) - \mathrm{II}_\xi(Y,Z)\mathrm{II}_\xi(X,W) \qquad \text{(H.12)}$$

and

$$g(\widetilde{R}(X,Y)Z, \xi) = \mathrm{II}_\xi(X, \nabla_Y Z) - \mathrm{II}_\xi(Y, \nabla_X Z) - \mathrm{II}_\xi([X,Y], Z) \qquad \text{(H.13)}$$
$$= g(\nabla_X S_\xi(Y) - \nabla_Y S_\xi(X) - S_\xi([X,Y]), Z), \qquad \text{(H.14)}$$

respectively.

EXERCISE H.5 Prove Gauss's *theorema egregium* when M is a surface in $\mathbb{R}^3$, by showing that the determinant of the shape operator is the Gaussian curvature. *Hint:* Choose X, Y, Z, and W appropriately in (H.12).

EXERCISE H.6 Assume that M is a hypersurface in Euclidean space. Choose local coordinates x^i on M such that the components of the metric tensor and second fundamental form of M are

$$g_{ij} = g(\partial_i, \partial_j) \qquad \text{and} \qquad b_{ij} = \mathrm{II}(\partial_i, \partial_j),$$

respectively. (As usual, $\partial_i := \partial/\partial x^i$.) Show that the Gauss and Codazzi–Mainardi equations can be written as

$$R_{ijk\ell} = b_{ik}b_{j\ell} - b_{i\ell}b_{jk} \qquad \text{(H.15)}$$

and

$$b_{\ell j,i} - b_{\ell i,j} = b_{jm}\Gamma^m{}_{\ell i} - b_{im}\Gamma^m{}_{\ell j}, \tag{H.16}$$

respectively. (The Christoffel symbols are those of the metric-compatible connection ∇ on M. Recall that the comma denotes a partial derivative. *Hint:* Show that $b_{ij} = S_{ij}$ where $S(\partial_i) = S^j{}_i \partial_j$, and indices are raised and lowered by using the metric. Recall the results of Exercises 8.31 and 8.34.

EXERCISE H.7 Let $M = \Sigma$ be a two-dimensional surface immersed in $N = \mathbb{R}^3$ via the parameterization

$$\sigma : (u, v) \rightarrow (x(u, v), y(u, v), z(u, v)), \tag{H.17}$$

and set $\sigma_u := \partial\sigma/\partial u$ and $\sigma_v := \partial\sigma/\partial v$. The unit normal vector to Σ is defined to be

$$n = \frac{\sigma_u \times \sigma_v}{\|\sigma_u \times \sigma_v\|}, \tag{H.18}$$

where "$\times$" denotes an ordinary cross product and $\|X\|$ is the ordinary Euclidean norm of X.

(a) Show that the second fundamental form of Σ can be written[3]

$$\text{II} = J\,du^2 + 2K\,du\,dv + L\,dv^2, \tag{H.19}$$

where

$$J = -g(n_u, \sigma_u) = g(n, \sigma_{uu}), \tag{H.20}$$

$$K = -g(n_v, \sigma_u) = g(n, \sigma_{uv}) = g(n, \sigma_{vu}) = -g(n_u, \sigma_v), \tag{H.21}$$

$$L = -g(n_v, \sigma_v) = g(n, \sigma_{vv}). \tag{H.22}$$

and the subscript variables are shorthand for differentiation with respect to that variable. For example, $\sigma_u = \partial\sigma/\partial u$ and $\sigma_{uu} = \partial^2\sigma/\partial u^2$.

Hint: The difficult part of this problem is just translating the formal differential geometric language into the traditional language. In other words, it is mostly a problem of notation. The surface Σ is the image of the map $\sigma : D \rightarrow \mathbb{R}^3$, where D is a domain in $\mathbb{R}^2$ with coordinates (u, v). The two vectors $\partial_u := \partial/\partial u$ and $\partial_v := \partial/\partial v$ span the tangent space to D, so their images $\sigma_*(\partial_u) =: X$ and $\sigma_*(\partial_v) =: Y$ span the tangent space of Σ. Note that if $f : \Sigma \rightarrow \mathbb{R}$ then

$$Xf = \sigma_*(\partial_u)f = (\partial_u)(\sigma^* f) = f(\sigma(u, v))_u$$

$$= \frac{\partial f}{\partial x}\frac{\partial x}{\partial u} + \frac{\partial f}{\partial y}\frac{\partial y}{\partial u} + \frac{\partial f}{\partial z}\frac{\partial z}{\partial u}$$

---

[3] The traditional notation is e, f, and g rather than J, K, and L, but we wish to reserve g for the metric. However, K is usually reserved for the Gaussian curvature but in this appendix we use K for a component of the second fundamental form.

$$= \left(\frac{\partial f}{\partial x}, \frac{\partial f}{\partial y}, \frac{\partial f}{\partial z}\right) \cdot \left(\frac{\partial x}{\partial u}, \frac{\partial y}{\partial u}, \frac{\partial z}{\partial u}\right)$$

$$= \text{grad } f \cdot \sigma_u = \sigma_u \cdot \text{grad } f,$$

where the dot denotes the usual Euclidean dot product. Dropping f gives

$$X = (\sigma_u)^i \partial_i.$$

Similarly, $Y^i = (\sigma_v)^i$ and $\xi = n^i \partial_i$. Finally, note that the second fundamental form appearing in this problem is really the pullback of the second fundamental form as we have defined it, even though, as is customary, we use the same notation for both. Thus, for example, we have

$$J = (\sigma^* \mathrm{II})(\partial_u, \partial_u) = \mathrm{II}(\sigma_*(\partial_u), \sigma_*(\partial_u)).$$

(b) Using Weingarten's formula (H.10), show that the shape operator (H.9) can be written

$$S = (EG - F^2)^{-1} \begin{pmatrix} GJ - FK & GK - FL \\ EK - FJ & EL - FK \end{pmatrix}, \tag{H.23}$$

where E, F, and G are the coefficients of the first fundamental form (metric) of Σ:

$$\mathrm{I} = E \, du^2 + 2F \, du \, dv + G \, dv^2. \tag{H.24}$$

Hint: You may wish to refer to Exercise 8.47.

(c) Show that the Gaussian curvature (the product of the principal curvatures) can be written as the ratio of the determinants of the second and first fundamental forms:

$$\lambda_1 \lambda_2 = \frac{\det \mathrm{II}}{\det \mathrm{I}} = \frac{JL - K^2}{EG - F^2}. \tag{H.25}$$

(d) The coordinates u and v on Σ are called **isothermal** if

$$g(\sigma_u, \sigma_u) = g(\sigma_v, \sigma_v) \qquad \text{and} \qquad g(\sigma_u, \sigma_v) = 0, \tag{H.26}$$

or equivalently, if the first fundamental form of the surface can be written

$$\mathrm{I} = \lambda^2 (du^2 + dv^2) \tag{H.27}$$

for some positive function $\lambda^2 = g(\sigma_u, \sigma_u) = g(\sigma_v, \sigma_v)$.[4] Show that if the surface admits an isothermal parameterization then

$$\sigma_{uu} + \sigma_{vv} = 2\lambda^2 H, \tag{H.28}$$

where $H = (\lambda_1 + \lambda_2)n/2$ (H is sometimes called the **mean curvature vector**). *Hint:* Differentiate the isothermal equations (H.26).

---

[4] More generally, a metric of the form $g_{ij} = \lambda^2 \delta_{ij}$ (in any dimension) is said to be **conformally flat**.

(e) A surface for which the mean curvature vanishes is called a **minimal surface**.[5]
Show that the catenoid, given parametrically by

$$\sigma(u, v) = (a \cosh v \cos u, a \cosh v \sin u, av), \quad 0 \leq u < 2\pi, \ -\infty < v < \infty,$$

is a minimal surface. *Hint:* Is the parameterization isothermal?

Remark The word "isothermal" comes from the word "isotherm" and means "at constant temperature". If φ is the scalar field representing the temperature at any point then the equation governing its behavior (the **heat equation**) is $\dot{\varphi} = \kappa^2 \nabla^2 \varphi$, where κ is the thermal diffusivity, the dot indicates a time derivative, and $\nabla^2 = \partial_x^2 + \partial_y^2 + \partial_z^2$ is the ordinary Laplacian. In the steady state the time derivative vanishes, so the temperature satisfies $\nabla^2 \varphi = 0$; this implies that φ is a harmonic function and that the level surfaces of φ are isotherms.

If we have a minimal surface, so that the mean curvature vanishes, then, if the parameterization is isothermal (and such a parameterization always exists; see [73], p. 31), by (H.28) each component $x_i(u, v)$ of σ is a harmonic function in parameter space ((u, v) space). The level curves of the coordinate functions, i.e., $x_i(u, v) =$ constant, are in some sense "coordinate isotherms", whatever that means. But this is not the reason for the terminology.

Instead, suppose that we have a *planar* two-dimensional object (with constant κ), described parametrically by $\sigma(u, v) = (x(u, v), y(u, v))$, and suppose that heat is flowing through the object. Gabriel Lamé showed [48] that the lines of constant u (respectively, v) are the isotherms and that the lines of constant v (respectively, u) are the adiabats (lines of heat flow through the system in the steady state) if and only if the coordinates u and v are "isothermal" in the sense that $\sigma_u \cdot \sigma_u = \sigma_v \cdot \sigma_v$ and $\sigma_u \cdot \sigma_v = 0$, where "·" is the usual Euclidean inner product.

EXERCISE H.8 A cylinder of radius a is given the following parameterization:

$$\sigma(u, v) = (a \cos u, a \sin u, v), \quad 0 \leq u < 2\pi, \ -\infty < v < \infty.$$

Show that the intrinsic curvature (as measured by the Gaussian curvature) is zero but that the extrinsic curvature (as measured by, say, the mean curvature) is nonzero. This confirms our intution that a cylinder embedded in $\mathbb{R}^3$ is intrinsically flat but extrinsically curved. *Hint:* Apply the results of Exercise H.7. *Remark:* A surface can have vanishing mean curvature, yet still be curved in the extrinsic sense, as long as

---

[5] The origin of this terminology comes from the Belgian physicist Joseph Plateau who, motivated by experiments with soap films on wire frames around 1850, asked the question whether there was a unique surface of minimal area whose boundary was some given closed curve (**Plateau's problem**). A surface of minimal area is a minimal surface, so that accounts for the interest in the subject. (Technically, a vanishing mean curvature only implies a zero first-order variation in the surface area when it is subjected to small normal displacements (see *e.g.* [22] or [73]), so the minimal surface condition does not necessarily imply minimal area although in practice it usually does.)

at least one principal curvature is nonvanishing. Moreover, in higher dimensions one must verify that the entire Riemann tensor is zero in order to verify that a manifold has no intrinsic curvature.

EXERCISE H.9 Enneper's surface (see the cover illustration) is a classic example of a minimal surface in $\mathbb{R}^3$. A standard parameterization is given by

$$\sigma(u, v) = \left(u - \frac{u^3}{3} + uv^2, v - \frac{v^3}{3} + vu^2, u^2 - v^2\right), \qquad (u, v) \in \mathbb{R}^2.$$

Remark: The image on the front cover reveals only a fragment of Enneper's surface; the entire surface has self-intersections. See [22], p. 205.

(a) Show that the coefficients of the first fundamental form are

$$E = G = (1 + u^2 + v^2)^2, \qquad F = 0.$$

(b) Show that the coefficients of the second fundamental form are

$$J = 2, \qquad K = -2, \qquad L = 0.$$

(c) Show that the principal curvatures are

$$\lambda_1 = \frac{2}{(1 + u^2 + v^2)^2} \qquad \text{and} \qquad \lambda_2 = -\frac{2}{(1 + u^2 + v^2)^2},$$

and thereby conclude that the surface is indeed minimal. *Remarks:* It's easy to see that the surface is minimal; this is so because the parameterization is isothermal and the coordinate functions are harmonic. Also, Osserman has shown ([73], p. 87) that, of all the complete regular (i.e., immersed) minimal surfaces in $\mathbb{R}^3$, only the catenoid and Enneper's surface have the property that their Gauss maps are injective.

EXERCISE H.10 Given a smooth function $f(x, y)$ on $\mathbb{R}^2$, its graph $z = f(x, y)$ determines a two-dimensional surface Σ in $\mathbb{R}^3$.

(a) Find the coefficients of the first fundamental form of Σ in terms of f and its derivatives.
(b) Find the coefficients of the second fundamental form of Σ in terms of f and its derivatives.
(c) Compute the Gaussian curvature of Σ.
(d) Write down a differential equation that must be satisfied by f in order for the surface to be minimal.

References

[1] R. Abraham and J. E. Marsden, *Foundations of Mechanics*, 2nd edn, updated 1985 printing (Addison-Wesley, Reading, 1978).

[2] C. Adams, *The Knot Book* (W. H. Freeman, New York, 1994).

[3] W. Ambrose and I. M. Singer, A theorem on holonomy, *Trans. AMS*, **75**, (1953), 428–443.

[4] G. Andrews, R. Askey, and R. Roy, *Special Functions* (Cambridge University Press, Cambridge, 1999).

[5] T. Apostol, *Calculus*, Vol. 2, 2nd edn (Wiley, New York, 1969).

[6] M. F. Atiyah and R. Bott, The Yang–Mills equations over Riemann surfaces, *Proc. Roy. Soc. A*, **308**, (1982), 523–615.

[7] J. Baez and J. Muniain, *Gauge Fields, Knots, and Gravity* (World Scientific, Singapore, 1994).

[8] P. Bamberg and S. Sternberg, *A Course in Mathematics for Students of Physics*, Vol. 2 (Cambridge University Press, Cambridge, 1990).

[9] N. Biggs, *Algebraic Graph Theory* (Cambridge University Press, Cambridge, 1993).

[10] R. Bishop and R. Crittenden, *Geometry of Manifolds* (American Mathematical Society, Providence, 1964).

[11] W. Boothby, *An Introduction to Differentiable Manifolds and Riemannian Geometry* (Academic Press, New York, 1975).

[12] R. Bott and L. Tu, *Differential Forms in Algebraic Topology* (Springer, New York, 1982).

[13] S. S. Cairnes, Triangulation of the manifold of class one, *Bull. Amer. Math. Soc.* **41**, (1935), 549–552.

[14] A. Candel and L. Conlon, *Foliations I* (American Mathematical Society, Providence, 2000).

[15] R. Carter, G. Segal, and I. Macdonald, *Lectures on Lie Groups and Lie Algebras*, London Mathematical Society Student Texts Vol. 32 (Cambridge University Press, Cambridge, 1995).

[16] S. S. Chern, "A simple intrinsic proof of the Gauss–Bonnet formula for closed Riemannian manifolds", *Ann. Math.*, **45:4**, (1944), 747–752. Reprinted in *Shing Shen Chern Selected Papers* (Springer-Verlag, New York, 1978).

[17] S. S. Chern, "Pseudo-Riemannian geometry and the Gauss–Bonnet formula", *An. da Acad. Brasileira de Ciências*, **35:1**, (1963), 17–26. Reprinted in *Shing Shen Chern Selected Papers* (Springer-Verlag, New York, 1978).

[18] S. S. Chern, Vector bundles with a connection, in *Global Differential Geometry*, Studies in Mathematics Vol. 27, ed. S. S. Chern (Mathematical Association of America, Washington DC, 1989).

[19] S. S. Chern, W. H. Chen, and K.S. Lam, *Lectures on Differential Geometry* (World Scientific, Singapore, 2000).

[20] L. Conlon, *Differentiable Manifolds*, 2nd edn (Birkhauser, Boston, 2001).

[21] G. de Rham, *Variétés Différentiables, Formes, Courants, Formes Harmoniques* (Herman, Paris, 1955).

[22] M. P. do Carmo, *Differential Geometry of Curves and Surfaces* (Prentice Hall, Englewood Cliffs, 1976).

[23] M. P. do Carmo, *Riemannian Geometry* (Birkhauser, Boston, 1992).

[24] A. Duval, C. Klivans, and J. Martin, "Simplicial matrix-tree theorems", *Trans. Amer. Math. Soc.*, **361:11**, (2009), 6073–6114.

[25] H. Flanders, *Differential Forms with Applications to the Physical Sciences* (Dover, New York, 1989).

[26] G. Francis and J. Weeks, "Conway's ZIP proof", *Amer. Math. Mon.* **106:5**, (1999), 393–399.

[27] T. Frankel, *The Geometry of Physics*, 3rd edn (Cambridge University Press, Cambridge, 2012).

[28] F. Gantmacher, *The Theory of Matrices*, Vol. 1 (Chelsea, New York, 1959).

[29] S. Goldberg, *Curvature and Homology* (Dover, New York, 1982).

[30] A. Granas and J. Dugundji, *Fixed Point Theory* (Springer-Verlag, New York, 2003).

[31] W. Greub, *Linear Algebra*, 4th edn (Springer-Verlag, New York, 1981).

[32] P. Gross and R. Kotiuga, *Electromagnetic Theory and Computation: A Topological Approach* (Cambridge University Press, Cambridge, 2004).

[33] V. Guillemin and A. Pollack, *Differential Topology* (Prentice Hall, Englewood Cliffs NJ, 1974).

[34] B. Hall, *Lie Groups, Lie Algebras, and Representations: An Elementary Introduction* (Springer, New York, 2003).

[35] F. Harary, *Graph Theory* (Addison-Wesley, Reading, 1969).

[36] J. Hart, R. Miller, and R. Mills, "A simple geometric model for visualizing the motion of the Foucault pendulum", *Amer. J. Phys.*, **55**, (1987), 67–70.

[37] A. Hatcher, *Algebraic Topology* (Cambridge University Press, Cambridge, 2002).

[38] A. Hatcher, *Vector Bundles and K-Theory*, unpublished manuscript, 2009.

[39] S. Hawking and G. Ellis, *The Large Scale Structure of Spacetime* (Cambridge University Press, Cambridge, 1973).

[40] F. Hehl and Y. Obukhov, Élie Cartan's torsion in geometry and in field theory, an essay, *Ann. Fond. Louis de Broglie*, **32:2–3**, (2007), 157–194.

[41] N. Hicks, *Notes on Differential Geometry* (Van Nostrand Reinhold, New York, 1965).

[42] W. V. D. Hodge, *The Theory and Applications of Harmonic Integrals* (Cambridge University Press, Cambridge, 1989).

[43] K. Hoffman and R. Kunze, *Linear Algebra* (Prentice-Hall, Englewood-Cliffs, 1961).

[44] A. Isidori, *Nonlinear Control Systems*, 3rd edn (Springer-Verlag, New York, 1995).

[45] L. Kauffman, *Knots and Physics* (World Scientific, Singapore, 1993).

[46] D. Klain and G-C. Rota, *Introduction to Geometric Probability* (Cambridge University Press, Cambridge, 1997).

[47] S. Kobayashi and K. Nomizu, *Foundations of Differential Geometry*, Vols. 1 and 2 (Wiley-Interscience, New York, 1996).

[48] G. Lamé, *Leçons sur les coordonnées curvilignes et leurs diverses applications* (Mallet-Bachelier, Paris, 1859).

[49] H. B. Lawson, *The Qualitative Theory of Foliations* CBMS Series Vol. 27 (American Mathematical Society, Providence, 1977).

[50] Jeffery Lee, *Manifolds and Differential Geometry*, Graduate Studies in Mathematics Vol. 107 (American Mathematical Society, Providence, 2009).

[51] John Lee, *Introduction to Smooth Manifolds* (Springer, New York, 2003).

[52] C. Livingston, *Knot Theory* (Mathematical Association of America, Washington, 1993).

[53] A. Lundell, A short proof of the Frobenius theorem, *Proc. Amer. Math. Soc.* **116:4**, (1992), 1131–1133.

[54] I. Madsen and J. Tornehave, *From Calculus to Cohomology* (Cambridge University Press, Cambridge, 1997).

[55] W. S. Massey, *Algebraic Topology: An Introduction* (Springer, New York, 1967).

[56] J. Matoušek, *Using the Borsuk–Ulam Theorem* (Springer, New York, 2003).

[57] J. Matoušek, *Lectures on Discrete Geometry* (Springer, New York, 2002).

[58] J. Milnor, *Topology from the Differentiable Viewpoint* (The University Press of Virginia, Charlottesville, 1965).

[59] J. Milnor, The Poincaré conjecture 99 years later: a progress report, unpublished, 2003.

[60] C. Misner, K. Thorne, and J. A. Wheeler, *Gravitation* (W. H. Freeman, San Francisco, 1973).

[61] H. Moffat and R. Ricca, Helicity and the Călugăreanu invariant, *Proc. Roy. Soc. Lond.* A **439**, (1992), 411–429.

[62] R. Montgomery, *A Tour of Subriemannian Geometries, Their Geodesics and Applications* (American Mathematical Society, Providence, 2002).

[63] J. Morgan and G. Tian, *Ricci Flow and the Poincaré Conjecture* (American Mathematical Society, Providence, 2007).

[64] S. Morita, *Geometry of Differential Forms* (American Mathematical Society, Providence, 2001).

[65] J. Munkres, *Topology: A First Course* (Prentice Hall, Englewood Cliffs, 1975).

[66] M. Nakahara, *Geometry, Topology, and Physics*, 2nd edn (IOP Publishing, Bristol, 2003).

[67] C. Nash and S. Sen, *Topology and Geometry for Physicists* (Dover, New York, 2011).

[68] E. Nelson, *Tensor Analysis* (Princeton University Press, Princeton, 1967).

[69] G. Naber, *Topology, Geometry and Gauge fields: Interactions*, 2nd edn (Springer, New York, 2011).

[70] B. O'Neill, *Semi-Riemannian Geometry (with Applications to Relativity)* (Academic Press, San Diego, 1983).

[71] J. Oprea, Geometry and the Foucault pendulum, *Amer. Math. Mon.* **102:6**, (1995), 515–522.

[72] D. O'Shea, *The Poincaré Conjecture: In Search of the Shape of the Universe* (Walker, New York, 2007).

[73] R. Osserman, *A Survey of Minimal Surfaces* (Van Nostrand Reinhold, New York, 1969).

[74] M. Reed and B. Simon, *Methods of Modern Mathematical Physics*, Vols. 1–4 (Academic Press, San Diego, 1980).

[75] J. Rotman, *An Introduction to Algebraic Topology* (Springer Verlag, New York, 1988).

[76] A. Shapere and F. Wilczek, *Geometric Phases in Physics* (World Scientific, Singapore, 1989).

[77] I. M. Singer and J. A. Thorpe, *Lecture Notes on Elementary Topology and Geometry* (Springer, New York, 1967).

[78] P. Slepian, *Mathematical Foundations of Network Analysis* (Springer-Verlag, Berlin, 1968).

[79] M. Spivak, *Calculus on Manifolds* (Addison-Wesley, Reading, 1965)

[80] M. Spivak, *A Comprehensive Introduction to Differential Geometry*, Vols. 1–5, 3rd edn (Publish or Perish, 1999).

[81] R. Stanley, *Enumerative Combinatorics*, Vol. 2 (Cambridge University Press, Cambridge, 1999).

[82] S. Sternberg, *Lectures on Differential Geometry* (Prentice Hall, Englewood Cliffs, 1964).

[83] G. Szpiro, *Poincaré's Prize: The Hundred-Year Quest to Solve One of Math's Greatest Puzzles* (Penguin, New York, 2008).

[84] L. Tu, *An Introduction to Manifolds*, 2nd edn (Springer, New York, 2011).

[85] V. A. Vassiliev, *Introduction to Topology* (American Mathematical Society, Providence, 2001).

[86] M. Visser, *Notes on Differential Geometry*, unpublished lecture notes (February 2011).

[87] T. Voronov, *Differential Geometry*, unpublished lecture notes (Spring 2009).

[88] F. Warner, *Foundations of Differentiable Manifolds and Lie Groups* (Scott Foresman, Glenview, 1971).

[89] D. West, *Introduction to Graph Theory* (Prentice-Hall, Upper Saddle River, 1996).

[90] J. H. C. Whitehead, On C^1 complexes, *Ann. Math.*, **41**, (1940), 809–824.

[91] J. Wolf, *Spaces of Constant Curvature* (McGraw-Hill, New York, 1967).

[92] G. Ziegler, *Lectures on Polytopes* (Springer, New York, 1995).

Index

Printed in the United States
By Bookmasters